Iterative Identification and Control

Springer
London
Berlin
Heidelberg
New York
Barcelona
Hong Kong
Milan
Paris
Singapore
Tokyo

Pedro Albertos and Antonio Sala (Eds.)

Iterative Identification and Control

Advances in Theory and Applications

With 130 Figures

Springer

Pedro Albertos, Professor
Antonio Sala, Doctor
Department of Systems Engineering and Control,
Polytechnic University of Valencia, C. Vera s/n Valencia, Spain

British Library Cataloguing in Publication Data
Iterative identification and control : advances in theory
 and control
  1.Automatic control 2.Iterative methods (Mathematics)
  I.Albertos Perez, P. II.Sala, Antonio
  629.8
  ISBN 1852335092

Library of Congress Cataloging-in-Publication Data
A catalog record for this book is available from the Library of Congress.

Apart from any fair dealing for the purposes of research or private study, or criticism or review, as permitted under the Copyright, Designs and Patents Act 1988, this publication may only be reproduced, stored or transmitted, in any form or by any means, with the prior permission in writing of the publishers, or in the case of reprographic reproduction in accordance with the terms of licences issued by the Copyright Licensing Agency. Enquiries concerning reproduction outside those terms should be sent to the publishers.

ISBN 1-85233-509-2 Springer-Verlag London Berlin Heidelberg
a member of BertelsmannSpringer Science+Business Media GmbH
http://www.springer.co.uk

© Springer-Verlag London Limited 2002
Printed in Great Britain

The use of registered names, trademarks etc. in this publication does not imply, even in the absence of a specific statement, that such names are exempt from the relevant laws and regulations and therefore free for general use.

The publisher makes no representation, express or implied, with regard to the accuracy of the information contained in this book and cannot accept any legal responsibility or liability for any errors or omissions that may be made.

Typesetting: Electronic text files prepared by editors
Printed and bound at The Cromwell Press, Trowbridge, Wiltshire
69/3830-543210 Printed on acid-free paper SPIN 10838293

Preface

This book deals with the design of high-performance feedback controls for physical systems ranging from industrial processes to aerospace systems. The approach is based on the relation between the modelling and experimental identification of the dynamics of the system and the subsequent model-based controller design. Because these two stages are coupled, that is, there is a strong dependency between them, an understanding of the joint approach allows higher performance controller design. The total control design effort is split into two phases and reduced.

The book presents a complete control design methodology that is applicable to a broad class of control problems. It is not limited to the approaches developed here and provides alternative approaches to get a final design. It outlines a general way of designing control systems for complex systems, where the process model is not well known and is difficult to be developed *a priori*.

The interplay between the modelling of dynamic systems and the design of feedback controllers based on these models is the main goal of this book. Combining both subjects into a cohesive development allows the consistent treatment of both problems to yield powerful new tools for system performance improvement. Central among the themes is that operation of a system in feedback with a controller exposes the areas in which the model accuracy is constraining the achieved controlled performance. As the model complexity needed to achieve a particular control objective may be unknown *a priori*, and getting complex models has a cost, the aim of iterative approaches is to develop methodologies to increase that complexity only when it is necessary, hence saving efforts in identification-oriented experiments and downtime.

The book presents new techniques in understanding the iterative improvement of performance through the successive fitting of models using closed-loop data and designing high-performance controllers using these models. It provides a comprehensive yet comprehensible introduction to the field. A number of different approaches are provided.

The subject matter includes:

- new approaches to understanding how to affect the fit of dynamical models to physical processes through the choice of experiments, data pre-filtering and model structure;
- connections between robust control design methods and their dependency on quality of model;
- experiment design in which data collected in operation under feedback can reveal areas limiting the achieved performance;
- iterative approaches to link these model fitting and control design phases in a cogent way to achieve improved overall performance.

This is an edited and coordinated volume of contributions from international leaders in the fields of system identification and robust control, who have developed both the industrial applications and the underpinning theories and are distinguished members of the most relevant automatic control societies (IFAC, IEEE). A great part of the content was discussed at a summer course held in Valencia, Spain, in the framework of an European Science Foundation project - COSY (control of complex systems). Nevertheless, a number of additional contributions, both theoretical and applied, have been included to present results from the most representative authorities in the field from different countries.

The book is interesting for a broad audience and offers comprehensive material for graduate courses in advanced control design. Control engineering students will find a review of the theory involved, valuable references for further study as well as suggestive applications of real control problems. Other than the main purpose of the book being a control design methodology, it also provides the tools for further research and development using alternative approaches, either for the modelling, the control design or the on-line tuning of any given control application. Last but not least, applied control engineers faced with actual control problems will find guidelines for improving their current controllers by combining the exploitation of the available models with their refinement by means of new experiments or using the already available data from the actual controlled plant.

We hope to deserve the confidence Springer-Verlag has shown in our project and to create a bridge between the challenging fields of modelling and control design.

We will appreciate any feedback from the readers.

Valencia,
November 2001

Pedro Albertos
Antonio Sala

Contents

List of Contributors xi

Introduction from the Editors xiii

Part I. Fundamentals and Theory

1 Modelling, Identification and Control 3
Michel Gevers
   1.1   A not-so-brief Historical Perspective 3
   1.2   A Typical Scenario and its Major Players 8
   1.3   Some Important New Insights 10

2 System Identification. Performance and Closed-loop Issues   17
Jesús Picó, Miguel Martínez
   2.1   Chapter Outline 17
   2.2   Modelling, Estimation and Identification 17
   2.3   Model Structures.. 19
   2.4   Implementation of the Model Structures: Predictors 22
   2.5   Open-loop Identification. What can be Achieved? 25
   2.6   Open-loop Identification Algorithms 27
   2.7   Closed-loop Identification 31

3 Data Preparation for Identification 41
Robert R. Bitmead, Antonio Sala
   3.1   Introduction 41
   3.2   Design Variables for Identification 43
   3.3   Data Preparation: Detrending and Filtering 55
   3.4   System Identification with an $\mathcal{H}_\infty$ Criterion 61
   3.5   Conclusions 62

4 Model Uncertainty and Feedback 63
Karl J. Åström
   4.1   Introduction 63
   4.2   Classical Control Theory 65
   4.3   State-space Theory 76
   4.4   Fundamental Limitations.................................. 80
   4.5 $\mathcal{H}_\infty$ Loop Shaping.. 86
   4.6   Summary 97

5 Open-loop Iterative Learning Control ... 99
Antonio Sala, Pedro Albertos
- 5.1 Introduction ... 99
- 5.2 Control with Two Degrees of Freedom ... 101
- 5.3 The Iterative Learning Paradigm ... 105
- 5.4 The Learning Algorithms ... 106
- 5.5 Examples ... 116
- 5.6 Conclusions ... 119

6 Model-based Iterative Control Design ... 121
Pedro Albertos
- 6.1 Introduction ... 121
- 6.2 Iterative Control Design ... 123
- 6.3 Application 1: Single Control Loop ... 126
- 6.4 Application 2: GPC Control Approach ... 136
- 6.5 Conclusions ... 142

7 Windsurfing Approach to Iterative Control Design ... 143
Brian D.O. Anderson
- 7.1 Introduction ... 143
- 7.2 Fundamental Problems in Adaptive Control ... 145
- 7.3 High-level Overview of the Windsurfer Approach to Adaptive Control ... 147
- 7.4 More Detailed Description of the Algorithm ... 150
- 7.5 Examples ... 155
- 7.6 Unstable Plants ... 161
- 7.7 Coping with the "Fundamental Problems" ... 162
- 7.8 Some Further Quantitative Directions ... 163

8 Iterative Optimal Control Design ... 167
Robert R. Bitmead
- 8.1 Introduction ... 167
- 8.2 Controller Tuning with Data ... 169
- 8.3 Iterative Identification and Control Design ... 172
- 8.4 Introducing Caution into Iterative Design ... 177
- 8.5 Embellishments and Conclusion ... 181

9 Identification and Validation for Robust Control ... 185
Michel Gevers
- 9.1 Introduction ... 185
- 9.2 The Identification/Control Setup and some Basic Formulæ ... 188
- 9.3 A Control-oriented Nominal Model ... 192
- 9.4 Caution in Iterative Design ... 196
- 9.5 Model Validation for Robust Control Design ... 200
- 9.6 Conclusions ... 208

Part II. Applications

10 Helicopter Vibration Control 211
Robert R. Bitmead
   10.1 Helicopter Vibration Control Example 211
   10.2 A Glimpse of the Solution 212
   10.3 Modelling Requirements 215
   10.4 Identification Experiments 215
   10.5 Data Analysis and Sanitizing 217
   10.6 Parametric System Identification 219
   10.7 Conclusions 222

**11 Control-relevant Identification and
Servo Design for a Compact Disc Player** 225
Paul M.J. Van den Hof, Raymond A. de Callafon
   11.1 Introduction 225
   11.2 The Compact Disc Mechanism 226
   11.3 Control-relevant Modelling of Radial Actuator 229
   11.4 Preliminaries and Notations 230
   11.5 Common Objectives in Identification and Control 232
   11.6 Estimation of Coprime Factorisations 233
   11.7 Application to CD Radial Actuator 236
   11.8 Control Design and Stability Robustness 238
   11.9 Summary 241

12 Control-relevant Identification and Robust Motion Control of a Wafer Stage 243
Raymond A. de Callafon, Paul M.J. Van den Hof
   12.1 Introduction 243
   12.2 The Wafer Stepper Positioning Mechanism 244
   12.3 Iterative Model Set Estimation and Control Design 251
   12.4 Elements of the Iterative Scheme 254
   12.5 Estimation of Model Set for the Wafer Stage 260
   12.6 Model Set-based Robust Control Design 266
   12.7 Summary 269

**13 Iterative Identification and Control:
a Sugar Cane Crushing Mill** 271
Robert R. Bitmead, Antonio Sala
   13.1 Introduction 271
   13.2 The Sugar Cane Crushing Process 272
   13.3 Identification Procedure 276
   13.4 Controller Design 288
   13.5 Iterative Approach 289
   13.6 Conclusions 294

References 297

Index 307

List of Contributors

Pedro Albertos (Ed.)

>Systems Engineering and Control Department,
>Universidad Politécnica de Valencia,
>P.O.Box 22012 , E-46071 Valencia, Spain
>Email: pedro@aii.upv.es

Brian D.O. Anderson

>The Australian National University,
>Canberra, ACT, 0200, Australia
>Email: brian.anderson@anu.edu.au

Karl J. Åström

>Department of Automatic Control,
>Lund University, Lund, Sweden
>Email: kja@control.lth.se

Robert R. Bitmead

>Department of Mechanical and Aerospace Engineering,
>University of California, San Diego,
>9500 Gilman Drive,
>La Jolla, CA 92093-0411, USA
>Email: rbitmead@ucsd.edu

Raymond A. de Callafon

>Department of Mechanical and Aerospace Engineering,
>University of California, San Diego,
>9500 Gilman Drive,
>La Jolla, CA 92093-0411, USA
>Email: callafon@ucsd.edu

Michel Gevers

>CESAME, Université Catholique de Louvain,
>Bâtiment Euler, B-1348 Louvain la Neuve, Belgium
>Email: gevers@auto.ucl.ac.be

Miguel Martínez

>Systems Engineering and Control Department,
>Universidad Politécnica de Valencia,
>P.O. Box 22012 , E-46071 Valencia, Spain
>Email: mami@isa.upv.es

Jesús Picó

>Systems Engineering and Control Department,
>Universidad Politécnica de Valencia,
>P.O. Box 22012 , E-46071 Valencia, Spain
>Email: jpico@isa.upv.es

Antonio Sala (Ed.)

>Systems Engineering and Control Department,
>Universidad Politécnica de Valencia,
>P.O. Box 22012 , E-46071 Valencia, Spain
>Email: asala@isa.upv.es

Paul M.J. Van den Hof

>Signals, Systems and Control Group,
>Department of Applied Physics,
>Delft University of Technology, 2628 CJ Delft, The Netherlands
>Email: p.m.j.vandenhof@tnw.tudelft.nl

Introduction from the Editors

In order to design a control system for a complex plant, there are basically two approaches: to use a good model of the plant and apply a model-based control design approach, or to algorithmically express the actions a skilled operator uses to control the plant and emulate this behaviour by means of an automatic controller. In the first case, a lot of effort should be devoted to obtain a good model of the plant, not knowing *a priori* if this model is covering all the interesting features to design the control. In the second case, the model of the plant is implicitly taken into account in the operator behaviour, but it is not considered for the controller design. In between these two extreme approaches, a number of solutions can be foreseen. The final goal of a control system is to provide a satisfactorily controlled system behaviour under normal operating conditions and to present a certain degree of robustness against unexpected changes in these conditions. What this means is that, at the end, the control engineer is not particularly interested in either the quality of the model being used to design the controller or in the quality of the theoretically designed controller before it is actually applied to the plant. Obviously, the final result strongly depends on both, but if one of the elements taking part in this procedure is significantly poor and the other one is very precise the global system will present an inappropriate behaviour. It should be a balance between the quality of the model and that of the designed controller. To solve a control problem, the control engineer should combine the efforts in using these techniques, which are not exclusive but complementary.

In this book, the iterative approach to designing a control system for a complex plant is presented and discussed. The iteration involves refinements of the model based on information gathered from the controlled plant, when it is performing at the edge conditions before violating some given constraints, and redesigning the controller based on the new models derived after treatment of the previous information. In fact, the iterative methodologies in this volume formalise the methods used by many control engineers in practice. The ideas in some chapters can be applied by using different identification and control design techniques, not being constrained to the ones given in the central chapters. A number of examples and industrial applications are outlined in order to illustrate and give insight into the versatility of the approach and the open challenges in design performing and powerful controllers.

For this purpose, the book is organized into two parts with altogether thirteen chapters. The first part deals with the basic ideas and tools required to develop the identification and control design approaches mainly used in this framework. Then, a number of approaches are outlined. In the second part, some industrial applications of these approaches are presented. On the basis of some of the applications, additional concepts are introduced.

The first chapter, *Modelling, Identification and Control*, written by Michel Gevers, is devoted to reviewing the state of the art in the field, giving a historical perspective of its evolution. The interplay between modelling, identification and robust control design, as well as the difficulties in coordinating these activities, are outlined and the progress and challenges of the field are discussed. This is a perfect introduction to get an overview of the aim of the book.

The second chapter, *System Identification. Performance and Closed-loop Issues*, developed by Jesús Picó and Miguel Martínez, summarises the techniques, results and forecasted limitations of experimental identification of real plants. The subject, which easily could be the matter of a whole book, is reduced to discrete time parametric models, and single-input single-output plants only are considered. The reader is referred to classical references for additional information. Much attention is devoted to closed-loop parameter estimation techniques, as they are at the core of many applications. It provides the basis for developing the identification stage of an iterative identification/control design approach.

One of the working ideas of the iterative approach is to use the data from the currently controlled plant to build an improved model of the plant, capturing the dynamic features that limit the design of higher performing controllers. A huge amount of data can be gathered from the plant, not all being interesting for that purpose, and some extra-exciting signals can be applied. The chapter on *Data Preparation for Identification*, written by Robert Bitmead and Antonio Sala, deals with these issues: what experiments should be carried out, how the data should be pre-processed before being used to fit a model and which structure should be used in the basis of such a model. Most of these tasks are performed off-line.

Prof. Karl Johan Åström presents in the fourth chapter, *Model Uncertainty and Feedback*, the basic concepts involved in the control design part, complementing the tools given in the previous chapters to construct a procedure for the iterative design of a control system. Without the aim of reviewing the basic control design techniques, the main goal of the chapter is to point out the limitations in achievable performances based on a given model, and the influence of the uncertainty of the model in the expected performances. In this framework, guidelines for comparing the dynamics of two systems, based on the concept of coprimeness and the use of appropriate metrics, such as Vinnicombe's metric, will be extensively used later on.

It appears that iterative control design techniques should be very appropriate for repetitive tasks, that is, control problems where the same output is expected in a number of successive operations of a given plant under similar reference signals, control requirements and constraints. This is also a broad field of both research and applications, mainly in the area of robotics and manufacturing. Antonio Sala and Pedro Albertos, in the fifth chapter on *Open-loop Iterative Learning Control*, introduce the concept of iterative improvement of the control action which, indirectly, leads to an implicit model of the controlled process. The point of the chapter is to stress that performance enhancement can be achieved with two-degree-of-freedom structures and that techniques and models for the feedback and feedforward blocks need not be the same. Adaptation of feedforward blocks is discussed, being a technique that is independent of the iterative improvement of the feedback path. Although the technique could be also applied in a closed-loop setting, in order to emphasise the basic concepts, attention is focused on open-loop controlled systems.

In Chapter 6, *Model-based Iterative Control Design*, Pedro Albertos illustrates the main ideas in iterative identification and control design by means of a number of academic examples. The main purpose is to emphasise the advantages and drawbacks of the approach without paying attention to the power or adequacy of the control design, identification or model parameter estimation algorithms. The common feature of the plants used in the designs is their apparent frequency decomposition due to either the presence of resonant modes or high-frequency components. This is quite usual in flexible manipulators or space distributed structures.

A deeper insight in the same problem is developed by Brian D.O. Anderson in the chapter on *Windsurfing Approach to Iterative Control Design*. The strong connection between the iterative control design approach and human learning in performing the control of an unstable plant, like the surfer application, is used as the main motivation. The internal model control design and closed-loop identification techniques are used as a basis for the discussion. Here, the limitations on the model uncertainty, the options in both the controller selection and the model reduction, and the validity of the designs are analysed. Also, the fundamental problems in applying this methodology are emphasised and some further investigations based on the measurement of the distance between two dynamic systems are outlined.

In Chapter 8, *Iterative Optimal Control Design*, Robert Bitmead explores the use of two optimal techniques in both phases: linear quadratic Gaussian optimal control (LQG) and least squares (LS) optimal identification. The iterative feature requires to develop methods in which the objective criterion for each phase is modified to reflect the overall control task at hand. The agreement between objectives in the solution of partial problems, *i.e.*, the model fitting, the control design and the controlled system behaviour, is a crucial issue to work in the right direction. The question of adding caution to the model and controller updates is also examined.

The last chapter of the first part of the book, *Identification and Validation for Robust Control*, contributed by Michel Gevers, analyses the influence of the model uncertainty in the robustness of the designed control, proposing the formulation of control-oriented identification criteria, whose minimisation yields a "control-oriented nominal model". Also, the distinction between identification (the construction of a nominal model) and validation (the construction of the uncertainty set around the model) are crucial for the controller design stage. In this way, as the validated model uncertainty set, obtained by prediction error identification, serves as the basis for robust control design, control-oriented guidelines should be used for the design of the experimental conditions under which identification should be performed.

A number of applications are presented and, in some detail, developed in the second part of the book. The chapter on *Helicopter Vibration Control*, worked out by Robert Bitmead and his team, tackles the problem of constructing a model from which to design a controller to reduce helicopter vibration. It is clear that the disturbance has a main frequency range of action and the model should cope with the helicopter dynamics in this region. In particular, what appears to be crucial in this application is the fitting of the model phase in the disturbance bandwidth, the frequency gain not being so important . This application shows the relevance of modelling for control, as well as the need to treat the data gathered from the controlled system before being used for the control design.

The second application, *Control Relevant Identification and Servo Design for a Compact Disc Player*, has been developed by Paul Van den Hof and Raymond de Callafon, and it has been well-reported and recognised. The identification approach, using normalised coprime factorisation, tries to get a reduced order model of the system as well as an estimate of the model uncertainty bounds. Thus, a model-based control design approach can be applied and the robust stability of the designed controlled system can be verified prior to the controller implementation.

The same authors, Raymond de Callafon and Paul Van den Hof, have worked out and written the application entitled *Control Relevant Identification and Robust Motion Control of a Wafer Stage*. The added complexity of a multivariable control problem is addressed in this application where, again, the interplay between (control-relevant) model identification, including model uncertainty bounding, and robust control design based on worst-case performance optimisation leads to an iterative approach. This also allows for the monitoring of the stability and performance robustness, enabling the possibility of guaranteeing an improvement of the controlled system performance in a single step of the iteration.

The last chapter outlines an application where many issues are put together: to acquire knowledge about the process, to select variables, to design experiments to get the suitable data, its pre-filtering, to build up a model, to design the control and to determine the options in the performance index. The *Iterative Identification and Control: a Sugar Cane Crushing Mill* chapter illustrates the most practical issues in designing a control system for a sugar cane factory. Here, the low *a priori* knowledge of both the process and the possible control solutions makes this application a real challenge for the control engineer. Worked out by Robert Bitmead and written and presented in collaboration with Antonio Sala, this chapter links the advanced control and identification theory with the "nuts and bolts" of an engineering control application.

As a final remark, after reviewing the content of the different chapters, it can be concluded that the book, including theoretical issues and practical applications, is suitable as a basis for a graduate course on advanced control as well as a reference for control engineers in their continuous updating of knowledge.

Part I

Fundamentals and Theory

1 Modelling, Identification and Control

Michel Gevers

Abstract. In this chapter, we first review the changing role of the model in control system design over the last fifty years. We then focus on the development over the last ten years of the intense research activity and on the important progress that has taken place in the interplay between modelling, identification and robust control design. The major players of this interplay are presented; some key technical difficulties are highlighted, as well as the solutions that have been obtained to conquer them. We end the chapter by presenting the main insights that have been gained by a decade of research on this challenging topic.

1.1 A not-so-brief Historical Perspective

There are many ways of describing the evolution of a field of science and engineering over a period of half a century, and each such description is necessarily biased, oversimplified and sketchy. But I have always learned some new insight from such sketchy descriptions, whoever the author. Thus, let me attempt to start with my own modest perspective on the evolution of modelling, identification and control from the post-war period until the present day.

Until about 1960, most of control design was based on model-free methods. This was the golden era of Bode and Nyquist plots, of Ziegler-Nichols charts and lead/lag compensators, of root-locus techniques and other graphical design methods.

From model-free to model-based control design
The introduction of the parametric state-space models by Kalman in 1960, together with the solution of optimal control and optimal filtering problems in a Linear Quadratic Gaussian (LQG) framework [90, 91] gave birth to a tremendous development of model-based control design methods. Successful applications abounded, particularly in aerospace, where accurate models were readily available.

From modelling to identification
The year 1965 can be seen as the founding year for parametric identification with the publication of two milestone papers. The paper [80] set the stage for state-space realisation theory which, twenty-five years later, became the major stepping stone towards what is now called *subspace identification*. The paper [12] proposed a Maximum Likelihood (ML) framework for the identification of input-output (*i.e.*, ARMAX) models that gave rise to the celebrated

prediction error framework that has since proven so successful. Undoubtedly – and as is so often the case – the advent of identification theory was spurred by a desire to extend the applicability of model-based control design to broader and broader fields of applications, for which no reliable models could be obtained, at least at a reasonable cost.

From the elusive true system to an approximate model

Most of the early work on identification theory was directed at developing more and more sophisticated model sets and identification methods with the elusive goal of converging to the "true system", under the assumption that the true system was in the model set. It was not until the 1980s that the effort shifted towards the goal of approximating the true system, and of characterising this approximation in terms of bias and variance error on the identified models. Once it was recognised that the identified model approximated the true system with some error, it made sense to tune the identification towards the objective for which the model was to be used. One of the main contributions of L. Ljung's book [108], first published in 1987, was to introduce the engineering concept of *identification design*, and to lay down some foundations for the formal design of goal-oriented identification. However, the specific contributions to control-oriented identification design were virtually non-existent until 1990.

From certainty equivalence to robust control design

A consequence of the early faith that the true system could be modelled almost perfectly and of the development of more and more sophisticated model-based control design methods was the application of the "certainty equivalence principle" for control design. Whether the model had been obtained by mathematical modelling or by identification from data, it was taken to represent the true system. It was as if the prevailing habits of adopting model-based control design techniques had almost completely obliterated our grandfathers' cautionary adoptions of gain and phase margins. There was an obvious need to introduce a formal way of injecting stability and performance safeguards in the model-based control design approaches. This is precisely what the robust control theory, initiated in the 1980s (see [158]), aimed to achieve: it was an attempt to preserve the obvious advantages of model-based control design while at the same time introducing robustness to model errors. Much of the effort in the development of robust control theory had to do with various ways of introducing model errors in the closed-loop system configuration: descriptions of additive, multiplicative, feedback errors, coprime factor perturbations, linear fractional transformations flourished.

From separate to synergistic design

It was not until the early 1990s that the identification community and the robust control community became aware of each other's work. Much of the successful robust control applications had been performed in situations where modelling techniques were able to deliver a fairly accurate model on the basis of first principles, and where it was reasonable to assume *a priori* bounds on the noise and modelling errors. Process control applications, in which identification methods are often the only path to a reasonable model and where these methods have also achieved many of their successes, were for the most part outside this realm. As a result of the lack of communication between these two research communities, the model uncertainty descriptions on which robust control design methods had been based (frequency domain descriptions, essentially) were very inconsistent with the tools delivered by prediction error (PE) identification. As a matter of fact, PE identification had very little to offer in terms of explicit quantification of the error in an estimated transfer function. There was very little understanding, *a fortiori*, of the interplay between the experimental conditions under which identification was performed and the adequacy of the resulting model (and model error) for control design. The prevailing philosophy was, "First estimate the best possible model, then design the controller on the basis of this estimated model and – possibly – of an estimate of the model error." Since the main stumbling block in applying robust control design techniques from identified models was the unavailability of adequate uncertainty descriptions, much of the early effort at combining identification with robust control theory went in the development of novel identification techniques that would deliver the kind of frequency domain uncertainty descriptions that the robust control theory of the 1980s required. The problem is that an identification method whose sole merit is to deliver an error bound may well produce a nominal model as well as an uncertainty set that are ill-suited for robust control design.

From model reduction to control-oriented low-order models
Of course, there is no fundamental objection to first spending a significant amount of effort in obtaining a very accurate model (including a model uncertainty set) for the unknown system by modelling and/or identification techniques, and then computing a robust controller from this model and its uncertainty set. This would typically lead to a high-order controller, which can of course later be reduced. However, there are both practical and theoretical reasons for adopting an identification method that directly leads to a low-order model and an uncertainty set that are tuned for robust control design. This will be one of the central themes of the present book.

A practical motivation is that there is no reason to waste enormous modelling and/or identification efforts in obtaining a highly accurate and complex model with fairly accurate error bounds if a simple model is obtained with much less effort and leads to a controller that achieves similar performance with the same stability safeguards. Examples abound to illustrate that extremely simple controllers, obtained from extremely simple models, often achieve high performance in complex systems. One such example will be presented for motivation in Section 1.3. A theoretical motivation is that a low-order model, obtained by model reduction techniques from a high-order model, may have higher variance error than a low-order model that has been identified directly for the purpose of control design [89].

Thus, much of the research effort of the 1990s focused on establishing synergies between identification and robust control design methods. These efforts have been directed at better matching the technical tools of both theories: new identification techniques have been developed that produce the kind of frequency domain model uncertainty descriptions that are prevalent in mainstream robust control theory, while new robust control analysis and design tools have been developed that are consistent with the parametric descriptions delivered by mainstream PE identification theory. More importantly, perhaps, relevant progress has been made at producing robust identification and control design procedures in which the experimental conditions and the identification criterion are designed to match the control performance criterion. Such matching must be achieved not only for the low-order nominal model (this is a design problem for the bias error distribution), but also for the model uncertainty set (this is a design problem for the variance error distribution). The bias and variance error are affected by different aspects of the experimental conditions and of the identification criterion.

From adaptive control to iterative control design
Dual control and adaptive control were two early attempts to address the issue of parametric uncertainty and model-based control design in a synergistic way. In dual control, the parameter estimation and the control design mechanism are obtained jointly as the result of a single but very complex optimisation problem. In adaptive control, the parameter adjustment scheme is subsidiary to the control objective. Both schemes were essentially developed for the case where the structure of the true system is known, and where the system is in the model set. The solution of the dual control problem proved to be computationally intractable, even in the simplest cases. As for adaptive control, after convergence mechanisms had been devised for the ideal case where the system is in the model set, attempts were made to robustify the adaptive control algorithms in order to take account of some modest degree of uncertainty. These attempts essentially consisted of introducing cautionary safeguards in the computation of the gain of the parameter adjustment scheme; see, *e.g.*, [5] for a representative example of these efforts. The major difficulty with adaptive control schemes is that the parameters of the feedback control system change at every sampling instant, making the closed-loop dynamics non-linear and the stability analysis of these dynamics extremely complex.

The study of the interplay between identification and control design, undertaken by various groups around 1990, led to the formulation of identification criteria that were a function of the underlying control performance criteria. As with all optimal experiment design results, the optimal solution is a function of the unknown system, and the only practical way to approach this optimal solution is then to attempt an iterative design. This has been pointed out by various authors (see, *e.g.*, [136]) and will be illustrated later in this chapter. In identification for control , this means that a succession of model updates and controller updates are intertwined. Each new model is identified from data obtained on the real plant on which the most recent controller is acting; each new controller is in turn computed from this most recent model. Representative examples of such iterative schemes, which first emerged in the early 1990s, can be found in [102, 139, 160].

A major difference between the iterative identification and control design schemes of the 1990s and the earlier adaptive control schemes is that in the former, the model and controller are kept constant in between two model and/or controller iterations. Thus, the closed-loop system performs in a batch-like mode, in which stationarity can be assumed – and hence asymptotic analysis can be applied – during the collection of each batch of data. This removes one of the fundamental problems of adaptive control schemes, namely the transient instability problem: see [7].Nevertheless, it was shown in [79] that, even with the simplest control performance criterion, such iterative schemes are not guaranteed to converge to a minimum of the control performance criterion over all models in the chosen restricted complexity model set. This may cause the iterative parameter adjustment scheme to drift to a controller parameter vector that makes the closed-loop system unstable.

As a result, important work has been undertaken to introduce prior stability checks into the iterative identification and control schemes. This work, which is still underway, has led to the inclusion of caution in the controller adjustment so that, from a presently operating stabilising controller and an updated model, stability and/or performance improvement guarantees can be established for the next controller [6, 46].

In the work on iterative identification and control design, the focus has been first on the formulation of control-oriented identification criteria, *i.e.*, the successive identification criteria are a function of the control performance criteria. As a result, the succession of nominal models have a bias error distribution that is "tuned for control design". This means that the (typically low-order) nominal models have a bias error that is small in the frequency areas where it needs to be small for the design of a better controller, typically around the present crossover frequency. Recent work has focused on the design of a control-oriented distribution of the variance error of the identified models, again leading to iterative designs [32, 62].

The idea is that, since one can manipulate the shape of the model uncertainty set by the choice of the experimental conditions under which the new model is identified, one should attempt to obtain model uncertainty sets for which the class of controllers achieve stability and the required performance with all models of that set is as large as possible.

1.2 A Typical Scenario and its Major Players

During the extensive research of the last decade on the interplay between identification and robust control, important progress has been made and some key lessons have been learned. In this section, we illustrate some of the salient features of the interplay. First, we present the typical scenario to which iterative identification and control design schemes are usually applied, as well as the major players.

In many applications, the system to be controlled is very complex and possibly non-linear, and it would therefore require a complex dynamical model to represent it with high fidelity. Any model-based control design procedure would therefore lead to a complex or high-order controller, since the complexity of a model-based controller is of the same order as that of the system. The practical situation considered here is where we want the to-be-designed controller to be linear and of low order.

Since we want to focus on ideas and concepts rather than on technically complicated issues, and for the sake of simplicity, we shall assume that the unknown true system can be represented with high fidelity by a single-input single-output linear time-invariant system. Thus, we assume that there is an unknown "true system" represented by

$$\mathcal{S} : y_t = G_0(z)u_t + v_t \tag{1.1}$$

where $G_0(z)$ is a linear time-invariant causal operator, y is the measured output, u is the control input, and v is noise, assumed to be quasi-stationary.

A typical situation is that we can perform experiments on this system with the purpose of designing a feedback controller. Most often, the system is already under feedback control, and the task is to replace the present controller by one that achieves better performance. This situation is representative of very many practical industrial situations. We denote the present controller by K_{id}:

$$u_t = K_{id}(z)[r_t - y_t] \tag{1.2}$$

where r_t is the reference excitation.

Using any set of N data collected on the unknown system, in open-loop or in closed-loop, we can apply prediction error identification and compute a model G_{mod} of the unknown G_0.[1] The model G_{mod} is typically a low-order approximation of the unknown G_0. Various validation methods have also been developed for the estimation of an uncertainty set $\mathcal{D}$ around G_{mod}, with the property that $G_0 \in \mathcal{D}$ with probability α, where α is any desired level close to 1 (e.g., $\alpha = 0.95$) [30, 64, 70].

---
[1] Most often, one would also compute a noise model for $v(t)$ in the form $v(t) = H_{mod}(z)e(t)$, with $e(t)$ white noise, but to keep things simple we shall only discuss here the interplay between the input-output model and the controller.

The traditional scenario in model-based robust control design was: *using the model G_{mod} and the uncertainty set $\mathcal{D}$ (if available), design a new controller $K(z)$ that achieves closed-loop stability and meets the required performance with all models in $\mathcal{D}$, and hence with the unknown true system G_0.* For this scenario to be successful, a very accurate model G_{mod} was typically required.

The present scenario based on the new insights gained on the interplay between identification and robust control is: *on the basis of the required performance, of any knowledge of the unknown system, and of the performance achieved with the present controller (if any), design a control-oriented identification experiment that produces a (new) G_{mod} and a (new) uncertainty set $\mathcal{D}$; then design a new controller $K(z)$ that achieves closed-loop stability and meets the required performance with all models in $\mathcal{D}$, and hence with the unknown true system G_0.* If necessary, repeat this design procedure, possibly with a more demanding performance criterion. In most versions of this new scenario, one first computes a class of controllers $\mathcal{K}(G_{mod}, \mathcal{D})$, which all achieve the required performance with all models in $\mathcal{D}$; the new controller K is then chosen within this class in such a way as to have some additional nice features (*e.g.*, low complexity).

The goal of the new scenario is to achieve the same or better performance based on models of lower complexity. In addition, the class of controllers $\mathcal{K}$ that achieve the required performance is larger because the model uncertainty set $\mathcal{D}$ is tuned towards that aim. All in all, the same or better performance is achieved with a controller that is easier to compute and of lower complexity than is possible with the traditional scenario.

The players within this (iterative) identification and robust control design scenario are therefore:

- the unknown plant G_0;
- the present controller K_{id} (if any);
- the present model G_{init} (if any);
- the identified model G_{mod};
- the uncertainty set of models $\mathcal{D}$ around G_{mod};
- the controller set $\mathcal{K}$ of controllers that achieve the prescribed performance;
- the new controller $K \in \mathcal{K}$.

Except for the unknown plant, the identification and control designer has some handle on all other players. It is the complexity of the interplay between all these players that makes the problem so challenging and interesting. A lot of progress has been accomplished and a lot of new insights gained, but it is no wonder that it has taken a decade so far to understand and conquer all the stumbling blocks along the way.

1.3 Some Important New Insights

The study of the interplay between the different players has led to some important new insights, and to significant progress on some key technical issues.

1.3.1 High-performance Control with Low-order Models

Experience shows that simple models often lead to high-performance controllers on complex processes. To illustrate this point, let us mention the modelling, identification and control application of the Philips compact disc (CD) player, taken from [47].

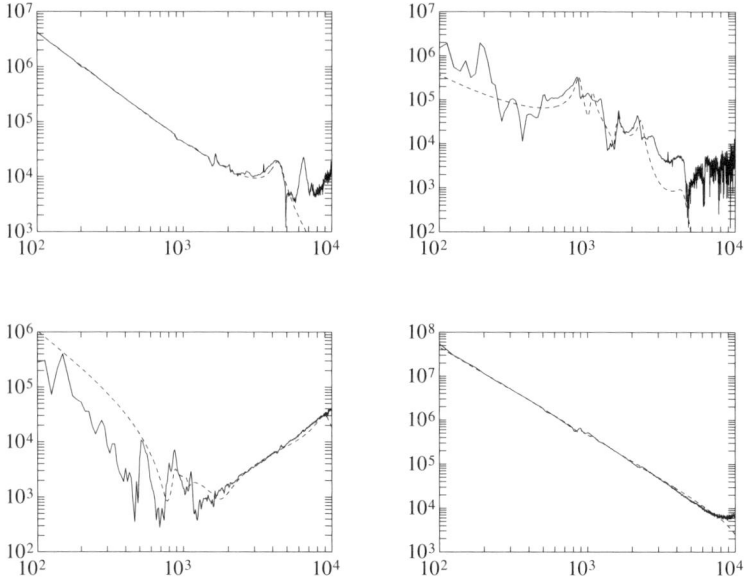

Fig. 1.1. Amplitude of spectral estimate (—) and of parametric model (- -)

Following the track on a CD involves two control loops. A first permanent magnet/coil system mounted on the radial arm positions the laser spot in the direction orthogonal to the track. A second permanent magnet/coil system controls an objective lens, which focuses the laser spot on the disc. The control system therefore consists of a 2-input/2-output system, with the spot position errors (in both radial and focus directions) as the variables to be controlled, and the currents applied to the magnet/coil actuators as the control variables. The modelling of this system using finite element methods or its estimation using spectral analysis techniques would lead to a 2-input/2-output model whose McMillan degree would be of the order of 150. However, by using an identification for control design criterion, a sixteenth order model has been identified that leads to excellent control performance: see [47] for details. A comparison between the spectral estimates and the identified models for the four input-output channels is presented in Figure 1.1. The spectral estimates have been obtained by taking 100 averages over 409600 time samples. The parametric models have been identified using 2000 closed-loop data samples.

A controller of degree 150, say, based on the full-order model obtained by modelling techniques, would be practically useless. Instead, data-based control-oriented identification has led to an approximate (and simplified) model, and to a reduced order controller that gives high performance. This application will be further discussed in detail in chapter 11.

1.3.2 The Role of Changing Experimental Conditions

It is now well recognised that one can find two plants whose Nyquist diagrams are practically indistinguishable, and a controller for the two plants for which the two closed-loop behaviours are enormously different. Thus, the quality of a plant model in a closed-loop setting can only be assessed in conjunction with the controller with which it must operate. This was beautifully pointed out in [141], where the following two principles were laid out.

Modelling Principle 1: arbitrarily small modelling errors can lead to arbitrarily bad closed-loop performance. The higher the performance sought of the controller, the more readily this phenomenon occurs. We illustrate this with the following example. Consider the two transfer functions

$$G_1(s) = \frac{1}{s+1} \quad \text{and} \quad G_2(s) = \frac{1}{(s+1)(0.1s+1)} \tag{1.3}$$

The left hand side of Figure 1.2 compares their open-loop step responses, while the right hand side compares the closed-loop step responses with a proportional feedback controller for two different values of the constant gain: $K = 1$ and $K = 100$.

Modelling Principle 2: larger open-loop modelling errors do not necessarily lead to larger closed-loop modelling errors. Stated otherwise, a very poor open-loop model may yield excellent matching in closed-loop. To illustrate this, we take the same plant $G_1(s)$ as above, and we now consider the model $G_2(s) = \frac{1}{s}$. Figure 1.3 illustrates that, even though G_2 would be rejected by any engineer as a model for G_1, the behaviours of the plants G_1 and G_2, in closed-loop with a proportional controller of gain $K = 100$ are almost indistinguishable.

The issue is that models, and the task of finding them (by modelling or identification), can only have their quality evaluated for a particular set of experimental conditions. Changing from open-loop operation to closed-loop operation with a specific controller is an obvious change of experimental conditions; but so is any change of controller.

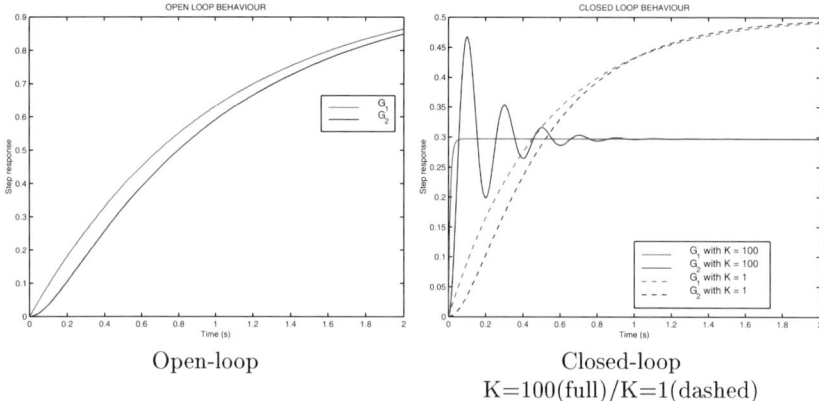

Fig. 1.2. Open loop (left) and closed-loop (right) step responses of G_1 and G_2 with two different controllers

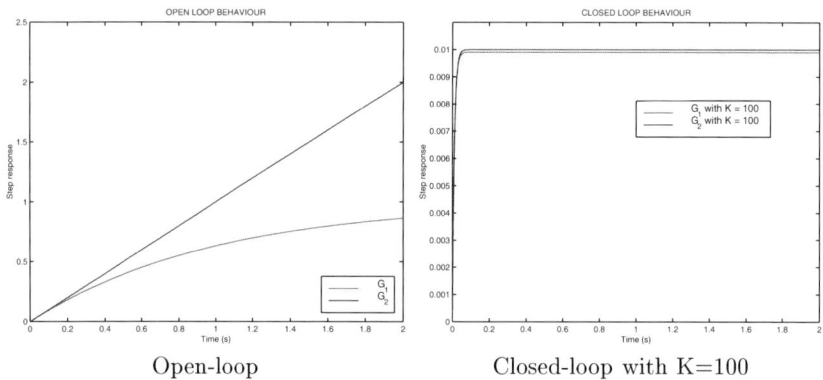

Fig. 1.3. Open-loop (left) and closed-loop (right) step responses of G_1 and $G_2 = \frac{1}{s}$

1.3.3 The Need for Iterative Identification and Control Design

To illustrate the need for iterative design, we take the simplest possible control design objective: model reference control. Thus, consider the true system (1.1) and suppose we have identified a model $\hat{G}(z) = G(z, \hat{\theta})$ of G_0 from some parameterised set of low-order models $\{G(z, \theta)\}$. Consider a control law

$$u_t = K(z)[r_t - y_t] \tag{1.4}$$

and assume that our control design objective is to design $K(z)$ such that the closed-loop transfer function from v_t to y_t is some pre-specified $S(z)$. Then, given a model $\hat{G}(z)$, the controller $K(z)$ is computed from[2]

$$\frac{1}{1+\hat{G}(z)K(z)} = S(z) \tag{1.5}$$

Compare the real closed-loop system of Figure 1.4 with the designed closed-loop system of Figure 1.5, with both loops driven by the same reference signal r_t.

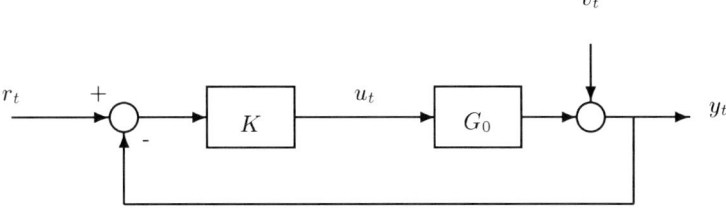

Fig. 1.4. Actual closed-loop system

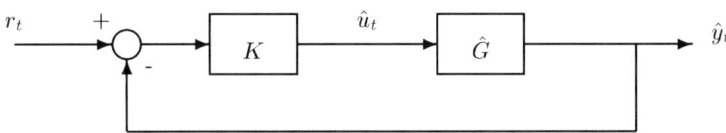

Fig. 1.5. Designed (or nominal) closed-loop system

Now, from Figures 1.4 and 1.5, one observes that:

$$y_t = \frac{G_0 K}{1+G_0 K} r_t + \frac{1}{1+G_0 K} v_t, \quad u_t = \frac{K}{1+G_0 K} r_t - \frac{K}{1+G_0 K} v_t,$$

$$\hat{y}_t = \frac{\hat{G}K}{1+\hat{G}K} r_t \tag{1.6}$$

The "control performance error" is defined as the error between the actual and the designed outputs:

$$y_t - \hat{y}_t = \left[\frac{G_0 K}{1+G_0 K} - \frac{\hat{G}K}{1+\hat{G}K}\right] r_t + \frac{1}{1+G_0 K} v_t \tag{1.7}$$

As observed in [9], this error can be rewritten as

$$y_t - \hat{y}_t = S(z)[y_t - G(z,\hat{\theta})u_t] \tag{1.8}$$

---

[2] We assume for simplicity here that a causal solution exists for $K(z)$, since this is not the focal point of our discussion.

Equation (1.8) can be seen as an equality between a control performance error on the left hand side (LHS) and a filtered identification prediction error (for an output error model) on the right hand side (RHS). Thus, it appears that if θ is obtained by minimising the mean square of the RHS of (1.8), *i.e.*, by closed-loop identification with a filter $S(z)$, then this will minimise the mean square control performance error. In other words, apparently there is a perfect match between control error and identification error. However, the controller $K(z)$ is also a function of the model parameter vector θ via (1.5). Since the data collected on the real closed-loop system of Figure 1.4 are a function of $K(z)$, they are also dependent on θ. Hence, a more meaningful and correct way to write (1.8) is as follows:

$$y_t - \hat{y}_t = S(z)[y_t(\theta) - G(z,\theta)u_t(\theta)] \tag{1.9}$$

Even though the RHS of (1.9) looks like a closed-loop prediction error, it cannot be minimised by standard identification techniques, because θ appears everywhere and not just in $G(z,\theta)$.

As a consequence, the approach suggested in most "identification for control" schemes is to perform identification and control design steps in an iterative way, whereby the i-th identification step is performed on filtered closed-loop data collected on the actual closed-loop system with the $(i-1)$-th controller operating in the loop, *i.e.*,

$$y_t - \hat{y}_t = S(z, \hat{\theta}_{i-1})[y_t(\hat{\theta}_{i-1}) - G(z,\theta)u_t(\hat{\theta}_{i-1})] \tag{1.10}$$

Although other variants exist, a typical iterative scenario is therefore as follows:

$$\hat{G}_1 \to K_1 \to \hat{G}_2 \to K_2 \to \ldots \to \hat{G}_i \to K_i \to \hat{G}_{i+1} \to K_{i+1} \to \ldots$$

We refer the reader to [60], [23] and [148] for details and for a survey on such iterative schemes.

An interesting question is whether by iteratively minimising over θ the mean square of the prediction errors defined by (1.10), one will converge to the minimum of

$$J(\theta) \triangleq E\{S(z,\theta)[y_t(\theta) - G(z,\theta)u_t(\theta)]\}^2 \tag{1.11}$$

This question was analysed in [79], where it was shown that the iterative identification and control schemes do not generically converge to the achievable minimum (within the model/controller set) of the control performance cost.

Despite this disappointing news from a theoretical point of view, the concept of iterative identification and control design was rapidly adopted in applications, in particular in process control applications. Representative examples can be found in [41, 47, 81, 125, 138]. One reason is that it is typical in such applications that large numbers of closed-loop data are flowing into the control computer, and it then makes a lot of sense to use these data to replace the existing controller by one that achieves better performance. The practical impact of iterative closed-loop identification and controller redesign has been assessed in [99], where some interesting observations are made on the distinction between this batch-like mode of operation and the more classical theory and methods of adaptive control.

1.3.4 The Need for Cautious Adjustments

As a result of the absence of convergence guarantees of the iterative schemes to a minimum of the control performance criterion, the procedure may well converge to a controller that makes the actual closed-loop system unstable. In addition, even in a situation where the procedure does converge to a minimum of the cost criterion, nothing guarantees that along the way the controller parameter vector will not take a transient value that destabilises the true system. In order to circumvent these difficulties, a lot of work has been devoted to the search for prior stability guarantees that can be checked before the new controller is implemented. The generic situation is as follows.

At some stage of the iterations a controller K_i, which was computed from a model $\hat{G}_i$, is operating on the true system G_0. This controller stabilises G_0 and achieves with G_0 a performance $J(G_0, K_i)$. With data collected on the closed-loop system (G_0, K), a new model $\hat{G}_{i+1}$ is identified, from which a new controller K_{i+1} is computed, which has a nominal performance $J(\hat{G}_{i+1}, K_{i+1})$. Before the new controller is actually applied to the plant G_0, one would like to have some prior guarantee that this new controller will stabilise and achieve a better performance with G_0. In the last few years, a significant amount of work has addressed this problem, either from a robust stability point of view, or from a robust performance point of view. Most of these results have led to the need to introduce some "caution" in the iterations, in that some measure of distance between the present and the new controller must be kept small. The origin of this need for caution is to be found in the observations made above about *the role of the changing experimental conditions*. As for the technical nature of these results, let it suffice here to say that the bound on the admissible change between two successive controllers is related to the distance between the successive closed-loop systems and to their corresponding performances. For details, see later chapters in this book or [6, 24, 46, 137].

1.3.5 The Benefits of Closed-loop Identification

One of the important lessons that has emerged from the study of the interplay between identification and control is the benefit of closed-loop identification when the model is to be used for control design. Until the late 1980s, it was commonly accepted within the identification community that closed-loop identification was preferably to be avoided. Optimality of closed-loop identification was first shown in the ideal context of optimal experiment design with full-order models (*i.e.*, where variance errors only are considered) [58, 63, 77], at least when the optimal controller contains a noise rejection objective. In the more practical case of identification with reduced order models, it is the required connection between the control performance criterion (obviously a closed-loop criterion) and the identification criterion, as described above, that establishes the need for closed-loop identification.

This observation triggered an important new activity in the design of special purpose closed-loop identification methods, the main goal pursued by these new methods being to obtain a better handle on the bias error in closed-loop identification [57, 73, 147, 150].

From a practical point of view, the newly-discovered benefits of closed-loop identification methods and the development of some of these special purpose methods came as welcome news to process control engineers who had never really liked the idea of opening the loop and applying special test signals to their systems in order to identify them.

Acknowledgements

The work reported herein is the result of collaborations with a large number of co-authors over the last ten years. I would like to acknowledge B.D.O. Anderson, P. Ansay, R.R. Bitmead, X. Bombois, B. Codrons, F. De Bruyne, G.C. Goodwin, S. Gunnarsson, H. Hjalmarsson, L. Johnston, C. Kulcsar, J. Leblond, O. Lequin, L. Ljung, B. Ninness, A.G. Partanen, M.A. Poubelle, G. Scorletti, L. Triest, P.M.J. Van den Hof, V. Wertz, and Z. Zang. Special thanks go to B.D.O. Anderson, X. Bombois and H. Hjalmarsson from whom I have learned so much.

This chapter presents research results of the Belgian Programme on Interuniversity Poles of Attraction, initiated by the Belgian State, Prime Minister's Office for Science, Technology and Culture. The scientific responsibility rests with its authors.

2 System Identification. Performance and Closed-loop Issues

Jesús Picó and Miguel Martínez

Abstract. Many of the most fruitful control design methods lie in the availability of a process model. Obtaining this model becomes, in many practical, cases a non-trivial task. Physical insight into the process may provide invaluable help. Yet, eventually, experimental identification has to be performed at some stage, either in open-loop or in closed-loop. A bunch of methods exist in both cases, each one with specific characteristics that make them more or less suitable for the given case. In this chapter, some of the most common parametric models and methods to identify their parameters are reviewed. Special emphasis is put on showing what can be achieved in practice. For the sake of simplicity, only SISO linear models have been considered.

2.1 Chapter Outline

The remainder of the chapter is organised as follows. After some general remarks on the definition of modelling, estimation and identification in Section 2.2, Section 2.3 is devoted to introducing parametric model structures based on discrete transfer functions. The implementation of one-step ahead process output predictors using these models is addressed in Section 2.4. Section 2.5 introduces the use of these predictors within an optimisation framework, to estimate the model parameters. Theoretical limits of performance are commented on. The practical implementation of the estimation algorithms, using derivative-based optimisation methods, is addressed in Section 2.6. Closed-loop identification is introduced in Section 2.7. Sections 2.8 and 2.9 are devoted to reviewing closed-loop identification methods relying on a conceptually open-loop approach. Finally, Sections 2.10 and 2.11 introduce methods based purely on a closed-loop approach.

2.2 Modelling, Estimation and Identification

A model is an abstract description of a system, relevant for some purpose, and consistent within a given experimental framework. Depending on the purpose (*e.g.*, prediction or control of a particular variable under given specifications over a given range), the model will take one out of several possible structures, and will focus on the relevant system feature to be described, disregarding others and, therefore, assuming different approximations and simplifications.

This translates into the desire that the modelled variable of interest y_m be approximately equal to the actual one y for inputs $u \in \mathcal{U}$, $\mathcal{U}$ being a valid set of input signals (see Figure 2.1), for instance, $\mathcal{U}$ being signals with spectrum limited under a given frequency or being signals of low amplitude.

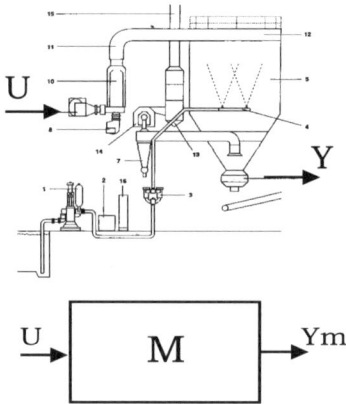

Fig. 2.1. System and model

The model can be obtained using different approaches:

- *physical modelling* (using natural laws) leads to so-called "grey box" models, where the model structure derives from the laws of behaviour implied, and only the values of the model parameters usually remain to be identified [29];
- *experimental modelling*, where minimum *a priori* knowledge about the process is assumed, leads to "black box" models. Two sub-problems arise:
  1. Obtain (estimate) the model structure.
  2. Obtain (identify) the model parameters.

Following [108, 109], in the identification problem we try to infer relationships between past process input-output data and future outputs. So, if there is an available observations vector $\phi(t)$ that is a function of a finite number of past process inputs and/or outputs, its relationship with the process output $y(t)$ is sought, where

$$y(t) = g_0(\phi(t)) \tag{2.1}$$

is assumed. A set of collected data $Z^N = \{[y(t), \phi(t)], t = 1 \ldots N\}$ (see Figure 2.2) is then used to estimate

$$\hat{y}(t) = \hat{g}_N(\phi(t)) \tag{2.2}$$

so that $\hat{g}_N$ is a good approximation of the actual mapping g_0.

2 System Identification. Performance and Closed-loop Issues   19

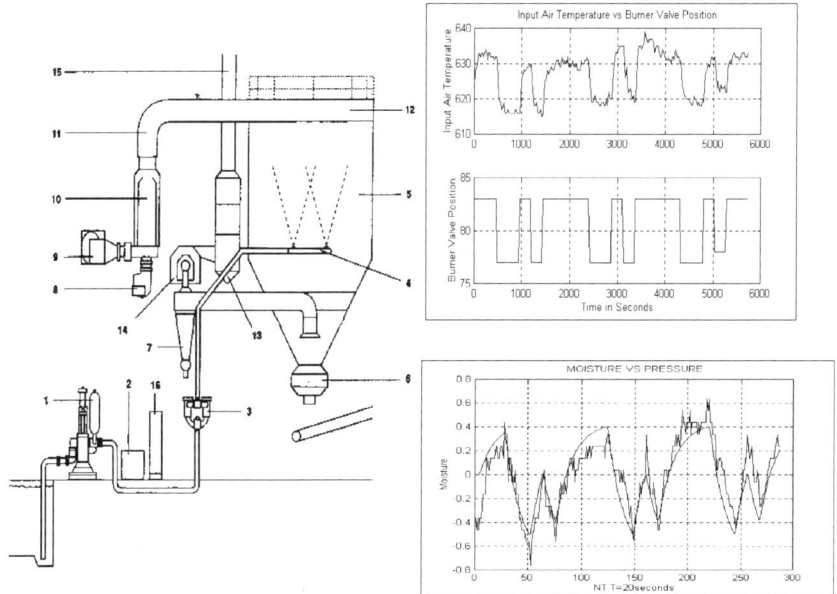

Fig. 2.2. Output samples for a real process

If the problem is approached from a black-box point of view, the identification problem consists of finding appropriate function expansions that will be used to build up $\hat{g}_N$.

2.3 Model Structures

The general framework considered here assumes that the real process $P(G, H)$ is described by:

$$y(t) = G(q)u(t) + v(t)$$
$$v(t) = H(q)e(t) \tag{2.3}$$

as depicted in Figure 2.3. Therefore, it is considered that what can be observed at the process output is the superposition of the unperturbed process output, and a perturbation (which, in general, cannot be measured). This perturbation is actually the outcome of all the internal and external perturbations generated in the system, or entering the system at some point, and affecting its dynamics from that point to the output. Most often, it is assumed (for the sake of modelling) that this perturbation is equivalent to a fictitious external white noise input signal $e(t)$ passing through a filter $H(q)$. This assumption is quite consistent with the kind of stochastic perturbations that usually affect the real processes.

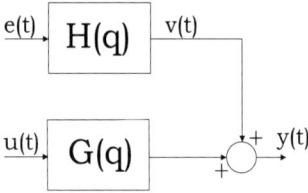

Fig. 2.3. Signals processing scheme for a real system

Deterministic perturbations (*e.g.*, a sudden change of an engine load, modelled as a step-like perturbation, 50 Hz interference from the supply net, effects of periodic batch-like incoming of material in a rolling mill, ...) are considered aside.

For modelling purposes, a family $\mathcal{M}(\theta) = \{\mathcal{G}(\theta), \mathcal{H}(\theta)\}$ of parameterised models is considered:

$$y_m(t) = G(q, \theta)u(t) + v_m(t)$$
$$v_m(t) = H(q, \theta)e(t) \qquad (2.4)$$

In the best case, it is expected that there exists a value θ_p of the model parameters such that $G(q) = G(q, \theta_p)$ (that is, $G(q) \in \mathcal{G}$) and $H(q) = H(q, \theta_p)$ ($H(q) \in \mathcal{H}$). The filter $H(q)$ is assumed to have direct input-output coupling. That is, $H(q) = 1 + \bar{H}$, where $\bar{H}$ is strictly proper. This general structure admits simplified variants, with less representational capability, but better conditioned for parameters tuning with less computational cost and better convergence rate. These variants try to capture the fact that the internal structure of $G(q, \theta)$ and $H(q, \theta)$ depends on the point through which it is considered that the perturbation affects the system. That is, where it enters or where it is internally generated. Thus:

1. If the perturbations enter, or are generated, at the process input, it is sensible to consider that the dynamics of the perturbation system H are the same as those of the unperturbed system G. Two alternatives may be considered:

   (a) The perturbation system H has no zeros. Physically, this usually corresponds to processes in which the perturbation at their output has a very low content in high frequencies (because of internal inertia, most physical systems behave as low-pass filters). The resulting model structure, depicted in Figure 2.4(a), is called *AutoRegresssive eXogenous* (ARX):

   $$A(q^{-1})y(t) = q^{-d}B(q^{-1})u(t) + e(t)$$

   (b) The perturbation system has zeros. This allows modelling of systems in which the perturbation "observed" at the output has a content in high frequencies bigger than that of the unperturbed system output (the zeros amplify the high-frequency signals), or systems where the output perturbation has a frequency content limited to medium and/or high frequencies (*e.g.*,between 1 kHz and 10 kHz). The resulting model structure, depicted in Figure 2.4(b), is called *AutoRegressive Moving Average eXogenous* (ARMAX):

$$A(q^{-1})y(t) = q^{-d}B(q^{-1})u(t) + C(q^{-1})e(t)$$

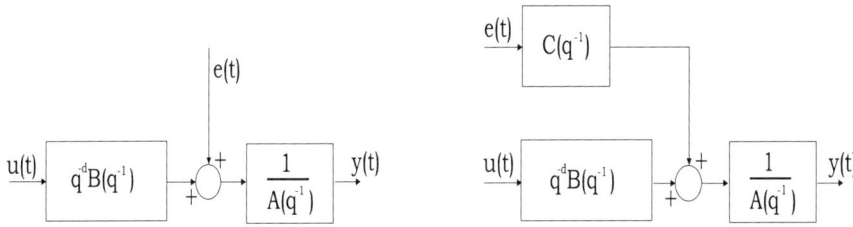

Fig. 2.4. (a) Scheme for an ARX model (b) Scheme for an ARMAX model

2. If the perturbation enters or originates at intermediate points of the process or at its output, it is sensible to consider different dynamics (at least partially) for the unperturbed system and the perturbation system. Two different situations may also be considered in this case:

   (a) The perturbation $e(t)$ acts directly upon the unperturbed process output. The signal $e(t)$ does not correspond necessarily to a white noise. The resulting structure is called the *output error* (OE) model (see Figure 2.5(a)). It is useful when the modelling focus is on the unperturbed process, and one is not interested in modelling the perturbation characteristics.

   $$y(t) = q^{-d}\frac{B(q^{-1})}{A(q^{-1})}u(t) + e(t)$$

   (b) If, for the perturbation, an autoregressive moving average model with dynamics different to that of the unperturbed system is considered, the resulting composite model is the so-called *Box-Jenkins* one, depicted in Figure 2.5(b):

   $$y(t) = q^{-d}\frac{B(q^{-1})}{A(q^{-1})}u(t) + \frac{C(q^{-1})}{D(q^{-1})}e(t)$$

Other models, resulting from mixing ideas extracted from the ones above, are also possible. Thus, for instance, to consider that the process input enters at some point before the noise and, hence, the signals $u(t)$ and

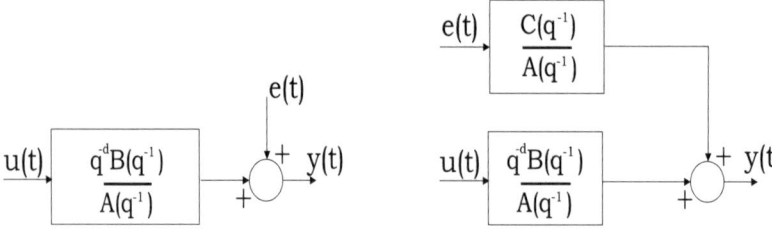

Fig. 2.5. (a) Output error model (b) Box-Jenkins model (BJ)

$e(t)$ excite partially common dynamics, the following structure can be assumed:

$$A(q^{-1})y(t) = q^{-d}\frac{B(q^{-1})}{F(q^{-1})}u(t) + C(q^{-1})e(t)$$

The basic ideas, shown for a SISO case, still hold for MIMO systems.

2.4 Implementation of the Model Structures: Predictors

The previous model structures can be used to construct a predictor that allows prediction of the process output. Later on, the prediction error will be used to determine the direction of parameter tuning. Next, the problem of implementing such a predictor is addressed.

Since $e(t)$ cannot be predicted (it is the source of variability, not correlated with any signal up to time t), then the unpredictable part of the process output signal $y(t)$ is just $e(t)$. Hence, noticing that

$$e(t) = H^{-1}(q)y(t) - H^{-1}(q)G(q)u(t)$$

the predictable part is

$$y(t) - e(t) = \left[1 - H^{-1}(q)\right]y(t) + H^{-1}(q)G(q)u(t)$$

Therefore, given data up to time $(t-1)$, the following predictor for $y(t)$ can be considered

$$\hat{y}(t \mid t-1, \theta) = \left[1 - H^{-1}(q, \theta)\right]y(t) + H^{-1}(q, \theta)G(q, \theta)u(t) \qquad (2.5)$$

where θ is a vector of parameters to be identified.

Once the structure of the predictor is clear, the next question is: how is the predictor implemented? For the models considered in the section above, the following results are obtained:

1. OE model:

$$H = 1$$
$$G = \frac{q^{-d}B(q^{-1})}{A(q^{-1})}$$
$$\hat{y}(t) = -a_1\hat{y}(t-1) - \ldots - a_n\hat{y}(t-n)$$
$$+ b_1 u(t-1-d) + \ldots + b_m u(t-m-d) \quad (2.6)$$

As easily seen, this predictor is not linear with respect to the parameters if these vary with t, because the predicted output depends on past predicted outputs. This fact will have practical implications when devising the algorithms to identify these parameters from experimental data.

2. ARX model:

$$H = \frac{1}{A(q^{-1})}$$
$$G = \frac{q^{-d}B(q^{-1})}{A(q^{-1})}$$
$$\hat{y}(t) = \left[1 - A(q^{-1})\right] y(t) + q^{-d}B(q^{-1})u(t)$$
$$\hat{y}(t) = -a_1 y(t-1) - \ldots - a_n y(t-n)$$
$$+ b_1 u(t-1-d) + \ldots + b_m u(t-m-d) \quad (2.7)$$

In this case, the predictor output is linear with respect to the parameters, even if these vary with t. Models with this characteristic are often referred to as *linear regressive*.

3. ARMAX model:

$$H = \frac{C(q^{-1})}{A(q^{-1})}$$
$$G = \frac{q^{-d}B(q^{-1})}{A(q^{-1})}$$
$$C(q^{-1})\hat{y}(t) = \left[C(q^{-1}) - A(q^{-1})\right] y(t) + q^{-d}B(q^{-1})u(t)$$
$$\hat{y}(t) = -c_1\hat{y}(t-1) - \ldots - c_c\hat{y}(t-c) + c_1 y(t-1) + \ldots$$
$$\ldots + c_c y(t-c) - a_1 y(t-1) - \ldots - a_n y(t-n)$$
$$+ b_1 u(t-1-d) + \ldots + b_m u(t-m-d) \quad (2.8)$$

As with the OE model, the predictor output is not a linear regression unless the parameters do not vary with time. Nevertheless, for slowly varying parameters, the model can be considered to be approximately linear, giving rise to a so-called *pseudolinear* regression.

4. Box-Jenkins model:

$$H = \frac{C(q^{-1})}{D(q^{-1})}$$

$$G = \frac{q^{-d}B(q^{-1})}{A(q^{-1})}$$

$$C(q^{-1})\hat{y}(t) = [C(q^{-1}) - D(q^{-1})]y(t) + q^{-d}\frac{D(q^{-1})B(q^{-1})}{A(q^{-1})}u(t)$$

$$y_u(t) \stackrel{\text{def}}{=} -a_1 y_u(t-1) - \ldots - a_n y_u(t-n) + u_f(t)$$

$$u_f(t) \stackrel{\text{def}}{=} b_1 u(t-1) \ldots + b_m u(t-m) + d_1 b_1 u(t-2) + \ldots$$
$$\ldots + d_1 b_m u(t-m-1) + \ldots + d_D b_1 u(t-D-1) + \ldots$$
$$\ldots + d_D b_m u(t-D-m)$$

$$\hat{y}(t) = -c_1 \hat{y}(t-1) - \ldots - c_c \hat{y}(t-c) + c_1 y(t-1) + \ldots$$
$$\ldots + c_c y(t-c) - d_1 y(t-1) - \ldots$$
$$\ldots - d_D y(t-D) + y_u(t) \qquad (2.9)$$

Its generality is an appealing characteristic, though at the price of having a complex predictor.

As seen, the signals appearing in the predictors are:

1. Past inputs, $u(t-k)$.
2. Past measured outputs, $y(t-k)$.
3. Past predicted outputs, $\hat{y}(t-k|\theta)$.
4. Past simulated outputs (generated using past inputs only), $\hat{y}(t-k|\theta)$

Therefore, the questions "How to select $\{g_n(\phi)\}$?", and "What variables should be chosen as regressors ($\phi(t)$)?" have been addressed. With respect to a third important question, "Which identification model structure do we select?", each model structure has its advantages and drawbacks. Thus, for instance:

1. ARX: $u(t-k), y(t-k)$. Also named the *series-parallel model*, is one of the most often used. In some cases, the model can become unstable even if the actual process is stable.

2. OE: $u(t-k), \hat{y}(t-k|\theta)$. Also called the *parallel model*, the use of past simulated outputs leads to recurrent configurations. If the actual process is stable, the model will also be stable.

3. ARMAX: $u(t-k), y(t-k), \hat{y}(t-k|\theta)$ includes past error predictions (and, hence, past predicted outputs). Allows for a good approximation of the perturbations, with a low number of regressors.

2.5 Open-loop Identification. What can be Achieved?

The problem of tuning the model parameters is addressed now, assuming that process data is obtained in open-loop. Given a real process

$$y(t) = G(q)u(t) + v(t)$$
$$v(t) = H(q)e(t)$$

for which a model

$$y(t) = G(q,\theta)u(t) + v(t)$$
$$v(t) = H(q,\theta)e(t)$$

is considered, the process output predictor (2.5), parameterised as

$$\hat{y}(t \mid t-1, \theta) = \left[1 - H^{-1}(q,\theta)\right] y(t) + H^{-1}(q,\theta) G(q,\theta) u(t) \qquad (2.10)$$

can be used to define the prediction error:

$$\epsilon(t,\theta) = y(t) - \hat{y}(t,\theta) = H^{-1}(q,\theta) \left[y(t) - G(q,\theta)u(t)\right] \qquad (2.11)$$
$$= H^{-1}(q,\theta) \left[(G(q) - G(q,\theta)) u(t) + v(t)\right] \qquad (2.12)$$

As seen in (2.12), the prediction error depends both on the perturbation $e(t)$, and on the mismatch between the true plant $P(q)$ and the modelled one $P(q,\theta)$. The prediction error can be filtered, using a filter $L(q,\theta)$, to enhance those frequency regions of interest.

$$\epsilon_f(t,\theta) = L(q,\theta)\epsilon(t,\theta) \qquad (2.13)$$

Now, given the set of experimental data

$$Z^N = \{[y(t), u(t)], t = 1 \ldots N\}$$

and the cost function

$$V_N\left(\theta, Z^N\right) = \frac{1}{N} \sum_{t=1}^{N} \epsilon_f^2(t,\theta) \qquad (2.14)$$

the goal is to obtain the set of parameters that minimise it, that is:

$$\hat{\theta} \stackrel{\text{def}}{=} \arg\min_{\theta} V_N(\theta, Z^N) \qquad (2.15)$$

Other cost functions different from (2.14) may be considered, for instance, the *maximum likelihood* [108].

Defining the optimal parameters as the ones minimising the average of the cost function calculated over infinitely many realisations:

$$\theta^\star \stackrel{\text{def}}{=} \arg\min_{\theta} E\left[\epsilon_f^2(t,\theta)\right] \qquad (2.16)$$

the difference between the true process and the estimated one can be written as:

$$G(q) - G(q,\hat{\theta}) = \{G(q) - G(q,\theta^\star)\} + \{G(q,\theta^\star) - G(q,\hat{\theta})\} \quad (2.17)$$

The first term on the right hand side describes the difference between the process and the best approximation of the process that can be obtained with the current model if the optimal parameters are found. Therefore, this approximation error corresponds to the *modelling error*, also called *bias*. The second term corresponds to the difference between the optimal and the current estimated model. Therefore, it corresponds to the *estimation error*.

The goal of the parameter estimation algorithms is to minimise the left hand side of the equality (2.17). Yet, in presence of modelling error (*i.e.*, unmodelled dynamics), this does not imply that the estimation error is minimised. Moreover, if the model parameters are estimated on-line, the estimation algorithm may not converge [14]. Robust identification techniques are used in this case [86].

Over infinitely many realisations of (2.15), different values of $\hat{\theta}_N$ are obtained. The variance of the estimates will depend on the number n of parameters to be estimated, the amount N of experimental data used in the estimation, the power spectrum of the noise v, and that of the input signals (u, e). The variance of the estimated $G(q,\hat{\theta})$ and $H(q,\hat{\theta})$ is given by [108]:

$$cov G(q,\hat{\theta}) \approx \frac{n}{N} \frac{\Phi_v}{\Phi_u} \quad (2.18)$$

$$cov H(q,\hat{\theta}) \approx \frac{n}{N} \frac{\Phi_v}{\sigma^2} \quad (2.19)$$

Thus, increasing the number of parameters decreases the bias, since the model has more representational capability, but increases the variance. On the other hand, the more experimental data the better if the complexity of the model allows adjustment of them.

On the other hand, applying Parseval's formula, the following expression can be obtained for $\theta^\star$:

$$\theta^\star = \arg\min_{\theta} \int_{-\pi}^{\pi} \frac{|L(q,\theta)|^2}{|H(q,\theta)|^2} \left\{ |G(q) - G(q,\theta)|^2 \Phi_u + \Phi_v \right\} \bigg|_{q=e^{j\omega}} d\omega \quad (2.20)$$

The expression (2.20) allows us to analyse under what conditions the bias term in (2.17) will vanish. Thus, denoting

$$\mathcal{G} \stackrel{\text{def}}{=} \bigcup_{\theta} \{G(q,\theta)\}$$

$$\mathcal{H} \stackrel{\text{def}}{=} \bigcup_{\theta} \{H(q,\theta)\}$$

the following cases arise:

1. If $G \in \mathcal{G}$ and $H \in \mathcal{H}$, there exists a value $\theta^\star$ such that $G = G(\theta^\star)$ and $H = H(\theta^\star)$.

2. If $G \in \mathcal{G}$ and $H \notin \mathcal{H}$ then there will be a bias, of which the magnitude will depend on that of the noise v. Thus, if Φ_v is high, the optimal value $\theta^\star$ will result from a trade-off between getting a suitable value for the noise model estimate $|H(q,\theta)|$ and a low value for $|G - G(q,\theta)|$.

3. If $G \in \mathcal{G}$ and $H \notin \mathcal{H}$, but the perturbation model $H(q,\xi)$ does not depend on the parameters θ of the process model, then a set of parameters $(\theta^\star,\xi^\star)$ can be obtained such that there is no bias for the estimation of G.

4. If $G \notin \mathcal{G}$, there will obviously be a bias.

2.6 Open-loop Identification Algorithms

The actual implementation of the algorithms devised to optimise the cost index (2.14) may introduce additional requirements or limitations to what can be achieved. Most of the optimisation methods used are based on a local iterative search in a downhill direction of (2.14) in the parameter space. Thus, the general expression for the estimated parameters is:

$$\hat{\theta}^{(i+1)} = \hat{\theta}^{(i)} - \mu_i R_i^{-1} \Delta \hat{f}_i \qquad (2.21)$$

where μ_i is a step size, $\Delta \hat{f}_i$ the current estimation of the gradient of V_N with respect to the parameters θ, and R_i is a matrix modifiying the search direction, typically some approximation of the hessian of V_N.

Determining how the model output varies with the parameters is the key point for the actual calculation of these terms. Eventually, most of the estimation algorithms can be cast under the same basic scheme. Thus, for recursive algorithms, the most common expressions are [108]:

$$\hat{\theta}^{(i+1)} = \hat{\theta}^{(i)} + \frac{P_i \varphi(i+1)}{1 + \varphi^T(i+1) P_i \varphi(i+1)} \left(y(i+1) - \hat{y}(i+1) \right) \qquad (2.22)$$

$$P_i^{-1} = \lambda_1 P_{i-1}^{-1} + \lambda_2 \varphi(i) \varphi^T(i) \qquad (2.23)$$

where $P_i \stackrel{\text{def}}{=} R_i^{-1}$ is the inverse of a data variance-covariance matrix, which defines the Gauss-Newton approximation of the hessian of V_N, $\varphi(t)$ is a vector with some estimation of the derivative of the process model with respect to its parameters, and λ_1, λ_2 are weighting terms.

In the next sections, some of the usual issues arising in the implementation of estimators for the most common models are addressed.

2.6.1 Estimation of an OE Model

Recall that for an OE model, its output is given by:

$$\hat{y}(t) = -a_1\hat{y}(t-1) - \ldots - a_n\hat{y}(t-n)$$
$$+ b_1 u(t-1-d) + \ldots + b_m u(t-m-d) \quad (2.24)$$

Since the model is recursive, a variation in any of its parameters will produce a direct variation at the model output, and an indirect one through the internal feedback. Thus, the corresponding derivatives used to obtain P_i and $\varphi(i)$ in (2.22)–(2.23) become:

$$\frac{\partial \hat{y}(t)}{\partial a_i} = -\frac{1}{A(q)} \hat{y}(t-i)$$
$$\frac{\partial \hat{y}(t)}{\partial b_i} = \frac{1}{A(q)} u(t-i-d) \quad (2.25)$$

Depending on whether a recursive or a non-recursive implementation of the estimation algorithm is sought, two possibilities arise for the predictor.

Given data up to time t, and having the parameters estimated in $(t-1)$, the model output for the recursive predictor becomes:

$$\hat{y}(t) = -\hat{a}_1(t-1)\hat{y}(t-1) - \ldots - \hat{a}_n(t-1)\hat{y}(t-n)$$
$$+ \hat{b}_1(t-1)u(t-1-d) + \ldots + \hat{b}_m(t-1)u(t-m-d) \quad (2.26)$$

Therefore, a time-varying predictor is the outcome. Then, the derivatives (2.25) are implemented using the current estimate of $A(q, t-1)$. In this case, the estimate of the derivative of V_N in the parameter space is built only from local data. Therefore, it is an approximation of the result achievable using the whole data set.

On the other hand, in a non-recursive implementation, the estimates are updated after each processing of all the input-output data set. In this case, the predictor and derivatives (2.25) are implemented using the estimates of the previous algorithm iteration.

If an optimisation of the cost function V_N was carried out with:

$$\theta^* = \arg\min_\theta \int_{-\pi}^{\pi} \frac{|L(q,\theta)|^2}{|H(q)|^2} \left\{ |G - G(q,\theta)|^2 \Phi_u + \Phi_v \right\} \Big|_{q=e^{j\omega}} d\omega$$

then no bias would appear in G (if $G \in \mathcal{M}$) even if $H \notin \mathcal{M}$.

An algorithm quite often used considers an approximated calculation of the derivatives based on taking into account only the steady state effect of variations of the parameters on the output, and neglecting the transient effect due to the internal feedback. Thus, in this case:

$$\frac{\partial \hat{y}(t)}{\partial a_i} \approx -\frac{\hat{y}(t-i)}{\hat{A}(1)\big|_{t-1}} \qquad \frac{\partial \hat{y}(t)}{\partial b_i} \approx \frac{u(t-i-d)}{\hat{A}(1)\big|_{t-1}} \quad (2.27)$$

In any case, the practical results obtained using any of the practical approaches are very similar in most cases. In Figure 2.6, the results for a first order system, whose output is affected by a coloured noise, are shown as an example. Estimating as if the process had an ARX structure yields clearly biased estimates, while doing so with an OE model does not.

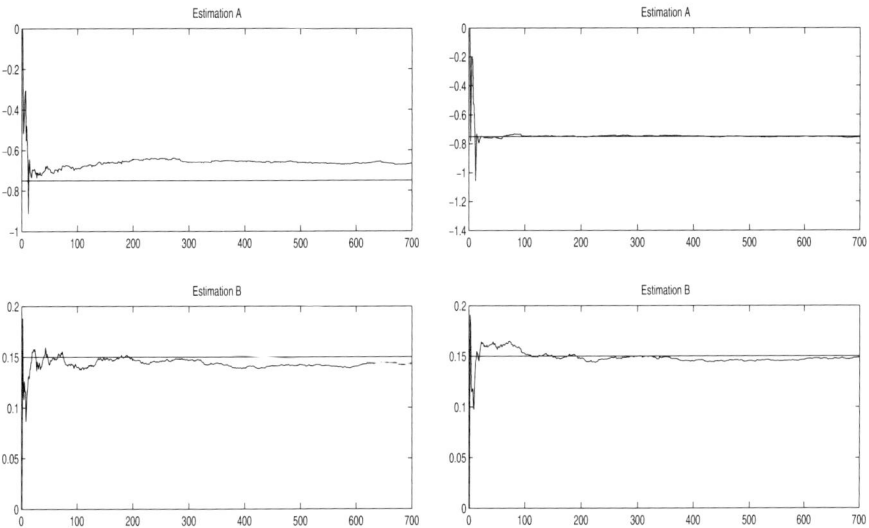

Fig. 2.6. Recursive estimation with (a) ARX model (b) OE model

2.6.2 Estimation of an ARX Model

Recalling that its output is given by:

$$\hat{y}(t) = -a_1 y(t-1) - \ldots - a_n y(t-n) \\ + b_1 u(t-1-d) + \ldots + b_m u(t-m-d) \quad (2.28)$$

The model is a linear regression, fed with input and output signals from the real process. The corresponding derivatives of the model output with respect to its parameters become equal to the regressors:

$$\frac{\partial \hat{y}(t)}{\partial a_i} = -y(t-i) \\ \frac{\partial \hat{y}(t)}{\partial b_i} = u(t-i-d) \quad (2.29)$$

Biased results are obtained if the model noise structure does not fit the real one, as seen in the example in Figure 2.6.

2.6.3 Estimation of an ARMAX Model

For the ARMAX model

$$A(q)y(t) = B(q)u(t) + C(q)e(t) \tag{2.30}$$

the following expressions are derived:

$$\hat{y}(t,\theta) = \left[1 - \frac{A(q,\theta)}{C(q,\theta)}\right] y(t) + \frac{B(q,\theta)}{C(q,\theta)} u(t) \tag{2.31}$$

$$\hat{y}(t) = -c_1 \hat{y}(t-1) - \ldots - c_c \hat{y}(t-c) + c_1 y(t-1) + \ldots$$
$$\ldots + c_c y(t-c) - a_1 y(t-1) - \ldots - a_n y(t-n)$$
$$+ b_1 u(t-1-d) + \ldots + b_m u(t-m-d) \tag{2.32}$$

The predictor is stable if $|C(q,\theta)| < 1$. Therefore, not all possible values for θ are allowed. Only those yielding a stable predictor are valid. To achieve this, some projection techniques are typically introduced in the basic estimation algorithm.

Defining:

$$\phi^T(t) \stackrel{\text{def}}{=} [-y(t-1), \ldots, -y(t-n_a), u(t-1), \ldots, u(t-n_b)] \tag{2.33}$$

$$\phi_f(t) \stackrel{\text{def}}{=} \frac{1}{C(q,\theta)} \phi(t) \tag{2.34}$$

$$y_f(t) \stackrel{\text{def}}{=} \frac{1}{C(q,\theta)} y(t) \tag{2.35}$$

then, the regressive expression

$$y_f(t) = \phi_f^T \theta_{A,B} + \hat{v}_f(t) \tag{2.36}$$

can be used to estimate the parameters of the polynomials A and B. Indeed, if the polynomial $C(q)$ was known, and $C(q,\theta) = C(q)$, the term $\hat{v}_f$ in (2.36) would be equal to the white noise signal $e(t)$. Therefore, Equation (2.36) is a linear regressive expression, and $\theta_{A,B}$ can be estimated by

$$\hat{\theta}_{A,B} = \arg\min_{\theta} \sum_{t=1}^{N} \left(y_f(t) - \phi_f^T(t)\theta_{A,B}\right)^2 \tag{2.37}$$

using, for instance, a Gauss-Newton algorithm [110].

Since $C(q)$ is not known, an estimate $C(q,\theta)$ has to be used to generate the filtered data. Then, from (2.30) and (2.36):

$$\hat{v}_f = \frac{1}{C(q,\theta)} v(t) \tag{2.38}$$

with $v(t) = C(q)e(t)$. Recall that $e(t)$ cannot be measured (in fact, as shown in Section 2.3, it is a fictitious signal useful for modelling the actual perturbations). Yet, the unfiltered residuals of the estimation (2.37)

$$\hat{v}(t) \stackrel{\text{def}}{=} y(t) - \phi^T(t)\theta_{A,B} \qquad (2.39)$$

give a useful approximation. Thus, assuming the approximation

$$\hat{v}(t) \approx C'\hat{v} + e = \psi^T(t)\theta_C + e(t) \qquad (2.40)$$

with

$$\psi(t) = [\hat{v}(t-1), \ldots, \hat{v}(t-n_c)] \qquad (2.41)$$
$$C' = c_1 q^{-1} + \ldots + c_{n_c} q^{-n_c} \qquad (2.42)$$

then, estimate with:

$$\hat{\theta}_C = \arg\min_\theta \sum_{t=1}^N \left(\hat{v}(t) - \psi^T(t)\theta_C\right)^2 \qquad (2.43)$$

and iterate through (2.37) and (2.43).

In this case, even if $G \in \mathcal{G}$ and $H \in \mathcal{H}$, the estimation will provide unbiased results only under the additional condition $\hat{v}(t) \approx C'\hat{v} + e$, which arises from the need to approximate unmeasurable signals for the actual implementation of the estimation algorithm. This condition can be seen to be equivalent to $C(q^{-1}) \approx 1$, and also to the one encountered in the literature [98, 110]:

$$Re\left\{C^{-1}(e^{i\omega}) - \frac{1}{2}\right\} > 0 \qquad (2.44)$$

As an example, the coefficients corresponding to second order polynomials, for which the previous condition holds, and the corresponding roots of the polynomial, are shown in Figure 2.7 in next page.

In Figure 2.8, the identification results for two systems, with only satisfying Condition (2.44), are shown. As clearly seen, the estimation of the coefficients of the polynomials $A(q)$ and $B(q)$ is good in both cases, while the one of $C(q)$ only in the first case.

2.7 Closed-loop Identification

The need for closed-loop identification has traditionally arisen when an existing controller cannot be switched off due to process restrictions (*e.g.*, the process is unstable in open-loop or has an integrator behaviour, cost of production interruption, ...) Yet, another important reason for using closed-loop identification instead of open-loop is that it may provide better results

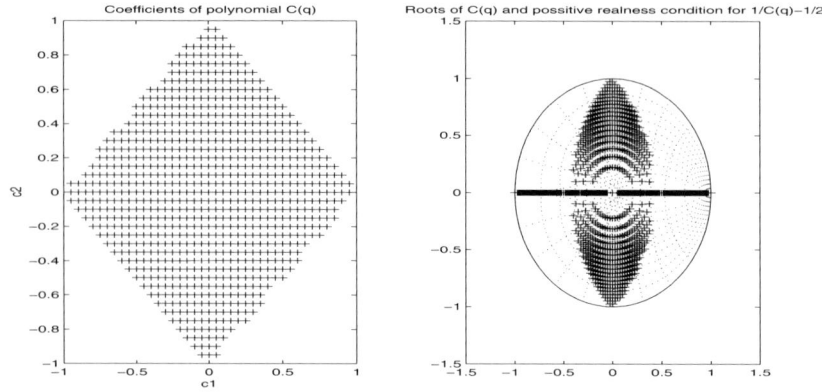

Fig. 2.7. (a) Coefficients for which $Re\left\{C^{-1}(e^{i\omega}) - \frac{1}{2}\right\} > 0$ (b) Roots

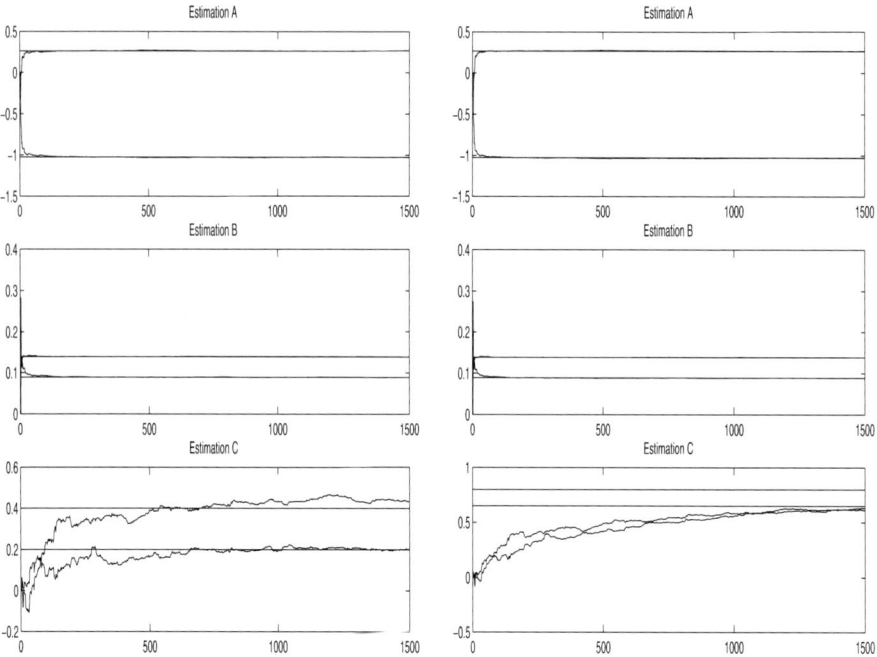

Fig. 2.8. Recursive estimation for (a) condition (2.44) satisfied and (b) not so

for controller design purposes [57, 148, 150]. A good open-loop model (*i.e.*, predicts open-loop response accurately) may not approximate correctly the process closed-loop behaviour, and *vice versa*, as shown in the first chapter (Section 1.3). Moreover, identification should be carried out only for those frequencies of interest for control, the ones where the closed-loop will operate (see Chapters 4 and 3). The need for closed-loop identification at the core of the resulting controller design strategies devised in the context of iterative identification and controller design has renewed interest in it in the last few years [3, 9, 100, 101, 124].

Two conceptually different approaches can be considered:

- identify the plant as if in open-loop. Depending on whether knowledge of the controller is assumed or not, two alternatives arise:
  - use of raw input-output data. Knowledge of the controller is not required, but some identifiability conditions must be met to ensure the correctness of the identified model [108];
  - use of filtered input-output data. The filter arises when the output prediction error is expressed as a function of past plant input and output signals, and past closed-loop prediction errors, and is a function of the controller parameters [9]. So to say, the identification structure tries to extract and use signals from the closed-loop to feed an open loop identification model;
- identify the plant as result of identifying the closed-loop. Again, two approaches are possible:
  - identify the closed-loop and, if the controller is known, obtain the plant model by deconvolution;
  - identify the plant model using a closed-loop identification model. The resulting identification structure leads to the use of output error algorithms [101].

2.7.1 Direct Closed-loop Identification Using Raw I/O Data

Consider the closed-loop depicted in Figure 2.9. The direct approach to identify the plant uses input-output data $\{u(t), y(t)\}$. The controller is not assumed to be known. Therefore, the open-loop predictor (2.12) is used.

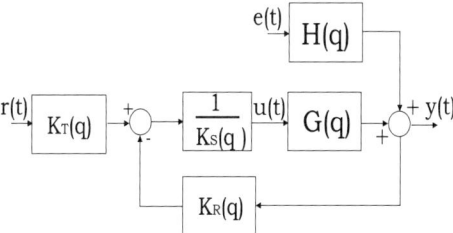

Fig. 2.9. Closed-loop with a two-degrees-of-freedom controller

Thus, the output and input are given by:

$$y(t) = K_T(q)G(q)S_p(q)r(t) + K_S(q)S_p(q)H(q)e(t) \qquad (2.45)$$

$$u(t) = \frac{1}{K_S(q)}[K_T(q)r(t) - K_R(q)y(t)] \qquad (2.46)$$

Defining the estimated and real sensitivity functions:

$$S_p(q, \theta) \stackrel{def}{=} \frac{1}{K_S(q) + K_R(q)G(q, \theta)} \tag{2.47}$$

$$S_p(q) \stackrel{def}{=} \frac{1}{K_S(q) + K_R(q)G(q)} \tag{2.48}$$

Then, the filtered prediction error becomes:

$$\epsilon(t, \theta) = L(q, \theta)\{y(t) - \hat{y}(t \mid t-1, \theta)\}$$
$$= \frac{L(q, \theta)}{H(q, \theta)} \left[\frac{K_T(q)}{K_S(q)} \frac{G(q)S_p(q) - G(q, \theta)S_p(q, \theta)}{S_p(q, \theta)} r(t) + \frac{S_p(q)}{S_p(q, \theta)} v(t) \right]$$
$$= \frac{L(q, \theta)}{H(q, \theta)} \left[\frac{K_T(q)(G(q) - G(q, \theta))}{K_S(q) + K_R(q)G(q)} r(t) + \frac{S_p(q)}{S_p(q, \theta)} v(t) \right] \tag{2.49}$$

If $r(t)$ is not a rich signal, and the main source for excitation in the system is the noise signal $e(t)$, then identification still may be possible. Yet, from (2.49), it can be seen that the estimated model may converge to the true one, or to the inverse of the controller $-K_S(q)/K_R(q)$. Generically,

1. If $G \in \mathcal{G}$ and $H \in \mathcal{H}$, then there exists a value $\theta^\star$ such that $G = G(\theta^\star)$ and $H = H(\theta^\star)$ only if [108]:
   - $r(t)$ is sufficiently rich, or
   - the controller is of sufficiently high order.
2. If $G \in \mathcal{G}$ and $H \notin \mathcal{H}$, then there will be a bias.

Let us consider a simple example to show the main ideas. Given a first order process and a proportional controller, as depicted in Figure 2.10, the regression vector

$$\phi_t = [-y_{t-1} u_{t-1}] \tag{2.50}$$

can be considered for estimation purposes.

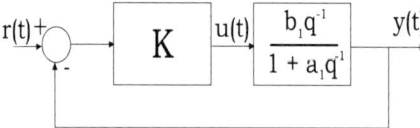

Fig. 2.10. Closed-loop with P controller

Taking into account the feedback

$$u_{k-1} = Kr_{k-1} - Ky_{k-1} \tag{2.51}$$

consider first that the reference signal r_t is assumed to be zero. This, in fact, corresponds to the worst case. Certainly, since in the usual case $\{r_t, u_t, y_t\}$

represent deviations with respect to a working point, the previous assumption is equivalent to considering a constant reference. As easily seen, in such a case, the two elements of the regression vector are linearly dependent. Consequently, the variance-covariance matrix

$$P_t^{-1} = \sum_{i=0}^{t} \phi_i \phi_i^T \tag{2.52}$$

becomes singular. Hence, the parameter estimation becomes ill-conditioned.

Therefore, to solve this problem, linear dependencies among the elements of the regression vector must be avoided. Possible approaches are:

1. Use a sufficiently rich reference signal $r(t)$. This may be unfeasible in many practical applications.
2. Addition of a dither signal on the controller output:

$$u_k^* = u_k + d_k$$
$$\phi_k = \begin{bmatrix} -y_{k-1}, u_{k-1}^* \end{bmatrix} = \begin{bmatrix} -y_{k-1}, -Ky_{k-1} + d_{k-1} \end{bmatrix}$$

such that the linear dependency is broken. This dither signal must have a small amplitude, so as not to spoil the real control signal, and zero mean, so as not to introduce offsets.

3. Use of a controller with an adequate structure: Thus, for the proposed example, the controller

$$u_{k-1} = -q_1 u_{k-2} - p_0 y_{k-1}$$
$$\phi_k = \begin{bmatrix} -y_{k-1}, -q_1 u_{k-2} - p_0 y_{k-1} \end{bmatrix}$$

adds the term u_{k-2}, which breaks the linear dependency.

In Figure 2.11, some experimental results are shown for this example. A recursive estimation was carried out.

This last idea can be easily generalised to ARX models of generic order. Thus, referring again to the scheme of Figure 2.9, if $K_T(q) = K_R(q) = K(q)$, the regression vector becomes:

$$\phi_k = \begin{bmatrix} -y_{k-1}, \ldots, -y_{k-n}, u_{k-d-1}, \ldots, u_{k-d-m} \end{bmatrix}$$
$$= [-y_{k-1}$$
$$\vdots$$
$$-y_{k-n}$$
$$k_0 y_{k-1} + \ldots + k_\nu y_{k-\nu-1} - s_0 u_{k-d-2} - \ldots - s_\mu u_{k-d-2-\mu}$$
$$\vdots$$
$$k_0 y_{k-m} + \ldots + k_\nu y_{k-\nu-m} - s_0 u_{k-d-m-1} - \ldots - s_\mu u_{k-d-m-1-\mu}]$$

To avoid linear dependencies, two conditions must be satisfied:

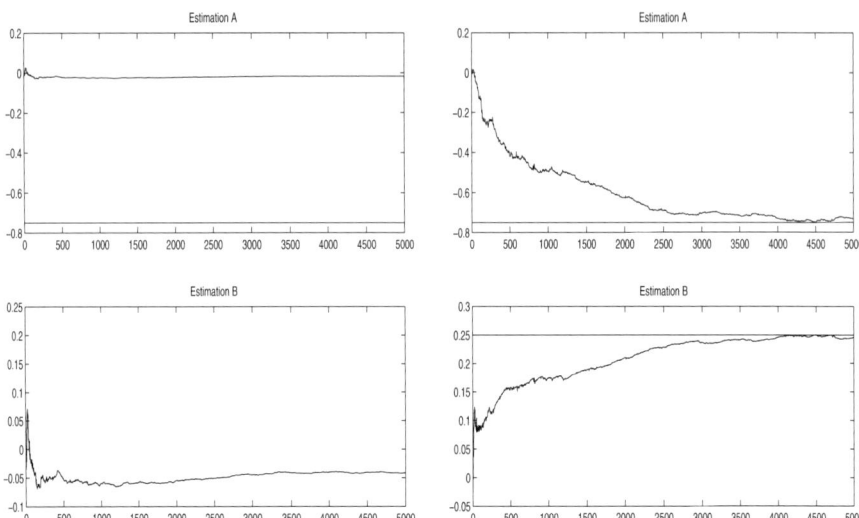

Fig. 2.11. Closed-loop recursive estimation for (a) proportional (b) dead-beat controller

1. $\mu \geq n$
2. $\nu + d \geq m$

along with $\mu \geq \nu$ for the controller to be realisable.

In the case of having a process with an ARMAX model, the condition

$$\max\{n + \mu, m + \nu + d\} \geq m + n + p \qquad (2.53)$$

arises [87], where p is the number of common roots between $A(q)K_S(q)q^d + B(q)K_R(q)$ and $C(q)$.

2.7.2 Direct Identification Using Filtered Data

As already mentioned, the use of filtered data is an alternative approach to direct identification using raw input-output data. The filter arises when the output prediction error is expressed as a function of past plant input and output signals, and past closed loop prediction errors, and is a function of the controller parameters. Therefore, the idea behind it is to use an open-loop identification model, but to somehow allow for the existence of feedback, thus enhancing the data used.

Consider again the closed-loop depicted in Figure 2.9. Recall that the open-loop predictor is given by:

$$\hat{y}_{OL}(t, \theta) = \left[1 - H^{-1}(q, \theta)\right] y(t) + H^{-1}(q, \theta)G(q, \theta)u(t) \qquad (2.54)$$

where prediction is made using input-output data from the process. If, instead, a closed-loop predictor is used, that is, one that uses simulated input-output data to make the prediction:

$$\hat{y}_{CL}(t,\theta) = \left[1 - H^{-1}(q,\theta)\right]\hat{y}_{CL}(t,\theta) + H^{-1}(q,\theta)G(q,\theta)\hat{u}(t) \quad (2.55)$$

$$K_S(q)\hat{u}(t) = K_T(q)r(t) - K_R(q)\hat{y}_{CL}(t,\theta) \quad (2.56)$$

Now, defining the open- and closed-loop prediction errors as

$$\epsilon_{OL}(t,\theta) \stackrel{\text{def}}{=} y(t) - \hat{y}_{OL}(t,\theta) \quad (2.57)$$

$$\epsilon_{CL}(t,\theta) \stackrel{\text{def}}{=} y(t) - \hat{y}_{CL}(t,\theta) \quad (2.58)$$

equation (2.55) can be rewritten, using (2.56), as follows:

$$\begin{aligned}\hat{y}_{CL}(t,\theta) &= \left[1 - H^{-1}(q,\theta)\right]\left[y(t) - \epsilon_{CL}(t,\theta)\right] \\ &\quad + H^{-1}(q,\theta)G(q,\theta)\left(\hat{u}(t) - u(t)\right) + H^{-1}(q,\theta)G(q,\theta)u(t) \\ &= \hat{y}_{OL}(t,\theta) + \left[H^{-1}(q,\theta) - 1\right]\epsilon_{CL}(t,\theta) \\ &\quad + H^{-1}(q,\theta)G(q,\theta)\frac{K_R(q)}{K_S(q)}\epsilon_{CL}(t,\theta) \end{aligned} \quad (2.59)$$

which yields easily the relationship:

$$\epsilon_{CL}(t,\theta) = \frac{H(q,\theta)K_S(q)}{K_S(q) + G(q,\theta)K_R(q)}\epsilon_{OL}(t,\theta) \quad (2.60)$$

Hence, the estimation is carried out using a usual open-loop prediction error (an open-loop predictor based on an OE model could also be used), but filtered. In practice, the filter has to be estimated initially. Therefore, some previous plant model is required. Recalling Equation (2.12), the following expression is obtained:

$$\epsilon_{CL}(t,\theta) = K_S(q)S_g(q,\theta)\left[(G(q) - G(q,\theta))u(t) + v(t)\right] \quad (2.61)$$

Unbiased estimates are obtained only under the condition of

$$K_S(q)S_g(q,\theta)H(q)e(t)$$

being white noise. To overcome this restriction, algorithms based on the asymptotic whitening of the prediction error, by means of using an extended disturbances model (e.g., ARMAX with $C_e(q)/A(q) \approx K_S(q)S_g(q,\theta)H(q)$), can be used [101]. A different approach to reach these results can be found in [9] and [101].

2.7.3 Indirect Method

Consider again the closed-loop depicted in Figure 2.9. The indirect approach to identify the plant consists basically of identifiying the closed-loop, and then extract the plant model from it.

Assume that the controller is known. Then, the output is given by:

$$y(t,\theta) = \frac{K_T(q)G(q,\theta)}{K_S(q) + K_R(q)G(q,\theta)} r(t) + \frac{K_S(q)H(q,\theta)}{K_S(q) + K_R(q)G(q,\theta)} e(t)$$
$$= T_{yr}(q,\theta)r(t) + T_{ye}(q,\theta)e(t) \qquad (2.62)$$

where (recall (2.47) and (2.48)):

$$T_{yr}(q,\theta) = K_T(q)G(q,\theta)S_p(q,\theta) \qquad (2.63)$$
$$T_{ye}(q,\theta) = K_S(q)H(q,\theta)S_p(q,\theta) \qquad (2.64)$$

A naive approach would consider the optimisation of some cost function related to the prediction error:

$$\epsilon(t,\theta) = L(q,\theta)\{y(t) - T_{yr}(q,\theta)r(t)\} \qquad (2.65)$$

Yet, Equation (2.62) can be seen as the one corresponding to a process in open-loop, with external input $r(t)$ and noise input $e(t)$. Therefore, the expression for the output predictor is:

$$\hat{y}(t\,|\,t-1,\theta) = \left[1 - T_{ye}^{-1}(q,\theta)\right] y(t) + T_{ye}^{-1}(q,\theta)T_{yr}(q,\theta)r(t) \qquad (2.66)$$

Hence, the filtered prediction error:

$$\epsilon_f(t,\theta) = L(q,\theta)T_{ye}^{-1}(q,\theta)\left[(T_{yr}(q) - T_{yr}(q,\theta))\,r(t) + K_S(q)S_p(q)v(t)\right]$$
$$= \frac{L(q,\theta)}{H(q,\theta)}\left[K_T(q)\left(G(q) - G(q,\theta)\right)S_p(q)r(t) + \frac{S_p(q)}{S_p(q,\theta)}v(t)\right] (2.67)$$

can be used instead of (2.7.3). Notice that the same expression that was obtained for the direct approach appears again. The important point in (2.67) is that the sensitivity function appears as a weighting one. This becomes more apparent if the expressions for variance and bias of the estimates are considered. Thus, for the first, the following expressions arise:

$$covT_{yr}(q,\theta) \approx \frac{n}{N}|S_p(q)|^2\frac{\Phi_v}{\Phi_r} \qquad covT_{ye}(q,\theta) \approx \frac{n}{N}|S_p(q)|^2\frac{\Phi_v}{\sigma^2} \quad (2.68)$$

And, for the bias:

$$\theta^* = \arg\min_\theta \int_{-\pi}^{\pi} F(q,\theta) \left\{|K_T(q)|^2\,|G - G(q,\theta)|^2\,\Phi_r + \left|\frac{1}{S_p(q,\theta)}\right|^2 \Phi_v \right\}\bigg|_{q=e^{j\omega}} d\omega$$
$$(2.69)$$

with

$$F(q,\theta) = \frac{|L(q,\theta)|^2\,|S_p(q)|^2}{|H(q,\theta)|^2}$$

Typically, the sensitivity function has high gain for low frequencies, and low gain for high frequencies. This implies that there will be less bias on the range of high frequencies, and less variance of the estimates at low ones.

2.7.4 Identification Using a Closed-loop Predictor

Consider again the closed-loop predictor given by Equations (2.55) and (2.56). The goal now is not to relate the corresponding closed-loop prediction error with the open-loop one, as in Section 2.7.2, but to use a fully closed-loop predictor. This will imply using simulated process input and output data, generated by the closed-loop predictor, instead of real data.

Rewrite (2.55) as:

$$\left[\left[1 - H^{-1}(q,\theta)\right], H^{-1}(q,\theta)G(q,\theta)\right] \left[\hat{y}_{CL}(t,\theta), \hat{u}(t)\right]^T \stackrel{\text{def}}{=} \hat{\theta}^T \Phi(t) \quad (2.70)$$

The process output and input are given by:

$$y(t) = \left[1 - H^{-1}(q)\right] y(t) + H^{-1}(q)G(q)u(t) + e(t) \stackrel{\text{def}}{=} \theta^T \varphi(t) + e(t) \quad (2.71)$$

$$u(t) = \frac{1}{K_S(q)} \left[K_T(q)r(t) - K_R(q)y(t)\right] \quad (2.72)$$

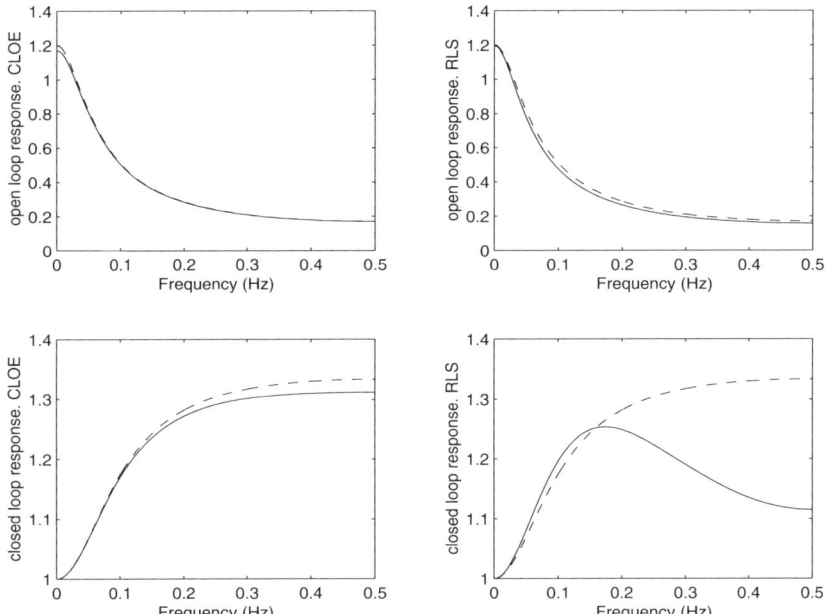

Fig. 2.12. Magnitude of the real and estimated open- and closed-loop responses

Adding and substracting the term $\theta^T \left[y(t), u(t)\right]^T$ to Equation (2.71), and making use of (2.70) and (2.72), the following expression is deduced for the closed-loop prediction error (2.58):

$$\epsilon_{CL}(t,\theta) = H(q)K_S(q)S_p(q)\left[\left(\theta - \hat{\theta}\right)^T \Phi(t) + e(t)\right] \quad (2.73)$$

Under conditions of signal richness and $H(q)K_S(q)S_g(q)e(t)$ being white noise (recall Section 2.7.2), unbiased estimates are obtained.

In Figure 2.12, the real and estimated magnitude of the open- and closed-loop responses are shown for a first order system. A coloured noise acts upon the system output. As seen, the estimated open-loop responses are very similar to the real ones, both using a RELS open-loop estimator, and a closed-loop output error one as described in this section. Yet, the estimated closed loop response is much closer to the real one when a closed-loop estimator is used.

Another version, using filtered closed-loop simulated data, can be found in [100, 101].

3 Data Preparation for Identification

Robert R. Bitmead and Antonio Sala

Abstract. System identification moves us from data to model and there is a modern tendency to automate as much of this process as is possible. In this chapter, we describe some important principles in practice for pre-processing the data for modelling and identification, unfortunately by hand. Three areas are considered that precede the application of identification software. These deal with: model structure selection, where one decides on the class of candidate models to be explored for a good fit to the plant and disturbance; experiment design, for the enhancement of parameter identifiability; and, data pre-processing through selection and filtering prior to identification. The order of these subjects is roughly that in which they need to be considered in applications. The study here is backed up by specific industrial examples in later chapters, notably Chapters 8 and 10.

3.1 Introduction

There is a strong element of approximation in system identification – a finite set of noisy experimental data is used to extract a linear finite-dimensional model of a non-linear, non-stationary, typically infinite-dimensional system. Further, the eventual application of the model that we have in mind could well take the system away from its current operating point and we need to understand how the approximation might be altered by changing the design variables of system identification prior to the use of the software.

There is in system identification, as in most other fields, a dichotomy between the circumstances pertaining in practice and those assumed to hold for the derivation of theoretical results. We shall rely for guidance on recent descriptions of the closed-loop modeling performance expressed as a frequency domain integral. This integral formula holds under quite weak assumptions and captures many of the issues concerned with managing the approximation. This is the design phase of the system identification and, not surprisingly, involves several interconnected decisions.

The first questions are raised before any system identification data is yet present – although one would hope to have a reasonable familiarity with the process and with the expectations of eventual performance. It is often the case that large volumes of operating plant data are available. While these typically are unsuited to system identification because they lack sufficient external excitation, they can help define the likely operating range for the equipment. One needs to consider what type of experiment might be able to display effectively the dynamics sought to be captured in the model. Before this can be well determined, it is necessary to have some concept of the nature of the model class to be fitted – this is the first entreé of *a priori* knowledge into the design.

Early decisions need to determine whether the collection of data from open-loop or closed-loop data is preferable. Clearly, for unstable systems, only the former is possible. An appreciation of the controller acting during the experiment and of the likely plant disturbances needs to be incorporated together with the ultimate objective. This is experiment design, where external signals to be injected into the loop need to be selected. These signals serve the purpose of exciting the dynamics that are regarded as important to be captured in the approximation. Once an initial model is identified, it is possible to make more informed choices for these variables.

One philosophical point of departure of theory and practice is that all data are equally valuable members drawn from the same stationary distribution. Experimental data can be very informative or completely uninformative. Here, we shall deal with questions such as *eyeballing* the data to determine whether it deserves inclusion into the identification, and with *massaging* the data by pre-processing to enhance its information about the target areas of the identification. This is typically done by filtering the data.

The standard approach to identification and modelling uses prediction error methods – a model is sought that minimises a measure of one-step-ahead prediction error. Thereafter, it is assumed that the identified model that is best according to this prediction criterion will also be the one that will produce the best achieved closed-loop control performance. Subsequently, modelling error can be dealt with at the control design phase via robust control techniques. That is, a controller can be synthesised so that it will produce consistent performance for all plants in a certain bounded set around the nominal model. The result is a controller detuning with some loss in nominal performance but an increased capacity to accommodate modelling errors. This detuning depends on both *model confidence* and *control objectives*.

Following the main point of this book, there is an interplay between modelling and control in the sense that successful feedback controllers can be designed based on crude models that capture the essential characteristics of the plant for control. Certain large modelling errors can lead to small closed-loop effects and *vice versa*. One method to accommodate this is to concentrate modelling effort to approximate closely the output of the plant when it is subjected to the signals it would eventually receive when operating in closed-loop with a good controller. Conversely, modelling effort that might stress accurate step-response or one-step-ahead prediction with pseudo-random inputs might not focus the modelling degrees of freedom to contribute to performance when the plant is placed under closed-loop control.

To illustrate this core issue, consider modelling a scalar plant to which we will apply proportional-plus-integral feedback control. We divide our modelling attention into three frequency regions. At very high frequencies, the fast-timescale response is captured well by one-step-ahead prediction error

[108]. Dynamics in this range are of less importance to achieved performance than knowledge at mid-range around the eventual gain- and phase-crossover values, because of controller roll-off at high frequency. Likewise, capturing the fine detail of the DC response (apart from the sign of the gain) is unimportant because the integral action will handle this. Thus, the critical values to capture with small model error are around the eventual crossover points, which are themselves determined by the controller.

Successful identification for control recognises this interconnection between control objective and modelling requirements, typically resulting in iterative approaches of successive model fitting and control design. We shall explore in this chapter the available mechanisms for affecting the fit of a model. Typically, we consider the measure of model quality being the frequency response fit between the true plant and the model. When we talk of redistributing the model error, we mean that we seek to identify a different model whose fit to the real system over frequency is tighter in some bands and is correspondingly more relaxed in others. A key to this study is the specification of the tools that allow us to manipulate the frequency fit. These include: the determination of experimental conditions (input signals, closed-loop or open-loop, *etc.*) for the collection of data, pre-filtering of the data, and selection of model structure.

3.2 Design Variables for Identification

Consider the process signals depicted in Figure 3.1. Here, $G(z)$ is the plant, $K(z)$ the controller, y_t is the plant output, u_t the plant input and r_t the external reference. We shall consider $\{(u_t, y_t)\}$ data collected from this loop

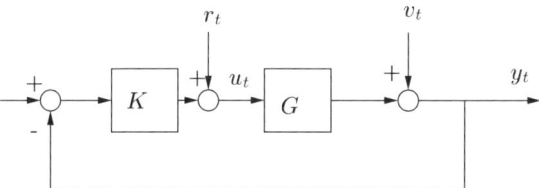

Fig. 3.1. Data generation loop

and their use for identifying a model.

System identification based on minimising the summed-squared filtered one-step-ahead prediction error between the measured input-output data and a parameterised model has the following objective function

$$V_N(\theta) = \sum_{t=1}^{N} \left(L(q)(y_t - \hat{y}_{t|t-1}(\theta))\right)^2$$

Here, θ is the vector of model parameters, N is the number of data, and $\hat{y}_{t|t-1}(\theta)$ is the one-step-ahead prediction of y_t using measured inputs and outputs up to and including time $t-1$ based on the model parameterised by θ. Using large numbers N, of data and appealing to Parseval's formula, we may rewrite this in the frequency domain. In open-loop, where the input sequence $\{u_t\}$ is statistically independent from the disturbance $\{v_t\}$, we have:

$$V_N(\theta) = \int_{-\pi}^{\pi} (|G(e^{j\omega}) - \hat{G}_\theta(e^{j\omega})|^2 \Phi_u(\omega) + \Phi_v(\omega)) \frac{|L(e^{j\omega})|^2}{|\hat{H}_\theta(e^{j\omega})|^2} d\omega \quad (3.1)$$

In closed-loop, the controller K causes correlation between the input and the disturbance. But the reference signal $\{r_t\}$ and $\{v_t\}$ remain uncorrelated, so we have

$$V_N(\theta) = \int_{-\pi}^{\pi} \Big[|G(e^{j\omega}) - \hat{G}_\theta(e^{j\omega})|^2 |K(e^{j\omega})|^2 \Phi_r(\omega)$$
$$+ |1 + \hat{G}_\theta(e^{j\omega}) K(e^{j\omega})|^2 \Phi_v(\omega) \Big] \frac{|L(e^{j\omega})|^2}{|1 + G(e^{j\omega}) K(e^{j\omega})|^2 |\hat{H}_\theta(e^{j\omega})|^2} d\omega \quad (3.2)$$

This material is well developed in [108]. Here: Φ_u, Φ_v and Φ_r are the spectra of the signals u_t, v_t and r_t; $G(z)$ and $\hat{G}_\theta(z)$ are the real plant and its parameterised model; $\hat{H}_\theta(z)$ is a parameterised noise model meant to capture the correlation properties of v_t; $L(z)$ is a data filter that is used to pre-filter the data sequences $\{u_t\}$ and $\{y_t\}$ prior to the identification.

The reason for the focus on these least squares one-step-ahead predictive methods is because of the existence of a broad class of applicable algorithms that have been widely studied, at least in the open-loop context. Algorithmically, they form the basis for much of the MATLAB® System Identification Toolbox, for example.

The following design choices can be made prior to the invocation of the identification algorithm:

- **Model structure:** the complexity and specific θ-parameterisation of the plant and disturbance models $(\hat{G}_\theta(z), \hat{H}_\theta(z))$.
- **Experiment design:** the determination of whether open-loop or closed-loop data will be used. In open-loop, by definition, the input u_t, and the disturbance v_t, are statistically independent.
- **Excitation:** the spectrum of the external reference input signal fed into the system in the data collection experiment, $\Phi_r(\omega)$.
- **Data selection:** the selection of well-excited, representative data away from non-linear limits, constraints *etc.*, required for both the linear model fitting and for subsequent model validation tests.
- **Data pre-filtering:** choice of the stable filter $L(z)$, which affects the complete identification objective function in both open- and closed-loop.

In practice, the selection of these variables is central to the inclusion of *a priori* information about the real system and the intended model application. For example, the model complexity needs to be managed in order to provide a subsequent tractable control design problem. When combined with the data filter, it becomes possible to concentrate model fitting in frequency regions of interest for the underlying control problem, such as disturbance rejection. Similarly, for many problems, the noise model $\hat{H}_\theta(z)$ can be selected as fixed, again reducing complexity and reflecting a knowledge of the disturbance process.

Specific examples of the selection of identification design variables will be presented in the later chapters. These will help to clarify the decisions needing to be made and to indicate the factors affecting these choices. Our aim in this part is to clarify what the variables are and how their roles in control design might differ from those in prediction error modelling.

3.2.1 Model Structure Selection

Model structure refers to a range of choices made prior to the explicit fitting of parameters, including:

- the degrees of the transfer functions being fitted for plant model and for disturbance model. For example, an ARMA model of the type

$$A(z)y_t = z^{-d}B(z)u_t + C(z)e_t$$

would require specification of the degrees of polynomials A, B and C, as well as the delay parameter d;

- the parameterisation of these transfer functions, so that the same system might be represented in a number of different ways. The ARMA parameterisation can be replaced by the linear-in-the-parameters MA model class

$$y_t = B(z)\left\{z^{-d}A^{-1}(z)u_t\right\} + e_t,$$

if the denominator polynomial $A(z)$ is known beforehand. Here, this has turned the identification into a standard regression problem between y_t and filtered versions of the input. Other parameterisations might seek to identify as explicit parameters the poles and zeros of the plant as opposed to the coefficients of the numerator and denominator polynomials;

- assumptions about the probabilistic nature of the disturbance process impinging upon the plant.

Ljung [108] develops these notions more fully in the context of prediction error model fitting – notably the model structure refers to a differentiable mapping from the parameter set to the model set so that the gradients of the predictors are stable. Further, there are a host of procedures and tests designed to assist in the determination of candidate model structures supported by experimental data. These include tests of order, delay and independence, which are usually applied to the analysis of residuals or prediction errors and are grounded in statistical procedure (see Chapter 11 of [144], for example).

In identification for control design, some extra comments are important and significantly different from the case for prediction error modelling:

- the *principle of parsimony* states that one ought not to use any more parameters for the description of system than are absolutely necessary. In the control context, the inclusion of more parameters typically leads to a more difficult control design problem. Thus, there is an added benefit to maintaining the dimension of θ as small. If one seeks specialised control designs, such as stable controllers, then the management of this constraint depends greatly on the degree of the plant model;

- for finite data sets and for non-stationary data, the confidence in each parameter reduces with increase of n, the dimension of θ. The parameter variance formulae of system identification [108, 144] establish that, asymptotically in data set size, the variance is proportional to n. For control design, this is an insidious effect because parameter variance leads to modelling error in a highly non-linear fashion. As we will see in Chapter 4, coping with modelling error usually requires detuning of the control objective;

- model structure selection is a primary means of injecting *a priori* knowledge about a system into the model fitting stage. The distinction drawn above between using an ARMAX model and a linear regression model with filtered inputs shows that lower-order (and therefore more accurate) parameterisations might be achieved through additional information about the plant – here the information was the position of the plant poles. Laguerre and Kautz parameterisations benefit from this observation. Another very common practical example arises when it is known that an integrator exists in the plant and can be included without requiring it to be identified, thereby reducing the complexity of θ;

- the nature of plant modelling and disturbance modelling is different in open-loop from in closed-loop. For prediction error modelling, the parameters of θ jointly are fitted to produce the minimising mean squared error. This happens as a global minimisation without regard to the distinction between model and plant. In identification for control, there is a clear distinction between the two – disturbance model errors without plant model errors cannot lead to instability of the control. Thus, there is a natural focus on fidelity of the plant model at the expense of the disturbance model. Indeed, one frequently chooses a high order for the disturbance model in order to separate this fit from that which concentrates on capturing a good low-order plant model.

These issues are both deep and difficult to formalise in the general context. Indeed, the systematic inclusion of *a priori* information about a process into system identification is called *grey-box modelling* [29]. Identification for control needs to consider not only these issues but also information about the control objective and design process. Euphemistically, one might say that this gives enormous scope for designers to demonstrate their skills.

3.2.2 Experiment Design and Excitation

Open-loop data versus closed-loop data

A critical component of the experimental conditions (and hence a major determinant of the ultimate data quality) is the nature of the correlation between the signals. Since, when it is all boiled down, least squares relies on correlation methods for computation, this appears first on our list.

In open-loop data, the disturbance is independent from the control input and so control inputs can be developed to expose many different modes of behaviour of the plant. The separation between input-related and disturbance-related outputs is made using correlation, provided the disturbance is truly random.

By contrast, the closed-loop case has the controller introducing correlation between the disturbance and the control. Without a separate reference signal r_t, there is complete correlation. Further, since a feedback controller is designed to suppress the sensitivity of the closed-loop to the particular plant, it is difficult to identify the details of the plant. However, when r_t is present, we may again resort to correlation to separate the signal components, although, this does not simply detangle the plant from the controller.

Our preference is firmly in favour of closed-loop data for three principle reasons: for many industrial systems, the data is only available in closed-loop for operational reasons; the closed-loop bias formula (3.2) incorporates weighting by the achieved sensitivity function into the model fit; and, the closed-loop experiment yields performance information about the current controller. Balanced against this view is the property that in open loop, one is able to separate by averaging the disturbance effects from the plant, thereby encouraging a higher fidelity plant model.

Reference signal selection

The frequency domain minimisation formula (3.2) reveals a wealth of information about the outcomes from system identification using prediction error methods with closed-loop data. In the absence of excitation signal r_t, the identified parameter θ will minimise

$$V_N(\theta) = \int_{-\pi}^{\pi} \frac{|1 + \hat{G}_\theta(e^{j\omega})K(e^{j\omega})|^2}{|1 + G(e^{j\omega})K(e^{j\omega})|^2} \times \frac{\Phi_v(\omega)|L(e^{j\omega})|^2}{|\hat{H}_\theta(e^{j\omega})|^2} \, d\omega. \qquad (3.3)$$

Ignoring for the moment model structure limitations, the obvious minimising value of θ produces $\hat{G}(z) = -K^{-1}(z)$, which results in a value of 0 for $V_N(\theta)$. That is, the identification selects $\hat{G}$ as the inverse of the controller rather than as the true plant G. Why is this?

There are two linear models that relate the data set $\{y_t, u_t : t = 1, \ldots, N\}$ when $r_t = 0$:

$$y_t = G(z)u_t + v_t$$
$$y_t = -K^{-1}(z)u_t \tag{3.4}$$

The former equation is the model of the plant. It is only ever approximately linear and is noisy. The latter equation is the controller $u_t = -K(z)y_t$, which is exactly linear if we implement it well and is not corrupted by noise. It is therefore not surprising that the latter solution is the one found by the minimisation.

Given that we already know the controller, there would appear to be little gained from this exercise of identifying with a reference signal. However, we should note that the model structure fixed earlier need not necessarily permit a perfect fit of $\hat{G}$ to $-K^{-1}$. What happens then? Turning again to (3.3), we see that if $\hat{G}$ cannot be made to coincide exactly with $-K^{-1}$, then the approximation of $\hat{G}$ to $-K^{-1}$ will be strongly weighted towards fitting well when $1 + GK$ is small. Note that this is where G is itself close to $-K^{-1}$, and so this is consistent with $\hat{G}$ forming a good approximation of G at these frequencies. The poor emphasis outside all but this (G, K) cross over band militates against this as a desirable approach in modelling.

The fact of the matter is that in order to fit a model $\hat{G}$ to the plant G we need to ensure that the term multiplying $G - \hat{G}$ has adequate excitation. In industrial environments, this has the unfortunate consequence of declaring much of the recorded plant operating data useless for model fitting purposes. A purpose-designed experiment needs to be conducted.

We note that (3.2) captures the open-loop case of (3.1) when K is set to 0, provided we take the reference input feeding directly into u_t. How ought the spectrum $\Phi_r(\omega)$ be chosen?

Expression (3.2) displays the two competing terms in the system identification. The desire to fit $\hat{G}$ to the plant G balances against the tendency to accept $\hat{G} = -K^{-1}$, with the relative weightings given by $\Phi_r(\omega)$ and $\Phi_v(\omega)$. This is akin to a frequency-dependent signal-to-noise ratio. Where the known excitation signal spectrum Φ_r dominates the unmeasurable noise spectrum Φ_v, the emphasis will be on fitting the model to the plant.

The excitation signals must have the appropriate frequency spectrum so that a good fit is achieved in the regions interesting for control design. Those regions are the frequencies over which significant plant disturbance occurs (robust performance), and other regions dictated by the robust stability requirement, such as around the gain and phase crossover points – or their multivariable equivalents.

In most plants, a natural limitation is placed on the excitation signal not to interfere too much with normal plant activity during experimentation. Industry management only allows for limited downtime for experimentation, if any.

The following list summarises the factors that determine the appropriate excitation to be applied to the closed-loop.

- **Disturbance spectrum:** Closed-loop control effort is needed at frequencies where disturbances are significantly present and one would expect that good plant models are needed where the control effort is large. Thus, an appropriate input spectrum (as a first guess) would be a multiple of the disturbance spectrum. To obtain that spectrum for a stable process, an FFT graph of the uncontrolled output behaviour might be used.
- **Ultimate control objective:** As high attenuation of disturbances across a wide frequency range is impossible in practice, the control objective may be often restricted to low frequencies even if disturbances in higher frequency ranges are present. In this way, a certain control bandwidth is assigned. It is notable that this decision about eventual control bandwidth is an opportunity to include *a priori* process information into the identification problem.
- **Independence from disturbance signals:** The capacity to distinguish the plant behaviour from the disturbance derives from the statistical independence of the excitation reference signal and the disturbance, effectively allowing decoupling.

We shall see in two examples of later chapters, Chapters 10 and 13, how specific choices of reference excitation are made. In each case, further aspects enter that limit the feasible excitation signal types. For the helicopter vibration control problem, we need to inject only smooth signals in order not to interact too strongly with non-linear rate limiters. In the sugar mill example of Chapter 13, the magnitude needs to be chosen so as not to cause saturation problems.

The bias formula (3.2) shows up quite clearly that the estimated model parameters depend on the frequency content of the input and the signal-to-noise ratio (amplitude). In non-linear systems, the estimated parameters additionally depend on the signal amplitudes due to the non-linearities, thus distorting the estimates from a Taylor series linearised model. We shall observe how data that exhibit too great an effect from non-linearities might need to be discarded.

3.2.3 Data Selection

The quandary of modelling followed by control design is that at the outset the plant is not being well regulated, so that disturbances dominate the measured data. Nevertheless, we add excitation signals to this, which take the operating point of the closed-loop systems still further afield. And yet we seek to extract a high fidelity model across some frequency range. Two underlying assumptions that enable this to be done are:

- the reference process r_t and the noise process v_t are statistically independent and so may be distinguished by correlation and averaging;
- the underlying processes are stationary and display dynamics largely captured by a linear model.

With practical operating data records, the first assumption dictates that we should seek to use large amounts of data in order to benefit fully from the (de)correlation properties. The second assumption, however, requires that we use only data that does not display obvious transient effects or non-linear effects, and this leads us to limit the size of acceptable trials.

In practice, neither of these assumptions is entirely valid and there is a need for a measure of caution in appealing to statistical limit theorems if the ground rules for their application are not demonstrated to hold.

If the true plant is not contained in the model set, then large data sets are not necessarily helpful to get a "better" model. Indeed, given that the bulk of the statistical analysis software tools, such as correlation analysis and order selection, are predicated in its calculations on an assumption that the estimated model is correct, these tools tend to encourage an overly optimistic view of the quality of fit to be expected with large data samples. For example, the penalty for a large order model structure diminishes with data sample size. When we know that the linear model we seek will necessarily be approximate, we need to select the most informative data subset of adequate, but not excessive, size.

Experimentally collected data sets may exhibit a series of features:

- there are periods in which the signals are dominated by the disturbances and regulation by the existing controllers is poor;
- there are other periods where the signals are well controlled into mid-range values and the input excitation can exert its effect;
- non-linear effects can be very evident when saturation is reached;
- some sensors may have a significant quantisation level;
- the signal environment is non-stationary, with the disturbance exhibiting sporadic presence (thus a stationary statistical model is also an approximation).

To improve a closed-loop identified model, we select only data where the disturbances and non-linearities do not dominate, so the process demonstrates excited behaviour capable of being captured by a linear model, with a reduced bias. If disturbance-dominated data had been used, then one would expect from an analysis of (3.2) that the identification with closed-loop data would concentrate its efforts on modelling the controllers rather than the plant itself.

3.2.4 Data Inspection

Once data are collected, it is important to inspect the raw results to ascertain their suitability for subsequent use in identification. Figure 3.2 shows some raw process data from the sugar mill example that is explored more fully later in Chapter 13. Note that the data appear to cover a reasonable range of the permitted values and have a broadband spectral content, as illustrated in Figure 3.3. (Note that we have suppressed the large DC or zero frequency value in this figure. We shall discuss this again shortly.)

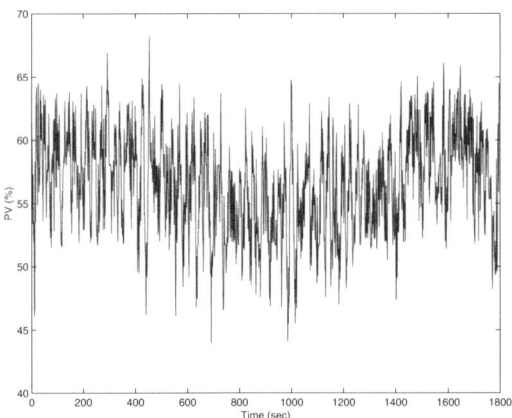

Fig. 3.2. Raw process variable data exhibiting wide-band frequency content

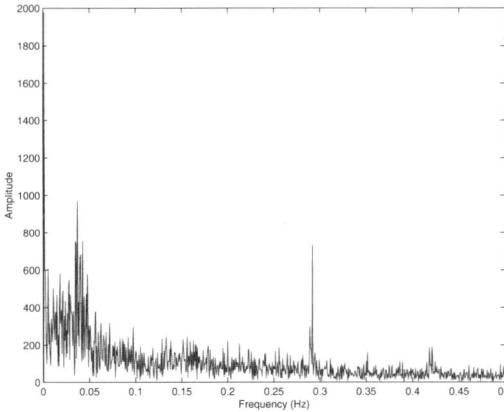

Fig. 3.3. Spectrum of raw process variable data – DC value suppressed

Now consider the heavily quantised variant of the same data displayed in Figure 3.4 on the following page. This has been artificially produced here by mapping the raw data. In practice, such signals arise from inappropriately setting the range of sensors or from failing to provide adequate amplification to signals. Quantisation is a non-invertible non-linear transformation of the data, which diminishes the information content. Its effect on system identification can be severe depending upon the correlation of the signal quantised. A zoomed version of the plot, Figure 3.5, containing both the raw and quantised data, illustrates the problems in loss of information content.

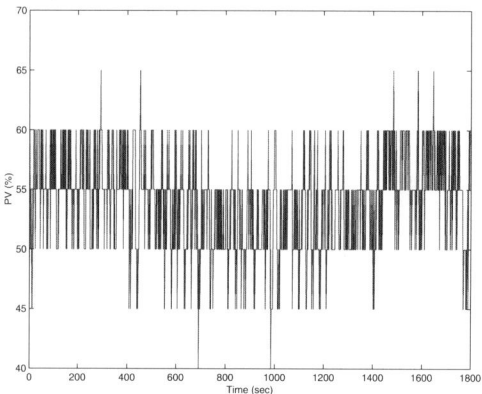

Fig. 3.4. Heavily quantised process variable data

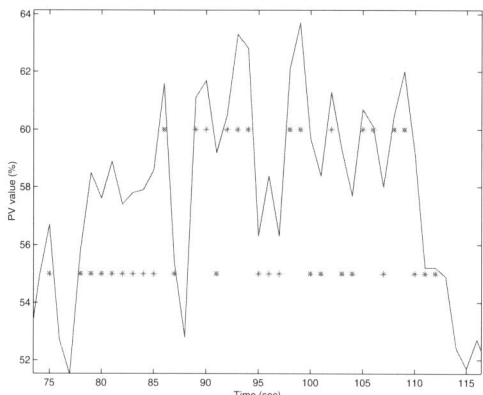

Fig. 3.5. Zoom of raw and quantised process variable data

How might one detect significant quantisation? The most obvious way is through inspection of the data via plots such as Figure 3.2. Alternatively, histograms can reveal the effects.

In a certain sense, this is a simple inspection test. However, if we were to rely on the spectrum to reveal the problem then detection becomes much more difficult. Figure 3.6 shows the spectrum of the quantised data. Compare this to Figure 3.3. Even if one plots both curves side by side, it is quite difficult to recognise the presence of quantisation effects from the spectrum alone. Naturally, this begs the question of whether the quantisation effects matter

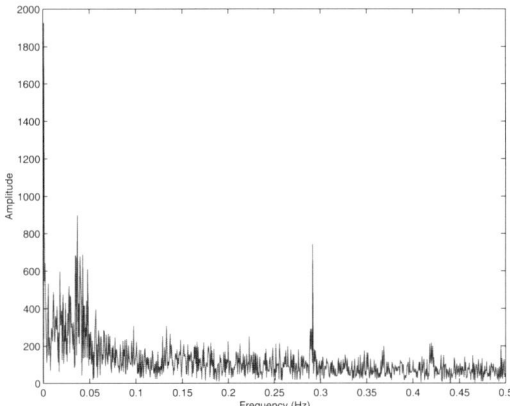

Fig. 3.6. Spectrum of quantised process variable data

for identification. Indeed they do when they occur on either input or output signals, and this effect is more pronounced with slowly-varying signals.

Other very common downstream processing of data can also mask the effects of quantisation and other non-linearities. Decimation is the process of removing data so that the dominant frequency range of interest occupies a larger proportion of the frequency domain. This is done by first low-pass filtering (to avoid aliasing effects) and then down-sampling the data. For example, if we seek to sub-sample the 1 Hz data x, shown in Figure 3.2, to yield 0.5 Hz sampled data, we might use the MATLAB® function y=decimate(x,2). This would first filter using an eighth order Chebyshev low-pass filter with cut-off at $0.8 \times \pi$ radians per s. Then, the filtered signal y is down-sampled (every second element is kept) to yield the new signal.

Figure 3.7 on the next page shows the data from Figure 3.2 decimated by a factor of two. Notice that the low-pass filtering stage of the decimation process has obscured the quantisation distortion. The histogram of this data fails to reveal the effect either. Other significant nonlinearities can be masked in a similar way by filtering procedures.

Other deleterious effects that can be observed by inspection include saturation, clipping and non-stationarity. The flap variable from the same data set as above is shown in Figure 3.8. Failing to recognise this non-stationarity could lead to a seriously compromised identification.

In this case, since the non-stationarity is a slow drift in the mean, we are able to use the data once it has been detrended to remove this effect – and

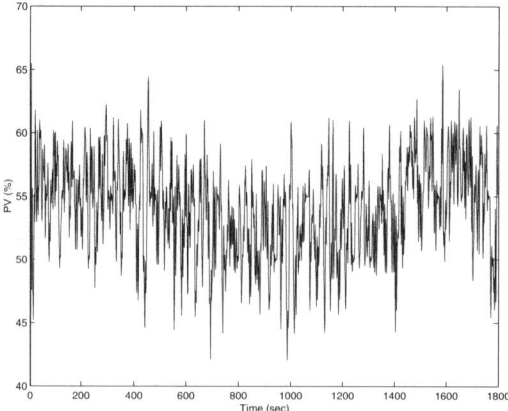

Fig. 3.7. Low-pass filtered and down-sampled quantised data

we perform the same processing on the other signals prior to identification. However, the saturation of the signal at zero (corresponding to a wide open flap) cannot be removed by processing and this data should not be used for fitting transfer functions dependent on the flap signal.

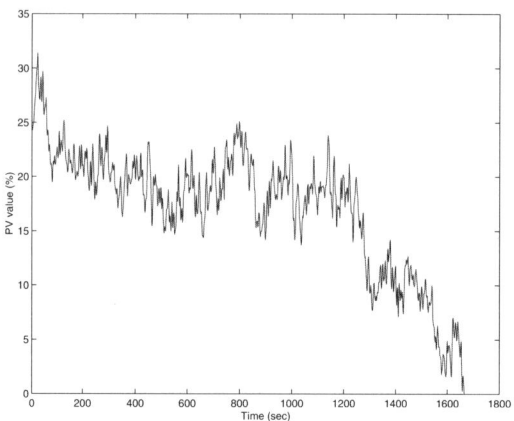

Fig. 3.8. Non-stationary process data

Training set and validation set

The principle of parsimony stated earlier attempts to limit the dimension of the parameter vector θ, fitted to a particular data set. With a single data set, the increased flexibility provided by more parameters causes the quality of model fit to improve with dimension. This dichotomy is reconciled by *validating* model fit, that is, evaluating its predictive performance on a statistically independent set. Parameters whose role is only to capture the random fluctuations of the noise rather than the repeatable variations due to systematic dynamic effects will add noise to the validation predictions and indicate over-parameterisation. Thus, the evaluation of model structure and quality of fit of a parameter vector identified in the training data set is performed using the validation data set. The selection of data for system identification therefore involves choosing two sets, the *training set* and the *validation set*. They should be statistically independent, which translates typically into their being separated in time.

Obviously, both sets can be extracted from a long experiment record if enough data is present to allow for transients to settle between the two subsets. This issue is a central one in the selection of the structure and size of the estimator. If only the training set is available, then the size selection is driven by the marginal improvement of new parameters (over the training set). If it is small, the decision to stop increasing model size is taken.

A natural problem arises in selecting the two data sets separated in time which exacerbates that of selecting just the one set. This is that, while being independent, the two sets should exhibit comparable dynamical effects and should be drawn from the same stationary population. Thus, non-stationarity between the two sets can be an issue.

3.3 Data Preparation: Detrending and Filtering

Recall (3.2), which relates the prediction error criterion to the frequency-domain formulation.

$$V_N(\theta) = \int_{-\pi}^{\pi} \left[|G(e^{j\omega}) - \hat{G}_\theta(e^{j\omega})|^2 |K(e^{j\omega})|^2 \Phi_r(\omega) \right.$$

$$\left. + |1 + \hat{G}_\theta(e^{j\omega})K(e^{j\omega})|^2 \Phi_v(\omega) \right] \frac{|L(e^{j\omega})|^2}{|1 + G(e^{j\omega})K(e^{j\omega})|^2 |\hat{H}_\theta(e^{j\omega})|^2} d\omega.$$

Our focus here will be the data filter $L(z)$, which forms a central part to very many system identification tasks because its magnitude shapes the entire model fit's distribution over frequency. Its selection provides one of the most powerful mechanisms with which to influence to the fit.

Note that the formula above derives from the prediction equations. If the model is of the form

$$y_t = \hat{G}(z)u_t + \hat{H}(z)e_t$$

then the associated one-step-ahead prediction and prediction error are

$$\hat{y}_{t|t-1} = \hat{H}^{-1}(z)\hat{G}(z)u_t + \left[1 - \hat{H}^{-1}(z)\right] y_t$$

$$\epsilon_t \stackrel{\triangle}{=} y_t - \hat{y}_{t|t-1}$$
$$= \hat{H}^{-1}(z) \left[y_t - \hat{G}(z)u_t \right]$$

Clearly then, the filtered prediction error is given by

$$\epsilon_t^f \triangleq L(z)\epsilon_t$$
$$= L(z)\hat{H}^{-1}(z)\left[y_t - \hat{G}(z)u_t\right] \tag{3.5}$$
$$= \hat{H}^{-1}(z)\left[\{L(z)y_t\} - \hat{G}(z)\{L(z)u_t\}\right] \tag{3.6}$$

That is, the weighting factor $|L(e^{j\omega})|^2$ may be introduced into the identification process by filtering *both* the input and output signal through the same linear filter $L(z)$ before going to the identification procedure.

We note also from (3.5) that achieving a weighting of the identification criterion by the factor $L(z)$ is totally equivalent, in this prediction error framework, to constraining the noise model $\hat{H}(z)$ to have specific properties. Indeed, one of the appealing properties of the filter L is that it may be used to accommodate known features of the real noise process, thereby reducing the complexity of estimation. For example, if we know that the disturbance contains steps, then this corresponds to knowledge that H contains an integrator, and we might obviate the need to estimate these known parameters by filtering the data by $L(z) = 1 - z^{-1}$, a differencer. Thus, if the true system is described by

$$y_t = G(z)u_t + v_t = G(z)u_t + \frac{\tilde{H}(z)}{1 - z^{-1}}e_t$$

then the differenced system has a description

$$\delta y_t = G(z)\delta u_t + \tilde{H}(Z)e_t$$

The plant between δy_t and δu_t remains the same as that between y_t and u_t, but the complexity of the noise model can be reduced.

The equivalence between weighting the fitting criterion, pre-filtering the data, and modifying the noise model is useful in providing methods for the inclusion of *a priori* process information into the identification. This will be brought out in specific examples later in Chapters 10 and 13.

3.3.1 Detrending Data

Plotting the spectrum of the raw chute signal in Figure 3.2, we see from Figure 3.9 that the content of the data is entirely dominated by the DC or constant component. This is no surprise, given that the ranging of the data between 0 and 100 per cent guarantees an offset from zero. When the data have their mean removed, so that the signal is expressed as a variation about zero, then we achieve the more informative plot of Figure 3.3. Had a system identification been performed using the raw, un-detrended data, then according to (3.2) the parameters would have been constrained to make the DC signal levels match almost perfectly. Often, particularly when seeking

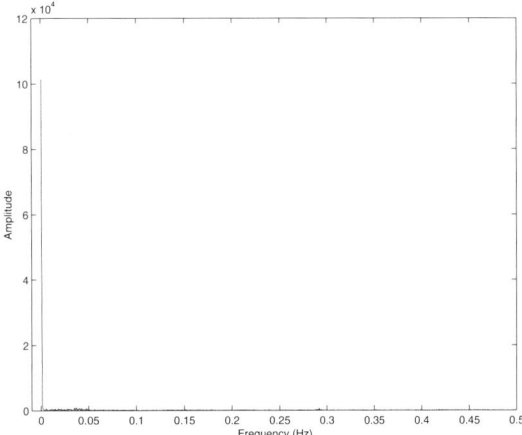

Fig. 3.9. Spectrum of raw chute data signal showing dominance of DC levels (note that the frequency axis has been shifted from zero)

linearised models, we do not really care to model the offsets and so it is important to detrend.

Detrending, since it corresponds precisely to a linear filtering of the data, can be interpreted as any other filter and included as part of L. However, it is important to apply identical filtering to both input and output signals. Equally, detrending only removes constant offsets. For a non-stationary signal such as that in Figure 3.8, more sophisticated pre-processing would be necessary. The detrend function of MATLAB® provides for the removal of piece-wise linear trends in addition to the DC value. Figure 3.10 shows the non-stationary data after processing through dtrend(nstat,'linear',[200 1200 1665]). While this has removed the visibly obvious trends in the data, it is uncertain what effect this might have in using this transformed data to fit a model. Specific process knowledge needs to be introduced in order to interpret the effects because the filtering effect of the detrending is time-varying and need not commute with other operations.

3.3.2 Data Filtering

We saw above that simple detrending can be used to eliminate DC offsets in the data when these are of little interest or, as in the case above, so dominate the fit as to jeopardise the identification. Other forms of linear, time-invariant filtering can also assist in concentrating the attention of the model fit to desirable frequency regimes. Indeed, one of the theses of this book is that this should be done to improve the quality of model fit to suit subsequent control design.

A differencer was also suggested earlier as a means of removing these offsets, since the transfer function $1 - z^{-1}$ has a zero at DC, and as a way

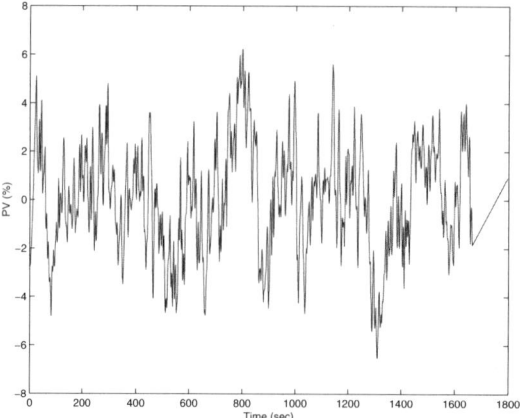

Fig. 3.10. Detrended non-stationary process data: *Caution!*

to avoid modelling the pole at $z = 1$ known to be in the noise model. The differencer also has a strong effect on other frequency regions – its frequency response magnitude increases roughly linearly with frequency and so it emphasises high frequencies at the expense of low frequencies. Figure 3.11 shows the frequency response magnitudes of three candidate detrenders. The dashed

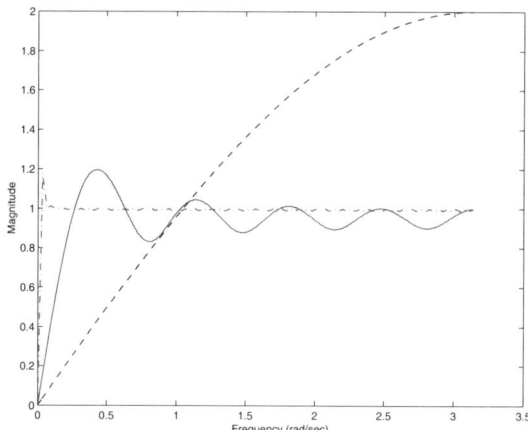

Fig. 3.11. Magnitude responses of several detrenders: differencer (dashed), 10-point detrender (solid), 100-point detrender (dot-dash)

curve is that of the differencer $1 - z^{-1}$, the other two are of mean-removing detrenders with transfer functions $1 - \frac{1}{N}\sum_{i=1}^{N} z^{-i}$ for $N = 10$ and $N = 100$. Note the zero at zero frequency and the level of distortion introduced into the other frequency bands.

If our intention in fitting a model is to develop a proportional-plus-integral controller, then we might not have much interest in the accuracy at low frequencies. So, a filter like the differencer or the 10-point detrender might suffice. However, if the sampling rate is well chosen, it is unlikely that good model quality will be useful at very high frequencies. Accordingly, we might wish to demerit this weighting by introducing a low-pass cutoff. There are numerous filter design tools in MATLAB® that might be used, such as butter(N,Wn), besself(N,Wn), cheby1(N,R,Wn), *etc.* Indeed, digital filter design is a vigorous area of activity itself. Figure 3.12 shows the magnitude response of a filter consisting of a cascade of the 10-point detrender with a fourth order Butterworth low-pass filter with a cut-off at 0.6 π rad/s. This might form a useful

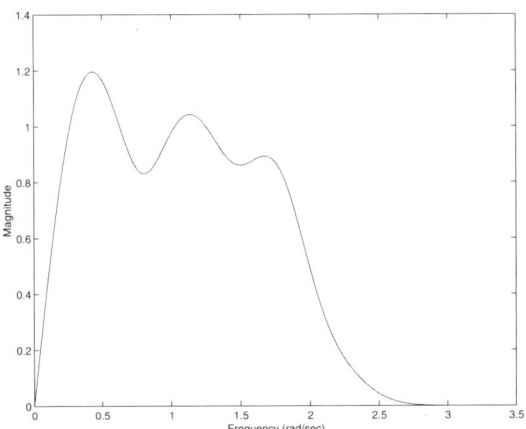

Fig. 3.12. Data pre-filter magnitude response

first-pass data pre-filter with which to commence a system identification.

Several broad comments can be made regarding this preparation phase of the data by pre-filtering:

- successive filtering of the data is possible, so that a differencer might be followed by a low-pass filter (as above) and this may then be followed by a filter which, say, emphasises the control cut-off frequencies. Because stable linear systems commute, the order of filtering is immaterial;
- it is important to document filterings of the data and to ensure that identical operations are performed on all data to be used to fit a linear model;
- the role of filtering in system identification is to implement a weighting function in the frequency domain performance measure. Its function is more persuasion and emphasis than complete redefinition of objective. Consequently, it can be handled fairly coarsely without undue difficulties

begin introduced. Thus, the pass-band ripple in Figures 3.11 and 3.12 is generally not of concern;
- system identification is based on least squares. Diminishing the frequency content of the filtered signals to zero can make this least squares problem ill-posed, like solving $0 = K \times 0$ for K. Thus, using filters that have very sharp cut-off and extreme attenuation in large bands can result in unreliable model fits in these bands. Some care is needed not to remove too much information from the signals;

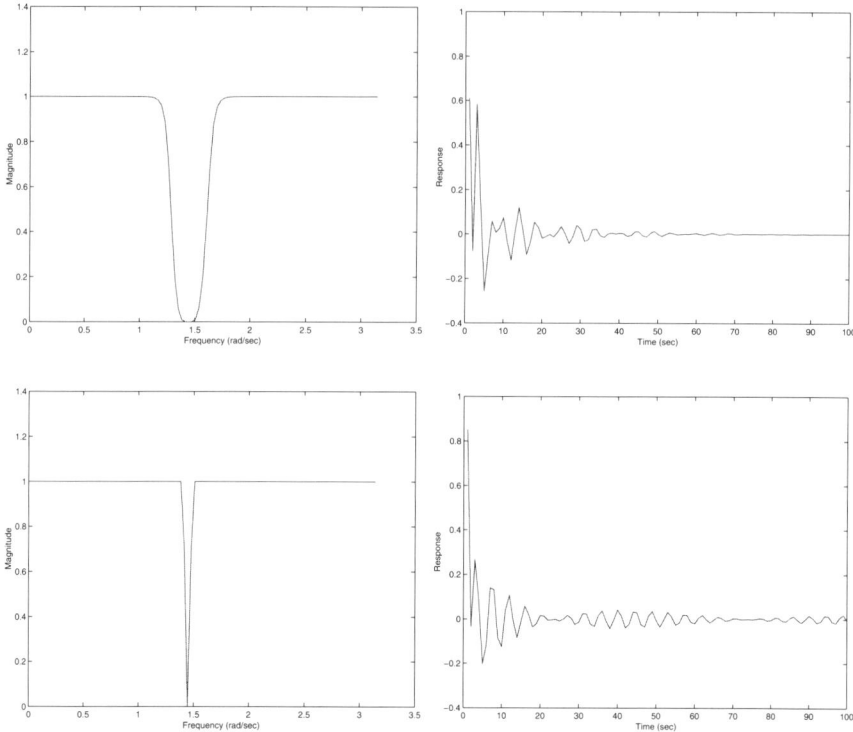

Fig. 3.13. Magnitude and impulse responses for wide-band (upper traces) and for narrow-band Butterworth bandstop filters

- note from Figure 3.11 that the 100-point detrender has a much sharper notch at DC than does the 10-point detrender. Note also that the response time of the former is 10 times longer than that of the latter. This is a manifestation of the time-bandwidth product relation – as the bandwidth of a filter is narrowed, correspondingly the decay time of its impulse response becomes longer, keeping the product roughly constant. This is illustrated in Figure 3.13, where the magnitude responses and impulse responses are displayed for two Butterworth bandstop filters. The wider band filter is fourth-order and has bandwidth of 0.12 rad/s. The narrower band filter is eighth order with a bandwidth of 0.02 rad/s. The importance of understanding the time–bandwidth product relation in system identification is that very narrow-band filters will require a long time (*i.e.*, a large amount of data) to forget their initial conditions. Some caution is needed in selecting appropriate bandwidths. This remark might be coupled with those above to counsel relatively mild filters without precipitous cut-offs or extreme nulls.

3.4 System Identification with an $\mathcal{H}_\infty$ Criterion

So far, our attention has been on the fitting of models that minimise a least mean squared prediction error criterion. The frequency domain formula (3.2) exhibits the fit to the frequency response of the real plant, $G(e^{j\omega})$. For robust control design, it is often desirable to have a model whose fit is bounded in terms of the maximum modulus of the frequency response error, *i.e.*, $\max_{\omega \in [0,\pi]} \left| G(e^{j\omega}) - \hat{G}(e^{j\omega}) \right|$. This is the $\mathcal{H}_\infty$-norm for stable systems. A bound on this error can translate into a constraint on the control design, the satisfaction of which *guarantees* robust stability [163]. Since fitting models with such a criterion involves estimation of frequency response values of G at specific points around the unit circle, which in turn requires estimation from filtered data, this is an opportune time to introduce the ideas.

Given frequency samples of G equally spaced around the unit circle,

$$\left\{ \hat{G}_k = G(e^{i2\pi k/N}) + \epsilon_k, \quad k = 0, \ldots, N-1 \right\}$$

with $|\epsilon_k| < d$, then it is possible using the methods of [67, 74] to compute an approximant, $\hat{G}$, of G with the following error bound.

$$\max_{\omega \in [0,\pi]} \left| G(e^{j\omega}) - \hat{G}(e^{j\omega}) \right| \leq \frac{2M}{\rho - 1} \mathcal{O}(\rho^{-N}) + 2d$$

where all the poles G lie inside a circle of radius $\rho^{-1} < 1$ and M is the maximum modulus of G on this circle. The interest in such problems focuses our attention on the accurate estimation of the frequency samples, in order to reduce the d term. How might this be done?

Considering (3.2), we see that if the data is filtered through a very narrow bandpass filter $L(z)$ with centre frequency $\frac{2\pi k}{N}$, and then a standard least squares estimation is performed on the complex gain, then this effectively matches the estimated gain to the frequency sample. The size of the error term ϵ_k depends on the narrowness of the filter and on the noise level in the data. Using the same set of data, this filtering and estimation may be performed at every frequency sampling point, provided the data contains sufficient energy to make the gain identifiable. Note that there appears a coupling between the accuracy with which one may estimate the frequency samples and the quantity of available data. Our preceding discussion is still pertinent in this case. This will be revisited in the practical example of helicopter vibration control in Chapter 10.

3.5 Conclusions

Our goals in this chapter have been to develop general principles, tools and techniques for the design of experiments and the preparation of data prior to the use of system identification algorithms to extract the best-fitting model. Practical experience suggests that efforts in this area can be exceptionally fruitful in understanding and managing the identification phase. Prior information about the plant, the disturbances and the ultimate design objective are most easily incorporated at this stage – hence the concentration of variables of identification.

In the following chapters, we shall see the development of specific industrial examples that demonstrate and benefit from these tools being applied to physical processes. Alongside this, we shall see the repeated appeal to the closed-loop frequency domain modelling formula (3.2) to underpin the development of theoretical approaches to encourage the identified model to be "good for control design". The purpose of this chapter has been to construct a bridge between considering the raw data and its processing and understanding on a conceptual level of the equivalent effects on the nature of the model fitted.

Acknowledgements

Research supported by US National Science Foundation under Grant ECS-0070146.

4 Model Uncertainty and Feedback

Karl J. Åström

4.1 Introduction

A key reason for using feedback is to reduce the effects of uncertainty that may appear in different forms as disturbances or as other imperfections in the models used to design the feedback law. Model uncertainty and robustness have been a central theme in the development of the field of automatic control. This paper gives an elementary presentation of the key results.

A central problem in the early development of automatic control was to construct feedback amplifiers whose properties remained constant in spite of variations in supply voltage and component variations. This problem was the key for the telephone industry that emerged in the 1920s. The problem was solved by [26]. We quote from this paper:

> " .. by building an amplifier whose gain is deliberately made say 40 decibels higher than necessary (10 000 fold excess on energy basis) and then feeding the output back on the input in such a way as to throw away excess gain, it has been found possible to effect extraordinary improvement in constancy of amplification and freedom from non-linearity."

Black's invention had a tremendous impact and it inspired much theoretical work. This was required both for understanding and for development of a design method. A novel approach to stability was developed in [118], fundamental limitations were explored by [27], who also developed methods for designing feedback amplifiers (see [28]). A systematic approach to design controllers that were robust to gain variations were also developed by Bode.

The work on feedback amplifiers became a central part of the theory of servomechanisms that appeared in the 1940s (see [71, 88]). Systems were then described using transfer functions or frequency responses. It was very natural to capture uncertainty in terms of deviations of the frequency responses. A number of measures, such as amplitude and phase margins and maximum sensitivities, were also introduced to describe robustness. Design tools such as the Bode diagram introduced to design feedback amplifiers also found good use in the design of servomechanisms. Bode's work on robust design was generalised to deal with arbitrary variations in the process by Horowitz [82]. The design techniques used were largely graphical.

The state-space theory that appeared in the 1960s represented significant paradigm shift. Systems were now described using differential equations. There was a very vigorous development that gave new insight, new concepts

[92, 94] and new design methods. Control design problems were formulated as optimisation problems that gave effective design methods (see [21, 22, 131]). Control of linear systems with Gaussian disturbances and quadratic criteria, the linear quadratic Gaussian (LQG) problem, was particularly attractive because it admitted analytical solutions (see [22, 90, 91, 93]). The design computations were also improved because it was possible to capitalise on advances in numerical linear algebra and efficient software. The controller obtained from LQG theory also had a very interesting structure. It was a composition of a Kalman filter and a state feedback.

The state-space theory became the predominant approach, [16]. Safonov and Athans [134] showed that the phase margin is at least 60° and the gain margin is infinite for an LQG problem where all state variables are measured. This result does, unfortunately, not hold for output feedback as was demonstrated in [50]. There were attempts to recover the robustness of state feedback using special design techniques called *loop transfer recovery*. The central issue is, however, that it is not straightforward to capture model uncertainty in a state variable setting. There was also criticism of the state-space theory [84].

The paper [158] represented a paradigm shift that brought robustness to the forefront. It started a new development that led to the so-called $\mathcal{H}_\infty$ theory. The idea was to develop systematic design methods that were guaranteed to give stable closed-loop systems for systems with model uncertainty. The original work was based on frequency responses and interpolation theory, which led to compensators of high order. The seminal paper [52] showed how the problem could be solved using state-space methods. Game theory is another approach to $\mathcal{H}_\infty$ theory. The game is to find a controller in the presence of an adversary that changes the process [17]. The $\mathcal{H}_\infty$ theory is now well described in books [51, 66, 164].

Major advances in robust design were made in the book [114], where the $\mathcal{H}_\infty$ control problem was regarded as a loop shaping problem. This gave effective design methods and it also re-established links with classical control. This line of research has been continued by [155], who has obtained definite results relating modelling errors and robust control. To do this he also had to invent a novel metric for systems [153]. This work brings $\mathcal{H}_\infty$ even closer to the classical results.

In this chapter, we will try to present the essence of the development in the simple setting of single-input single-output systems. We start with a presentation of some aspects of classical control theory in Section 4.2. Robustness issues for state-space theory is reviewed briefly in Section 4.3, where we also present an example that illustrates that a blind use of state-space theory can lead to closed-loop systems with very poor robustness properties. This can be overcome by analysing the robustness and modifying the design criteria. In Section 4.4, we discuss fundamental limitations on performance due to time delays and right half plane poles and zeros. This does not appear naturally in state-space theory, where the major requirements are observability and controllability. Section 4.5 presents some key results from $\mathcal{H}_\infty$ loop shaping. To do this, we have to discuss the important problem of determining if two systems are close from the point of view of feedback. We also have to introduce better stability concepts and ways to characterise model uncertainty. We end the section with a discussion of Vinnicombe's theory, which gives much insight, necessary conditions and very nice connections to classical control theory.

4.2 Classical Control Theory

The fields of automatic control emerged in the mid 1940s, when it was realised that it was a common framework for problems associated with feedback control from a wide range of fields such as telecommunications, industrial processes, vehicles, power systems, *etc.* An essential component came from telecommunications, where a key problem had been to design accurate reliable amplifiers from components with variable properties.

The feedback amplifier

A schematic diagram of the amplifier is shown in Figure 4.1. Let the raw gain of the amplifier be A. The feedback amplifier has very remarkable properties as can be seen from its input-output relation. It follows from Figure 4.1 that

$$\frac{V_2}{V_1} = -\frac{R_2}{R_1} \frac{1}{1 + \frac{1}{A}\left(1 + \frac{R_2}{R_1}\right)} \tag{4.1}$$

Notice that the gain V_2/V_1 is essentially given by the ratio R_2/R_1. If the raw amplifier gain A is large, the gain is virtually independent of A. Assume, for example, that $R_2/R_1 = 100$ and that $A = 10^4$. With a 10 per cent change of A the variation gives only a gain variation of 0.1 per cent. Feedback thus has the amazing property of drastically reducing the effects of uncertainty. The linearity is also increased significantly.

The risk of instability is a drawback of feedback. Nyquist developed a theory for analysing stability of feedback amplifiers [118]. Systematic methods

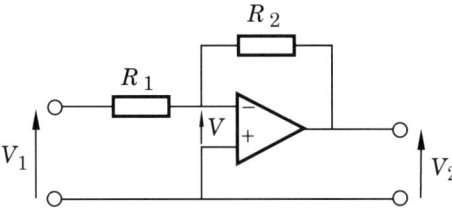

Fig. 4.1. Schematic diagram of a feedback amplifier

for designing feedback systems were developed in [27] and further elaborated on in [28]. These ideas formed one of the foundations of automatic control.

In today's terminology, we could say that Black used feedback to design an amplifier that was robust to variations in the gain of the process. As a side-effect, he also obtained a closed-loop system that was extremely linear.

Generalisation

The ideas of feedback are applicable to a wide range of systems. This is illustrated in Figure 4.2, which shows a basic feedback loop consisting of a process and a controller. The purpose of the system is to make the process

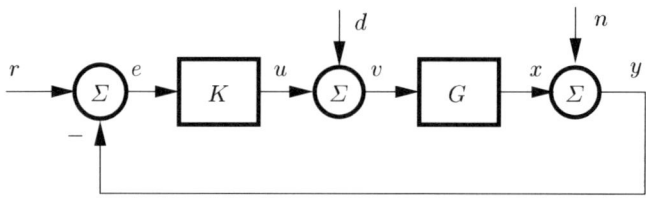

Fig. 4.2. Block diagram of a feedback system

variable x follow the set point r in spite of the disturbances d and n that act on the system. The properties of the closed-loop system should also be insensitive to variations in the process. There are two types of disturbances. The load disturbance d drives the system away from its desired state and the measurement noise n, which corrupts the information about the system obtained from the sensors.

The system in Figure 4.2 has three inputs r, d and n, and four interesting signals x, y, e and u. There are thus 12 relations that are of potential interest. Assume that the process and the controller are linear time-invariant systems that are characterised by their transfer functions G and K respectively. The relations between the signals are given by the transfer functions:

$$G_{xr} = \frac{GK}{1+GK} \qquad G_{xd} = \frac{G}{1+GK} \qquad G_{xn} = -G_{xr}$$

$$G_{yr} = G_{xr} \qquad G_{yd} = G_{xd} \qquad G_{yn} = \frac{1}{1+GK}$$

$$G_{er} = 1 - G_{xr} = G_{yn} \qquad G_{ed} = -G_{xd} \qquad G_{en} = -G_{yn}$$

$$G_{ur} = \frac{K}{1+GK} \qquad G_{ud} = -G_{xr} \qquad G_{un} = -G_{ur}$$

Here, G_{ij} denotes the transfer function from signal j to signal i. Notice that there are only four independent transfer functions.

$$\begin{aligned} G_{xr} &= \frac{GK}{1+GK} \\ G_{xd} &= \frac{G}{1+GK} \\ G_{yn} &= \frac{1}{1+GK} \\ G_{ur} &= \frac{K}{1+GK} \end{aligned} \tag{4.2}$$

We call these "the gang of four". The transfer functions $G_{xr} = T$ and $G_{yn} = S$ have special names, S is called the *sensitivity function*, and T is called the *complementary sensitivity function*. Notice that both S and T depend only on the loop transfer function $L = GK$. The sensitivity functions are related by

$$S + T = 1 \tag{4.3}$$

This explains the term complementary sensitivity function. These functions have interesting properties as is discussed in the following.

Stability and stability margins

Many properties of the system in Figure 4.2 can be derived from the loop transfer function $L = GK$. The stability of the system can be investigated by Nyquist's stability criterion, which says that the closed-loop system is stable if

$$\frac{1}{2\pi} \Delta \arg_\Gamma (1 + L(s)) = -\mathcal{P}_{rhp}(L) \tag{4.4}$$

where $\Delta \arg$ is the argument variation when s traverses a contour Γ that encloses the right half plane (RHP) and $\mathcal{P}_{rhp}(L)$ is the number of poles of L in the right half plane.

Stability is normally investigated by analysing the Nyquist curve, see Figure 4.3. To achieve stability, the Nyquist curve must be sufficiently far away from the critical point -1. The distance from the critical point can also be used as a measure of the degree of stability. In this way, the notions of gain margin A_m and phase margin φ_m are defined (see Figure 4.3). A gain margin A_m implies that the gain can be increased with a factor less than A_m without making the system unstable. Similarly, for a system with a phase margin φ_m, it is possible to increase the phase shift in the loop by a quantity less than φ_m without making the system unstable.

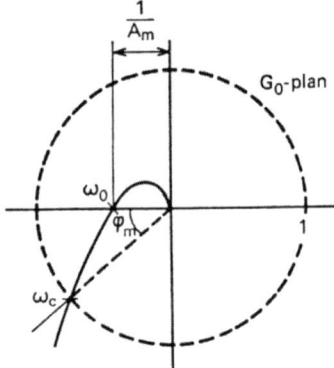

Fig. 4.3. Nyquist curve with phase and gain margins

4.2.1 Small Process Variations

The effects of small variations in the process will now be investigated. The signal transmission of the closed-loop system from set point r to output y is given by the complementary sensitivity function

$$T = \frac{GK}{1 + GK}$$

We have

$$\frac{dT}{T} = \frac{dG}{G} - \frac{K dG}{1 + GK} = \frac{1}{1 + GK} \frac{dG}{G} = S \frac{dG}{G} \tag{4.5}$$

The function S tells how the closed-loop properties are influenced by small variations in the process. The maximum sensitivities

$$M_s = \max_\omega |S(i\omega)|, \qquad M_t = \max_\omega |T(i\omega)| \tag{4.6}$$

are also used as robustness measures. The variable $1/M_s$ can be interpreted as the shortest distance between the Nyquist curve and the critical point -1, see Figure 4.3. The maximum complementary sensitivity is also denoted by M_p in classical literature.

4.2.2 Disturbance Rejection and the Sensitivity Function

The sensitivity function also has other physical interpretations. Consider the system in Figure 4.2. If there is no feedback, the Laplace transform of the output is Y_{ol}. The output under closed-loop control is given by

$$Y_{cl} = \frac{1}{1 + GK} Y_{ol}$$

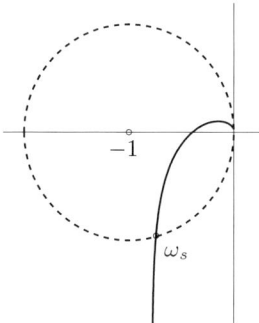

Fig. 4.4. Nyquist curve with circle for constant sensitivity $S(i\omega) = 1$. Disturbances with frequencies outside the circle are attenuated and frequencies inside the circle are amplified by the feedback

It thus follows that

$$\frac{Y_{cl}}{Y_{ol}} = \frac{1}{1+GK} = S \tag{4.7}$$

The sensitivity function thus tells how the disturbances are influenced by feedback. Disturbances with frequencies such that $|S(i\omega)|$ is less than 1 are reduced by an amount equal to the distance to the critical point and disturbances with frequencies such that $|S(i\omega)|$ is larger than 1 are amplified by the feedback. This is illustrated in Figure 4.4.

4.2.3 Large Process Variations

Equation 4.5 gives the sensitivity for small perturbations of the process. It is also possible to get expressions for large variations in the process. To see how much the process can change without making the closed-loop system unstable, we will use the Nyquist diagram. Consider a point on the Nyquist curve in Figure 4.5 on the following page. The distance to the critical point is $|1 + L|$. If the process changes by ΔG, the point changes by $K\Delta G$. The system will remain stable as long as

$$|K\Delta G| < |1 + GK|$$

and the number of right hand poles of GK does not change. This implies that the perturbations must have the property that ΔG does not have any poles in the right half plane.

The admissible variation in process dynamics is thus given by

$$|\frac{\Delta G}{G}| < \left|\frac{1+GK}{GK}\right| = \left|\frac{1}{T}\right| \tag{4.8}$$

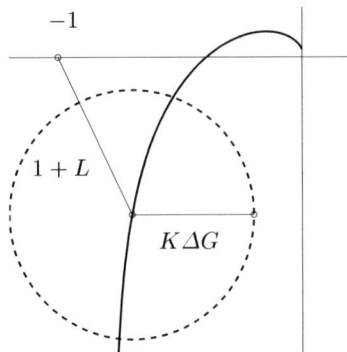

Fig. 4.5. Nyquist curve with circle for the loop transfer function $L = GK$. With process variations of magnitude $|\Delta G|$, the point in the middle of the circle can move anywhere in the circle

which can also be expressed as

$$|\Delta G| < \left|\frac{G}{T}\right| \tag{4.9}$$

It follows from Equation 4.8 that the largest relative precision $1/M_t$ is required at the frequency ω_t where the complementary sensitivity has its maximum. It also follows that the requirements on model precision are very modest when T is small.

Equation 4.9 implies that allowable modelling error is governed by the ratio $|G/T|$. This means that large errors can be allowed when $|G/T|$ is large and high model precision is required when $|G/T|$ is small.

A similar estimate based on the maximum sensitivity is that

$$|\Delta G| < \left|\frac{1}{SK}\right| \tag{4.10}$$

It follows from this equation that large modelling errors are allowed when the sensitivity and the controller gains are large.

Bode's integrals

It follows from Equations (4.5) and (4.8) that it would be highly desirable to make the sensitivity functions S and T as small as possible. This is, unfortunately, not possible because it follows from Equation (4.3) that $S + T = 1$. There are also other constraints on the sensitivities. It was shown in [28] that if the loop transfer function $L(s)$ goes to zero faster than $1/s$ for large s, then

$$\int_0^\infty \log|S(i\omega)|\,d\omega = \int_0^\infty \log\left|\frac{1}{1+L(i\omega)}\right|d\omega = \pi\sum p_i$$
$$\int_0^\infty \log|T(1/i\omega)|\,d\omega = \int_0^\infty \log\left|\frac{L(1/i\omega)}{1+L(1/i\omega)}\right|d\omega = \pi\sum \frac{1}{z_i} \quad (4.11)$$

where p_i are the right half plane poles of L and z_i are the right half plane zeros of L, see also [145]. These equations imply that the sensitivities can be made small at one frequency only at the expense of increasing the sensitivity at other frequencies. This phenomena is sometimes called the *water bed effect*. It also follows from the equations that the presence of poles in the right half plane increase the sensitivity and that zeros in the right half plane increase the complementary sensitivity. A fast RHP pole gives higher sensitivity than a slow pole, and a slow RHP zero gives higher sensitivity than a fast zero.

4.2.4 Bode's Relations

The amplitude and the phase curves are also related. It is not possible to achieve high-phase advance without using high gains and it is not possible to obtain transfer functions that decrease rapidly without having large phase lags. These facts are expressed analytically by some relations derived in [27].

Consider a transfer function $G(s)$ with no poles or zeros in the right half plane. Introduce

$$\log G(i\omega) = A(\omega) + i\Phi(\omega) \quad (4.12)$$

a logarithmic frequency scale $u = \log \omega/\omega_0$, $\omega = \omega_0 e^u$, and the functions

$$a(u) = A(\omega_0 e^u), \quad \phi(u) = \Phi(\omega_0 e^u)$$

Assume that $(\log G(s))/s$ goes to zero as s goes to infinity, then

$$A(\omega_0) - A(\infty) = -\frac{2}{\pi}\int_0^\infty \frac{v\Phi(v) - \omega_0\Phi(\omega_0)}{v^2 - \omega_0^2}dv$$
$$= -\frac{1}{\omega_0\pi}\int_{-\infty}^\infty \frac{d(e^u\phi(u))}{du}\log\coth\left|\frac{u}{2}\right|du$$
$$\Phi(\omega_0) = \frac{2\omega_0}{\pi}\int_0^\infty \frac{A(v) - A(\omega_0)}{v^2 - \omega_0^2}dv = \frac{1}{\pi}\int_{-\infty}^\infty \frac{da(u)}{du}\log\coth\left|\frac{u}{2}\right|du$$
$$(4.13)$$

an approximate version is that

$$\Phi(\omega) \approx \frac{2}{\pi}\frac{da(u)}{du} \quad (4.14)$$

This means that if the slope $n = da(u)/du$ of the magnitude curve is constant the phase is $n\pi/2$. This relation appears in practically all elementary courses in feedback control.

Bode's relations impose fundamental limitations on the performance that can be achieved. A simple observation is that even if it is desirable that the loop gain decreases rapidly at the crossover frequency, it is not possible to have a steeper slope than -2 without violating stability constraints.

An interesting question is if the limitations imposed by Bode's relations can be avoided by using non-linear systems. The Clegg integrator [40] is a non-linear system where the magnitude curve has the slope -1 and the phase lag is only $38°$.

4.2.5 Bode's Ideal Loop Transfer Function

In his work on design of feedback amplifiers Bode suggested an ideal shape of the loop transfer function. He proposed that the loop transfer function should have the form

$$L(s) = \left(\frac{s}{\omega_{gc}}\right)^n \tag{4.15}$$

The Nyquist curve for this loop transfer function is simply a straight line through the origin with $\arg L(i\omega) = n\pi/2$, see Figure 4.6. Bode called (4.15)

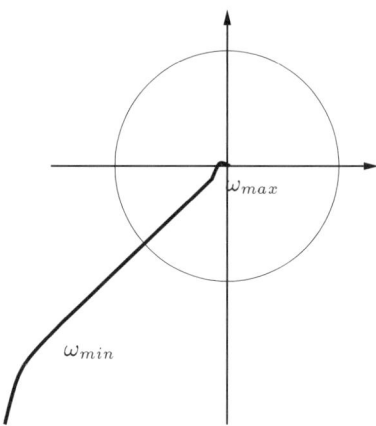

Fig. 4.6. Nyquist curve for Bode's ideal loop transfer function

the ideal cut-off characteristic. In the terminology of automatic control, we will call it *Bode's ideal loop transfer function*. One reason why Bode made the particular choice of $L(s)$ given by Equation (4.15) is that it gives a closed-loop system that is insensitive to gain changes. Changes in the process gain will change the crossover frequency but the phase margin is $\varphi_m = \pi(1+n/2)$ for all values of the gain. The gain margin is infinite. The slopes $n = -1.333$, -1.5 and -1.667 correspond to phase margins of 60°, 45° and 30°. Bode's idea to use loop shaping to design controllers that are insensitive to gain variations were later generalised by [82] to systems that are insensitive to other variations of the plant, culminating in the QFT method [83].

The transfer function given by Equation (4.15) is an irrational transfer function for non-integer n. It can be approximated arbitrarily close by rational frequency functions. Bode also suggested that it was sufficient to approximate L over a frequency range around the desired crossover frequency ω_{gc}. Assume, for example, that the gain of the process varies between k_{min} and k_{max} and that it is desired to have a loop transfer function that is close to (4.15) in the frequency range $(\omega_{min}, \omega_{max})$. It follows from (4.15) that

$$\frac{\omega_{max}}{\omega_{min}} = \left(\frac{k_{max}}{k_{min}}\right)^{1/n}$$

With $n = -5/3$ and a gain ratio of 100, we get a frequency ratio of about 16 and with $n = -4/3$ we get a frequency ratio of 32. To avoid having too large a frequency range, it is thus useful to have n as small as possible. There is, however, a compromise because the phase margin decreases with decreasing n and the system becomes unstable for $n = -2$.

4.2.6 Fractional Systems

It follows from Equation (4.15) that the loop transfer function is not a rational function. We illustrate this with an example.

Example.
Consider a process with the transfer function

$$G(s) = \frac{k}{s(s+1)} \tag{4.16}$$

Assume that we would like to have a closed-loop system that is insensitive to gain variations with a phase margin of 45°. Bode's ideal loop transfer function that gives this phase margin is

$$L(s) = \frac{1}{s\sqrt{s}} \tag{4.17}$$

Since $L = GK$ we find that the controller transfer function is

$$K(s) = \frac{s+1}{\sqrt{s}} = \sqrt{s} + \frac{1}{\sqrt{s}} \tag{4.18}$$

This is called a *fractional transfer function*, or a *fractional system*, because the transfer function is not a rational function. To implement a controller,

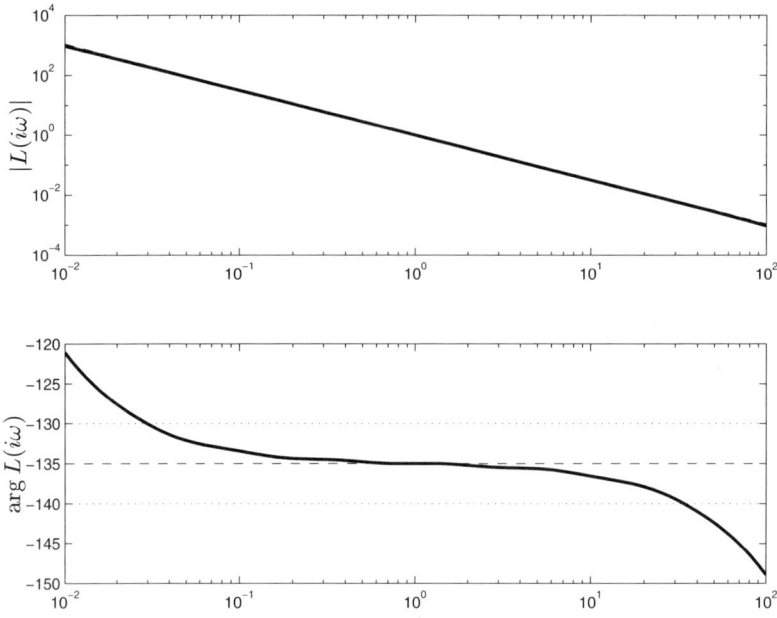

Fig. 4.7. Bode diagram of the loop transfer function obtained by approximating the fractional controller (4.18) with the rational transfer function (4.19). The fractional transfer function is shown in full lines and the approximation in dashed lines

the transfer function is approximated with a rational function. This can be done in many ways. One possibility is the following

$$\hat{K}(s) = k \frac{(s+1/64)(s+1/16)(s+1/4)(s+1)^2(s+4)(s+16)(s+64)}{(s+1/128)(s+1/32)(s+1/8)(s+1/2)(s+2)(s+8)(s+32)(s+128)} \quad (4.19)$$

where the gain k is chosen to equal the gain of $\sqrt{s} + 1/\sqrt{s}$ for $s = i$. Notice that the controller is composed of sections of equal length having slopes 0, +1 and −1 in the Bode diagram.

Figure 4.7 shows the Bode diagrams of the loop transfer functions obtained with the fractional controller (4.18) and its approximation (4.19). The differences of the gain curves are not visible in the graph. Figure 4.7 shows that the phase margin is close to 45° even if the gain changes substantially. The differences in phase is less than 2° for a gain variation of three orders of magnitude. With a tolerance of 5°, we can even allow a gain variation of four orders of magnitude. The range of gains can be extended by increasing the order of the approximate controller (4.19). Even if the closed-loop system has the same phase margin when the gain changes, the response speed will change with the gain.

The example shows that robustness is obtained by increasing controller complexity. The range of gain variation that the system can tolerate can be increased by increasing the complexity of the controller.

Fractional systems did not receive much attention after Bode's work. In the 1990s there was, however, a resurgence in interest in fractional systems [18, 120, 121, 129, 130]. Oustaloup coined the acronym CRONE (Commande Robuste d'Ordre Non Entier), which translates as Robust Control of Fractional Order, for his controller.

4.2.7 Systems with Two Degrees of Freedom (2-DOF)

The system in Figure 4.2 is said to have error feedback because control is based on the error $e = r - y$, which is the difference between the reference and the output signals. There are significant advantages in having control systems with other configurations. An example of such a system is shown in Figure 4.8. In this system, the set point is fed through a model before it is

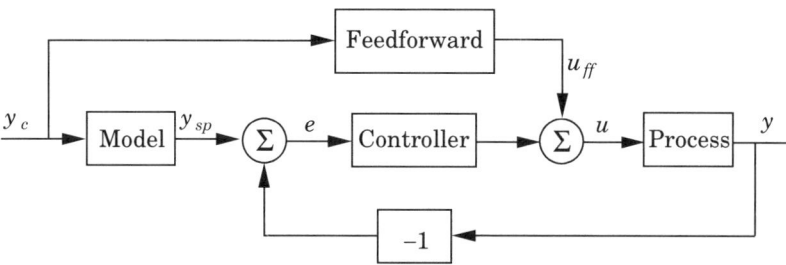

Fig. 4.8. Block diagram of a system with two degrees of freedom

compared with the process output. There is also a feedforward link, which essentially is a combination of the model and the inverse process model or an approximate inverse. Ideally, the feedforward signal u_{ff} generates a signal which, when applied to the process, produces the ideal output in response to set point changes. The feedback controller that acts on the error will only make corrections if the output y deviates from the ideal output y_{sp} generated by the model.

A system with error feedback only is called a system with one degree of freedom. The system in Figure 4.8 is called a system with two degrees of freedom (2-DOF) because the signal path from the set point to the control signal is different from the signal path from the output to the control signal. This terminology was introduced by [82], who analysed these systems carefully.

A very nice property of systems with two degrees of freedom is that the problem of set point response can be separated from the problems of robustness and disturbance rejection. Referring to Figure 4.2, we will first design a feedback by compromising between disturbance attenuation and robustness. When this is done, we will then design a model and a feedforward that gives the desired response to the setpoint.

There are many variations of systems with two degrees of freedom, the following quote from [82] is still valid provided that the system is linear.

> "Some structures have been presented as fundamentally different from the others. It has been suggested that they have virtues not possessed by others, and have been given special names ... all 2-DOF configurations have basically the same properties and potentials ..."

4.2.8 Quantitative feedback theory (QFT)

Bode's technique of dealing with gain variations is both elegant and effective. A limitation of Bode's work was that it was limited to gain variations only. A very nice generalisation of Bode's work was done by Horowitz, who extended it to arbitrary variations of a process transfer function. He characterised model uncertainty by sets of amplitudes and phase for each frequency, called *templates*. Horowitz also developed a graphical design technique to design feedback systems that were robust to these types of disturbances. He used a system configuration with two degrees of freedom to deal with set point responses. Horowitz's design technique, called quantitative feedback theory, is described in several books, see [82, 83]. It has been applied successfully to a wide range of problems.

Summary

In this section, we have reviewed classical control theory with a focus on model uncertainty and robustness. It is worthwhile to note that model uncertainty was a key motivation for introducing feedback and that classical control theory had very effective ways of dealing with uncertainty, both qualitatively and quantitatively. Process uncertainty could be described very easily as a variation in the process transfer function with the caveat that the disturbances do not change the number of right half plane poles of the system. The theory has given important concepts and tools such as the transfer function, Nyquist's stability theory, the Nyquist curve, Bode diagrams, Bode's integrals and Bode's ideal loop transfer function. Robustness measures such as amplitude and phase margins and the maximum sensitivities were also introduced. Bode's ideal loop transfer function is probably the first design method that addressed robustness explicitly. Horowitz's quantitative feedback theory is a continuation of this idea.

4.3 State-space Theory

The state-space theory represented a paradigm shift, which led to many useful system concepts and new methods for analysis and design. The systems were represented by differential equations instead of transfer functions. For linear systems, the standard model used was

$$\frac{dx}{dt} = Ax + Bu + v$$
$$y = Cx + e$$
(4.20)

where u is the input, y the output and x is the state. The uncertainty is represented by the disturbances v and e and by variations in the elements of the matrices A, B and K. The disturbances e and v have been typically described as stochastic processes, see [59] and [15].

The control problem was formulated as to minimise the criterion

$$J = E \lim_{T \to \infty} \frac{1}{T} \int_0^T (x^T Q_1 x + u^T Q_2 u) dt$$
(4.21)

Since the equations are linear with stochastic disturbances and the criterion is quadratic, the problem was called the linear quadratic Gaussian control problem (LQG). The solution to the control problem is given by

$$u = L(x_m - \hat{x}) + u_{ff}$$
$$\frac{d\hat{x}}{dt} = A\hat{x} + Bu + K(y - C\hat{x})$$
(4.22)

This control law has a very nice interpretation as feedback from the error $x_m - \hat{x}$, which is the difference between the ideal states x_m and the estimated states $\hat{x}$. The estimated states are given by the Kalman filter. Controllability and observability are key conditions for solving the problem.

There are many other design methods based on the state-space formulation that give controllers with the structure (4.22), for example, pole placement. They differ from the LQG method in the sense that other techniques are used to obtain the matrices K and L.

In Figure 4.9, we show a block diagram of the controller obtained from LQG theory. In the figure, we have also used a system configuration with two degrees of freedom. The system has a very attractive structure. The observer or the Kalman filter delivers an estimate of the state based on a model of the system and the input and output signals of the system. Notice that the state may also have components that represent the disturbances. There is a feedback from the deviations of the estimated state from its desired value x_m. Set point following is obtained by the usual two-degree-of-freedom configuration.

Robustness

In the model (4.20), it is natural to describe model uncertainties as variations in the elements of the matrices A, B and K. This is, however, a very restricted class of perturbations, which does not cover neglected dynamics or small time

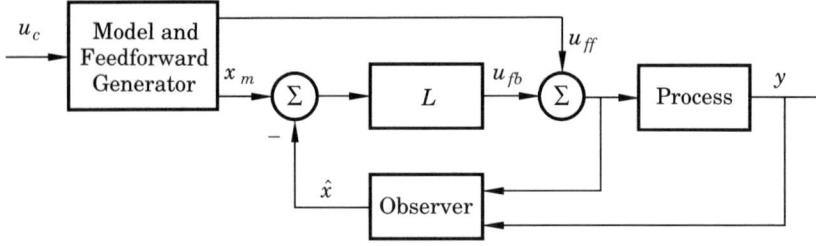

Fig. 4.9. Block diagram of a system state feedback having two degrees of freedom

delays. Such uncertainties are easier to describe in the frequency domain. The LQG theory was also criticised heavily by classic control theorists because it did not take robustness into account [84]. The deficiency was remedied by the $\mathcal{H}_\infty$ theory, which will be discussed later.

Very strong robustness properties could be established for systems with LQG control, when all states are measured. In this case, the system has a phase margin of 60° and infinite gain margin [134], which indicates a very good robustness.

The nice robustness properties of systems with state feedback do not hold for systems with output feedback. Nice counter-examples were given in the paper [50]. For systems with output feedback, it was attempted to recover the robustness of full state feedback making very fast observers. This approach led to a design technique called *loop transfer recovery* [53].

The only formal requirements on the system to be controlled in state-space theory is that the system is observable and controllable. There are no consideration of right half plane poles and zeros or time delays. Because of this, it is necessary to investigate the robustness of the design and to make appropriate modifications to achieve good robustness. We use an example to illustrate what happens if this is not done.

Example 4.1 (a fast system with a low bandwidth). Consider a system that is described by

$$\frac{dx}{dt} = \begin{pmatrix} -1 & 1 \\ 0 & 0 \end{pmatrix} x + \begin{pmatrix} a \\ 1 \end{pmatrix} u$$
$$y = \begin{pmatrix} 1 & 0 \end{pmatrix} x$$

The system is controllable if $a \neq 1$. We will assume that $a = -10$. The system is of second order and one state variable is measured directly. The system can be controlled with an observer of first degree. The closed-loop system is then of order 3. We assume that a state feedback and an observer are designed so that the closed-loop system is

$$(s + \alpha\omega_0)(s^2 + \omega_0 s + \omega_0^2) \qquad (4.23)$$

The transfer function of the system is

$$G(s) = \frac{as+1}{s(s+1)} \quad (4.24)$$

To obtain a fast closed-loop system, we choose $\omega_0 = 10$ and $\alpha = 1$. Straightforward calculations show that the controller has the transfer function

$$K(s) = \frac{s_0 s + s_1}{s+r} \quad (4.25)$$

with $r = 9274.5$, $s_0 = 925.5$ and $s_1 = 1000$. The loop transfer function is

$$L(s) = G(s)K(s) = \frac{(as+1)(s_0 s + s_1)}{s(s+1)(s+r)} = -9255 \frac{(s-0.1)(s+1.0805)}{s(s+1)(s+9274)}$$

The process pole at $s = -1$ is almost cancelled by the controller zero at $s = -s_1/s_0 = -1.0805$. The Bode diagram of the loop transfer function is shown in Figure 4.10.

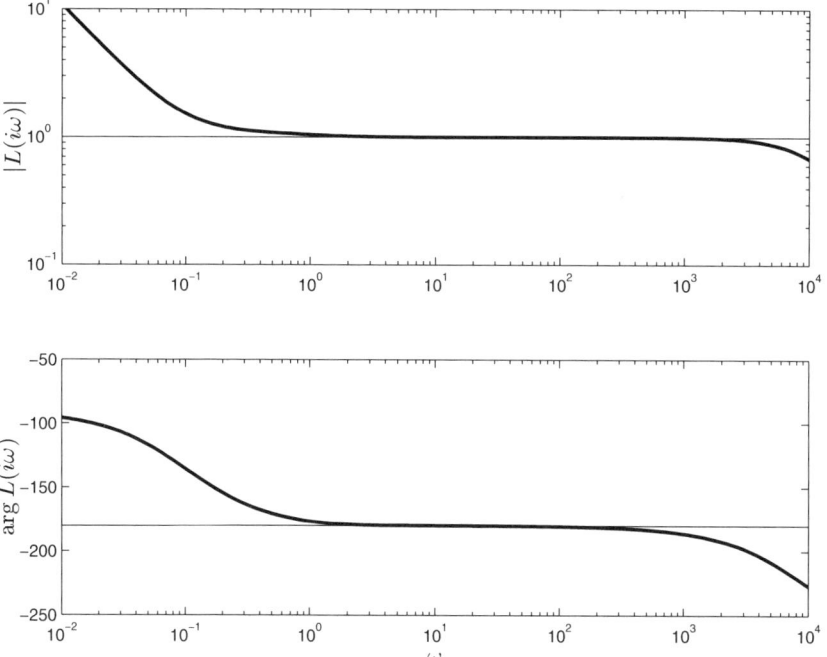

Fig. 4.10. Bode diagram for the loop transfer function of a system with a slow unstable zero at $s = 0.1$, where the specifications are $\omega_0 = 10$ and $\alpha = 1$

The loop transfer function has a low frequency asymptote that intersects the line $\log|L(i\omega)| = 0$ at $\omega = -1/a$, i.e., at the slow unstable zero. The magnitude then becomes close to 1 and it remains so until the break point at $\omega = r \approx \alpha a \omega_0^3$, i.e., the controller pole. The phase is also close to $-180°$ over that frequency range, which means that the stability margin is very poor. The crossover frequency is 6.58 and the phase margin is $\varphi_m = 0.15°$. The maximum sensitivities are $M_s = 678$ and $M_t = 677$, which also shows that the system is extremely sensitive. The slope of the magnitude curve at crossover is also very small, which is another indication of the poor robustness of the system.

The example illustrates clearly the danger of using a design method in a routine manner. It also shows that it is not sufficient to check controllability and observability or the closed-loop poles. In this particular case, the system is extremely sensitive even if the system is both controllable and observable and closed-loop poles are well damped. For this particular problem, there are severe limitations because of the slow right half plane zero combined with fast closed loop poles. In trying to make designs that violate these limitations by making a closed-loop system that is too fast, we obtain a closed-loop system that has very poor stability margins even if the closed-loop poles are quite well damped. Also notice that even if the gain crossover frequency is 6.58 rad/s, the sensitivity becomes larger than 1 for $\omega = 0.107$, which is close to the right half plane zero. Feedback is thus not effective for disturbances having higher frequencies than 0.107, because disturbances will be amplified by the feedback. It is, of course, possible to obtain sensible control designs, for example, by choosing smaller values of ω_0.

The example shows that it is important to be aware of the limitations when designing control systems.

Summary

State-space theory is an elegant way to approach a control problem. A nice aspect is that it naturally deals with multivariable systems. The theory has given important concepts such as observability and reachability, It has also given several design methods, such as linear quadratic control, loop transfer recovery and pole placement. It has also introduced powerful computational methods based on numerical linear algebra. A severe drawback is that robustness is not dealt with properly. This means that it is possible to formulate control problems which give solutions that have very poor robustness properties. It is easy to avoid the difficulties when we are aware of them, simply by evaluating the robustness of a design and to reduce the requirements until a suitable compromise is reached.

4.4 Fundamental Limitations

It is very useful to determine the performance that can be achieved without sacrificing robustness. Such estimates will be provided in this section. In particular, we will consider limitations that arise from poles and zeros in the right half plane and time delays. The results are based on [10, 140, 142] and [11], where many more details are given.

Consider a system with the transfer function $G(s)$. Factor the transfer function as

$$G(s) = G_{mp}(s)G_{nmp}(s) \tag{4.26}$$

where G_{mp} is the minimum-phase part and G_{nmp} is the non-minimum-phase part. Let $\Delta G(s)$ denote the uncertainty in the process transfer function. It is assumed that the factorisation is normalised so that $|G_{nmp}(i\omega)| = 1$ and the sign is chosen so that G_{nmp} has negative phase. The achievable bandwidth is characterised by the gain crossover frequency ω_{gc}.

4.4.1 The Crossover Frequency Inequality

We will now derive an inequality for the gain crossover frequency. The loop transfer function is $L(s) = G(s)K(s)$. Requiring that the phase margin is φ_m, we get

$$\arg L(i\omega_{gc}) = \arg G_{nmp}(i\omega_{gc}) + \arg G_{mp}(i\omega_{gc}) + \arg K(i\omega_{gc}) \geq -\pi + \varphi_m \tag{4.27}$$

Assume that the controller is chosen so that the loop transfer function $G_{mp}K$ is equal to Bode's ideal loop transfer function given by Equation (4.15), then

$$\arg G_{mp}(i\omega) + \arg K(i\omega) = n\frac{\pi}{2} \tag{4.28}$$

where n is the slope of the loop transfer function at the crossover frequency. Equation (4.28) is also a good approximation for other controllers because the amplitude curve is typically close to a straight line at the crossover frequency. The parameter n in (4.28) is then the slope n_{gc} at the crossover frequency. It follows from Bode's relations (4.13) that the phase is $n_{gc}\pi/2$. It follows from Equations (4.27) and (4.28) that the crossover frequency satisfies the inequality

$$\arg G_{nmp}(i\omega_{gc}) \geq -\alpha \tag{4.29}$$

where

$$\alpha = \pi - \varphi_m + n_{gc}\frac{\pi}{2} \tag{4.30}$$

This *crossover frequency inequality* gives the limitations imposed by non-minimum-phase factors. A straightforward method to determine the crossover frequencies that can be obtained is to plot the left hand side of Equation (4.29) and determine when the inequality holds. The following example gives a simple rule of thumb.

82   K.J. Åström

Example 4.2 (a simple rule of thumb). To see the implications of (4.29), we will make some reasonable design choices. With a phase margin of 45° ($\varphi_m = \pi/4$), and a slope of $n_{gc} = -1/2$ we get $\alpha = \pi/2$ and Equation (4.29) becomes

$$\arg G_{nmp}(i\omega_{gc}) \geq -\frac{\pi}{2} \tag{4.31}$$

This gives the simple rule that the phase lag of the minimum-phase components should be less than 90° at the gain crossover frequency.

4.4.2 A Zero in the Right Half Plane

We will now discuss limitations imposed by right half plane zeros. We will first consider systems with only one zero in the right half plane. The non-minimum-phase part of the plant transfer function then becomes

$$G_{nmp}(s) = \frac{z-s}{z+s} \tag{4.32}$$

Notice that G_{nmp} should be chosen to have unit gain and negative phase. We have

$$\arg G_{nmp}(i\omega) = -2\arctan\frac{\omega}{z}$$

and (4.29) gives the following upper bound on the crossover frequency:

$$\frac{\omega_{gc}}{z} \leq \tan \alpha/2 \tag{4.33}$$

and, using the simple rule of thumb (4.31), we get $\omega_{gc} < z$.

A right half plane zero gives an upper bound to the achievable bandwidth. The bandwidth decreases with decreasing frequency of the zero. It is thus more difficult to control systems with slow zeros.

4.4.3 Time Delays

The transfer function for such systems has an essential singularity at infinity. The non-minimum-phase part of the transfer function of the process is

$$G_{nmp}(s) = e^{-sT} \tag{4.34}$$

We have $\arg G_{nmp}(i\omega) = -\omega T$ and the crossover frequency inequality, Equation (4.29) becomes

$$\omega_{gc}T \leq \pi - \varphi_m + n_{gc}\frac{\pi}{2} = \alpha \tag{4.35}$$

The simple rule of thumb (4.31) gives $\omega_{gc}T \leq \frac{\pi}{2} = 1.57$.

Time delays thus give an upper bound on the achievable bandwidth.

4.4.4 A Pole in the Right Half Plane

Consider a system with one pole in the right half plane. The non-minimum-phase part of the transfer function is thus

$$G_{nmp}(s) = \frac{s+p}{s-p} \tag{4.36}$$

where $p > 0$. Notice that the transfer function is normalised so that G_{nmp} has unit gain and negative phase. We have

$$\arg G_{nmp}(i\omega) = -2\arctan\frac{p}{\omega}$$

and the crossover frequency inequality (4.29) gives

$$\omega_{gc} \geq \frac{p}{\tan\alpha/2} \tag{4.37}$$

The simple rule of thumb (4.31) gives $\omega_{gc} \geq p$.

Unstable poles give a lower bound on the crossover frequency. For systems with right half plane poles, the bandwidth must thus be sufficiently large.

4.4.5 Poles and Zeros in the Right Half Plane

Consider a system with

$$G_{nmp}(s) = \frac{(z-s)(s+p)}{(z+s)(s-p)} \tag{4.38}$$

For $z > p$, we have

$$\arg G_{nmp}(i\omega) = -2\arctan\frac{\omega}{z} - 2\arctan\frac{p}{\omega} = -2\arctan\frac{\omega/z + p/\omega}{1 - p/z}$$

The right hand side has its maximum for $\omega = \sqrt{pz}$ and the inequality (4.29) becomes

$$\frac{z}{p} \geq \tan^2\alpha/2 = \tan^2\left(\frac{\pi}{4} - \frac{\varphi_m}{4} + n_{gc}\frac{\pi}{8}\right) \tag{4.39}$$

The simple rule of thumb (4.31) gives $z \geq 25.3p$. Table 4.1 gives the phase margin as a function of the ratio z/p for $\varphi_m = \pi/4$ and $n_{gc} = -1/2$. The phase margin that can be achieved for a given ratio p/z is

$$\varphi_m < \pi + n_{gc}\frac{\pi}{2} - 4\arctan\sqrt{\frac{p}{z}} \tag{4.40}$$

When the unstable zero is faster than the unstable pole, i.e., $z > p$, the ratio z/p thus must be sufficiently large in order to have the desired phase margin. The largest gain crossover frequency is the geometric mean of the unstable pole and zero.

Table 4.1. Achievable phase margin for $n_{gc} = -1/2$ and different zero-pole ratios z/p.

z/p	2	2.24	3.86	5	5.83	8.68	10	20
φ_m	-6.0	0	30	38.6	45	60	64.8	84.6

Example 4.3 (the X-29). Considerable design effort has been devoted to the design of the flight control system for the X-29 aircraft [39, 133]. One of the design criteria was that the phase margin should be greater than 45° for all flight conditions. At one flight condition, the model has the following non-minimum-phase component:

$$G_{nmp}(s) = \frac{s-26}{s-6}$$

Since $z = 4.33p$, it follows from Equation (4.40) that the achievable phase margins for $n_{gc} = -0.5$ and $n_{gc} = -1$ are $\varphi_m = 32.3°$ and $\varphi_m = -12.6°$. It is interesting to note that many design methods were used in a futile attempt to reach the design goal. A simple calculation of the type given in this section would have given much insight.

Example 4.4 (Klein's unridable bicycle). An interesting bicycle with rear wheel steering, which is impossible to ride, was designed by Professor Klein in Illinois [96]. The theory presented in this paper is well suited to explain why it is impossible to ride this bicycle. The transfer function from steering angle to tilt angle is given by

$$\frac{\theta(s)}{\delta(s)} = \frac{m\ell V}{Jc} \frac{V - as}{s^2 - mg\ell/J}$$

where m is the total mass of the bicycle and the rider, J the moment of inertia for tilt with respect to the contact line of the wheels and the ground, h the height of the centre of mass from the ground, a the vertical distance from the centre of mass to the contact point of the front wheel, V the forward velocity, and g the acceleration of gravity. The system has a RHP pole at $s = p = \sqrt{mg\ell/J}$, caused by the pendulum effect. Because of the rear wheel steering, the system also has a RHP zero at $s = z = V/l$. Typical values $m = 70$ kg, $\ell = 1.2$ m, $a = 0.7$, $J = 120$ kgm^2 and $V = 5$ m/s, give $z = V/a = 7.14$ rad/s and $p = \omega_0 = 2.6$ rad/s. The ratio of the zero and the pole is thus $p/z = 2.74$, with $n_{gc} = -0.5$. The inequality (4.29) shows that the phase margin can be at most $\varphi_m = 10.4$.

The reason why the bicycle is impossible to ride is thus that the system has a right half plane pole and a right half plane zero that are too close together. Klein has verified this experimentally by making a bicycle where the ratio z/p is larger. This bicycle is, indeed, possible to ride.

So far, we have only discussed the case $z > p$. When the unstable zero is slower than the unstable pole, the crossover frequency inequality (4.29) cannot be satisfied unless $\varphi_m < 0$ and $n_{gc} > 0$.

4.4.6 A Pole in the Right Half Plane and Time Delay

Consider a system with one pole in the right half plane and a time delay T. The non-minimum-phase part of the transfer function is thus

$$G_{nmp}(s) = \frac{s+p}{s-p} e^{-sT} \tag{4.41}$$

The crossover frequency condition (4.29) gives

$$2\arctan\frac{\omega_{gc}}{p} - \omega_{gc}T \geq \varphi_m - n_{gc}\frac{\pi}{2} \tag{4.42}$$

The system cannot be stabilised if $pT > 2$. If $pT < 2$, the left hand side has its smallest value for $\omega_{gc}/p = \sqrt{2/(pT) - 1}$. Introducing this value of ω_{gc} into (4.42), we get

$$2\arctan\sqrt{\frac{2}{pT} - 1} - pT\sqrt{\frac{2}{pT} - 1} > \varphi_m - n_{gc}\frac{\pi}{2}$$

The simple rule of thumb with $\varphi_m = \pi/4$ and $n_{gc} = -0.5$ gives

$$pT \leq 0.326 \tag{4.43}$$

Example 4.5 (pole balancing). To illustrate the results, we can consider the balancing of an inverted pendulum. A pendulum of length ℓ has a right half plane pole $\sqrt{g/\ell}$. We assume that the neural lag of a human is 0.07 s. The inequality (4.43) gives $\sqrt{g/\ell}\, 0.07 < 0.326$, hence $\ell > 0.45$. The calculation thus indicates that a human with a lag of 0.07 s should be able to balance a pendulum whose length is 0.5 m. To balance a pendulum whose length is 0.1 m, the time delay must be less than 0.03 s. Pendulum balancing has also been done using video cameras as angle sensors. The limited video rate imposes strong limitations on what can be achieved. With a video rate of 20 Hz, it follows from (4.43) that the shortest pendulum that can be balanced with $\varphi_m = 45°$ and $n_{gc} = -0.5$ is $\ell = 0.23$ m.

4.4.7 Other Criteria

The phase margin is a crude indicator of the stability margin. By carrying out detailed designs the results can be refined. This is done in [11], which gives results for designs with $M_s = M_t = 2$ and $M_s = M_t = 1.4$.

- A RHP zero z

$$\frac{\omega_{gc}}{z} \leq \begin{cases} 0.5 & \text{for } M_s, M_t < 2 \\ 0.2 & \text{for } M_s, M_t < 1.4 \end{cases}$$

- A RHP pole p

$$\frac{\omega_{gc}}{p} \geq \begin{cases} 2 & \text{for } M_s, M_t < 2 \\ 5 & \text{for } M_s, M_t < 1.4 \end{cases}$$

- A time delay T

$$\omega_{gc} T \leq \begin{cases} 0.7 & \text{for } M_s, M_t < 2 \\ 0.37 & \text{for } M_s, M_t < 1.4 \end{cases}$$

- A RHP pole-zero pair with $z > p$

$$\frac{z}{p} \geq \begin{cases} 6.5 & \text{for } M_s, M_t < 2 \\ 14.4 & \text{for } M_s, M_t < 1.4 \end{cases}$$

- A RHP pole p and a time delay T

$$pT \leq \begin{cases} 0.16 & \text{for } M_s, M_t < 2 \\ 0.05 & \text{for } M_s, M_t < 1.4 \end{cases}$$

A time delay or a zero in the right half plane gives an upper bound of the bandwidth that can be achieved. The bound decreases when the zero z decreases and the time delay increases. A pole in the right half plane gives a lower bound on the bandwidth. The bandwidth increases with increasing p. For a pole-zero pair, there is an upper bound on the pole/zero ratio. Notice that the inequalities given above are rules of thumb and not guaranteed hard bounds.

4.5 $\mathcal{H}_\infty$ Loop Shaping

A consequence of the introduction of state-space theory was that interest shifted from robustness to optimisation. New developments that started by the development of $\mathcal{H}_\infty$ control by George Zames in the 1980s gave a strong revival of robustness [158]. This led to a very vigorous development that has given new insight and new design methods. These results will be discussed in this section. To keep the presentation simple, we will only deal with systems having one input and one output, but techniques as well as results can be generalised to systems with many inputs and many outputs. For more extensive treatments, we refer to [51, 66, 75, 155, 164].

Problem formulation

If a system structure with two degrees of freedom is used, the problems of setpoint response can be dealt with separately and we can therefore focus on robustness and attenuation of disturbances. The set point will not be considered and Figure 4.2 can be simplified to the system shown in Figure 4.11. The

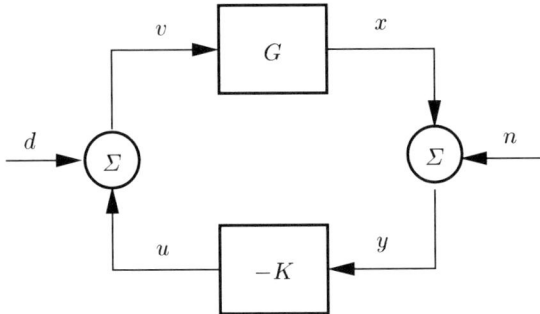

Fig. 4.11. Block diagram of a simple feedback system

system has two inputs: the measurement noise n and the load disturbance d. The problem we will consider is the design of a controller with the following properties:

- insensitive to changes in the process properties;
- ability to reduce the effects of the load disturbance d;
- does not inject too much measurement noise into the system.

Stability

Stability is a primary robustness requirement. There is one problem with the stability concept introduced in Section 4.2. Because the stability is based on the loop transfer function, only there may be cancellations of poles and zeros in the process and the controller. This does not pose any problems if the cancelled factors are stable. The results will, however, be strongly misleading if the cancelled factors are unstable because there will be internal signals in the system that will diverge. We illustrate this with an example.

Example 4.6 (pole zero cancellation). Consider the system in Figure 4.11 with

$$K(s) = \frac{s-1}{s} \qquad G(s) = \frac{1}{s-1}$$

The loop transfer function is $L = 1/s$ and the system thus appears stable. Notice, however, that the transfer function from disturbance d to output is

$$G_{yd} = \frac{s}{(s+1)(s-1)}$$

A load disturbance will thus make the output diverge and it does not make sense to call the system stable.

The problem illustrated in Example 4.6 is well known. Classically, it is resolved by formally introducing the rule that cancellation of unstable poles is not permissible. This can also be encapsulated in an algebra for manipulating systems that do not permit division by factors having roots in the right half plane [126, 127].

Another way is to introduce a stability concept that takes care of the problem directly. It follows from the analysis in Section 4.2 that the closed-loop system is completely characterised by the transfer functions given by Equation (4.2). Based on this, we can say that a system is stable if all these transfer functions (4.2) are stable. This is sometimes called *internal stability*. The transfer functions (4.2) can be conveniently combined in the matrix

$$F(s) = \begin{pmatrix} \dfrac{1}{1+GK} & -\dfrac{K}{1+GK} \\ \dfrac{G}{1+GK} & -\dfrac{GK}{1+GK} \end{pmatrix} \tag{4.44}$$

Notice that this transfer function (4.44) represents the signal transmission from the disturbances d and n to the signals v and x in the block diagram in Figure 4.2. Let the transfer functions of the process and the controller be represented as

$$G(s) = \frac{B_c}{A_c} \qquad K(s) = \frac{B_c}{A_c}$$

respectively. The matrix (4.44) can then be represented as

$$F(s) = \begin{pmatrix} \dfrac{A_c A_p}{A_c A_p + B_c B_p} & -\dfrac{A_p B_c}{A_c A_p + B_c B_p} \\ \dfrac{A_c B_p}{A_c A_p + B_c B_p} & -\dfrac{B_p B_c}{A_c A_p + B_c B_p} \end{pmatrix} \tag{4.45}$$

and the stability criterion is that the equation

$$C_{pol} = A_c A_p + B_c B_p \tag{4.46}$$

should have all its roots in the left half plane. This is also called internal stability. We will simply say that the system (G, K) is stable.

Example 4.7 (pole zero cancellation). Applying the result to the example with pole-zero cancellation, we find that

$$F(s) = \begin{pmatrix} \dfrac{s}{s+1} & -\dfrac{s-1}{s+1} \\ \dfrac{s}{(s-1)(s+1)} & -\dfrac{1}{s+1} \end{pmatrix}$$

The characteristic polynomial is

$$C_{pol} = (s-1)s + s - 1 = (s-1)(s+1)$$

which has a root in the right half plane.

How to compare two systems

A fundamental problem when discussing robustness is to determine when two systems are close. This seemingly innocent problem is not as simple as it may appear. For feedback control, it would be natural to claim that two systems are close if they have similar behavior under a given feedback [141, 152]. The fact that two systems have similar open-loop characteristics does not mean that they will behave similarly under feedback.

Example 4.8 (similar open-loop, different closed-loop). The stable systems with transfer functions

$$G_1(s) = \frac{1000}{s+1}, \quad G_2(s) = \frac{1000a^2}{(s+1)(s+a)^2}$$

have very similar open-loop responses for large values of a. This is illustrated in Figure 4.12, which shows the step responses for $a = 100$. The differences between the step responses is barely noticeable in the figure. The closed-loop

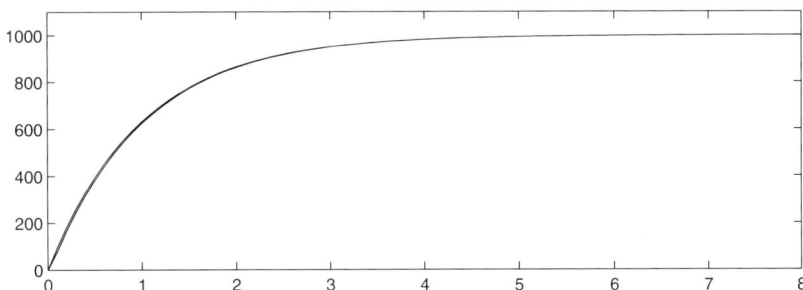

Fig. 4.12. Step responses for systems with the transfer functions $G_1(s) = 1000/(s+1)$ and $G_2(s) = 10^7/((s+1)(s+100)^2)$

systems obtained with unit feedback have the transfer functions

$$G_{1cl} = \frac{1000}{s+1001}, \quad G_{2cl} = \frac{10^7}{(s-287)(s^2+86s+34879)}$$

The closed-loop systems are *very* different because the system G_{2cl} is unstable.

Example 4.9 (different open-loop, similar closed-loop). The systems with the transfer functions

$$G_1(s) = \frac{1000}{s+1}, \qquad G_2(s) = \frac{1000}{s-1}$$

have very different open-loop properties because one system is unstable and the other is stable. The closed-loop systems obtained with unit feedback are, however,

$$G_{1cl}(s) = \frac{1000}{s+1001} \qquad G_{2cl}(s) = \frac{1000}{s+999}$$

which are very close.

There are many examples of this fact in the literature of adaptive control, where the importance of considering the closed-loop properties of a model has been recognised for a long time [14].

The examples given above show that the naive way of comparing two systems by analysing their responses to a given input signal is not appropriate for feedback control. The difficulty is that it does not work when one or both systems are unstable as in Example 4.7 and 4.8.

One approach is to compare the outputs when the inputs are restricted to the class of inputs that give bounded outputs. This approach was introduced in [56, 159] using the notion of gap metric. Another approach was introduced in [151, 152]. To describe this approach, we assume that the process is described by the rational transfer function

$$G(s) = \frac{B(s)}{A(s)}$$

where $A(s)$ and $B(s)$ are polynomials. We next introduce a stable polynomial $C(s)$ whose degree is not smaller than the degrees of $A(s)$ and $B(s)$. The transfer function $G(s)$ can then be written as

$$G(s) = \frac{B(s)/C(s)}{A(s)/C(s)} = \frac{D(s)}{N(s)} \qquad (4.47)$$

Vidyasagar proposed to compare two systems by comparing the stable rational transfer functions D and N. This is called the *graph metric*. A difficulty was that the graph metric was difficult to compute.

Coprime factorisation

The polynomial K in (4.47) can be chosen in many different ways. We will now discuss a convenient choice.

We start by introducing a suitable concept. Two rational functions D and N are called coprime if there exist rational functions X and Y that satisfy the equation

$$XD + YN = 1$$

The condition for coprimeness is essentially that $D(s)$ and $N(s)$ do not have any common factors. The functions $D(s)$ and $N(s)$ will now be chosen so that

$$DD^* + NN^* = 1 \qquad (4.48)$$

where we have used the notation $D^*(s) = D(-s)$. A factorisation (4.47) of G where N and D satisfy (4.48) is called a *normalised coprime factorisation* of G. When using such a factorisation, the polynomials A and B in (4.47) do not have common factors.

4.5.1 Vinnicombe's Metric

A very nice solution to the problem of comparing two systems that is appropriate for feedback was given by Vinnicombe [153, 155]. Consider two systems with the normalised coprime factorisations

$$G_1 = \frac{D_1}{N_1} \qquad G_2 = \frac{D_2}{N_2}$$

To compare the systems, it must be required that

$$\frac{1}{2\pi} \Delta \arg_\Gamma (N_1 N_2^* + D_1 D_2^*) = 0 \qquad (4.49)$$

where Γ is the Nyquist contour. In the polynomial representation, this condition implies that

$$\frac{1}{2\pi} \Delta \arg_\Gamma (B_1 B_2^* + A_1 A_2^*) = \deg A_2 \qquad (4.50)$$

If the winding number constraint is satisfied, the distance between the systems is defined as

$$\delta_\nu(G_1, G_2) = \sup_\omega \frac{|G_1(i\omega) - G_2(i\omega)|}{\sqrt{(1 + |G_1(i\omega)|^2)(1 + |G_2(i\omega)|^2)}} \qquad (4.51)$$

We have $|\delta_\nu(G_1, G_2)| \leq 1$. If the winding number condition is not satisfied, the distance is defined as $\delta_\nu = 1$. Vinnicombe showed that δ_ν is a metric, which he called it the *ν-gap metric*.

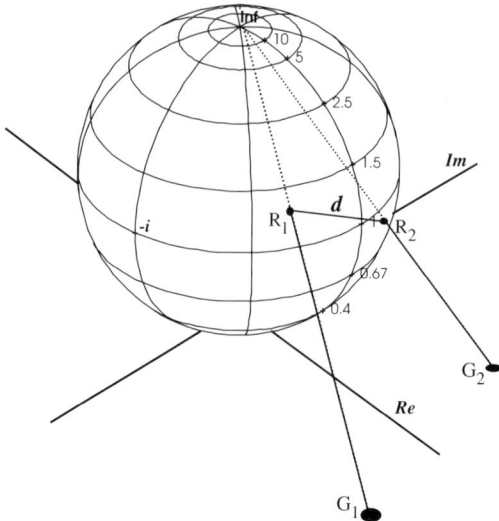

Fig. 4.13. Geometric interpretation of the Vinnicombe metric

Geometric interpretation

Vinnicombe's metric is easy to compute and it also has a very nice geometric interpretation. The expression

$$d = \frac{|G_1 - G_2|}{\sqrt{(1 + |G_1|^2)(1 + |G_2|^2)}}$$

can be interpreted graphically as follows. Let G_1 and G_2 be two complex numbers. The Riemann sphere is located above the complex plane. It has diameter 1 and its south pole is at the origin of the complex plane. Points in the complex plane are projected onto the sphere by a line through the point and the north pole, as in Figure 4.13. Let R_1 and R_2 be the projections of G_1 and G_2 on the Riemann sphere. The number d is then the shortest chordal distance between the points R_1 and R_2, as in Figure 4.13.

Coprime factor uncertainty

Classical sensitivity results such as (4.8) were obtained based on additive perturbations. The system G was perturbed to $G + \Delta G$, where ΔG is a stable transfer function. These types of perturbations are not well suited to deal with feedback systems as is illustrated by Example 4.8. A more sophisticated way of describing perturbations is required for this. The development of metrics for systems gave good insight into what should be done. Uncertainty will be described in terms of the normalised coprime factorisation of a system.

Consider a system described by

$$G + \Delta G = \frac{N + \Delta N}{D + \Delta D} = ND^{-1} = D^{-1}N \tag{4.52}$$

where the pair (N, D) is a normalised coprime factorisation of G, and the perturbations ΔN and ΔD are stable proper transfer functions. Figure 4.14 shows a block diagram of the closed-loop system with the perturbed plant.

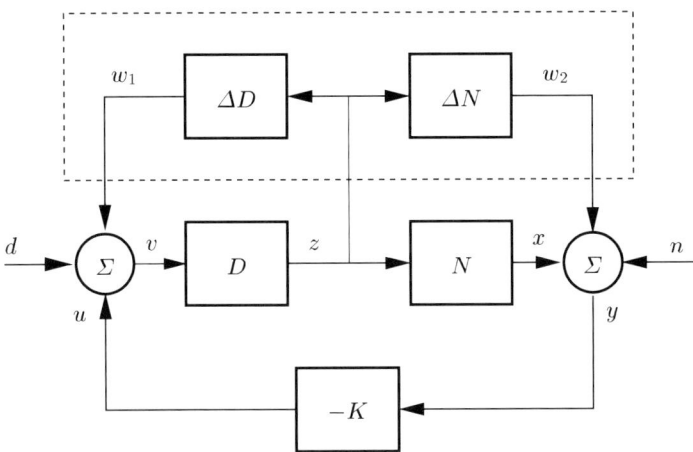

Fig. 4.14. Block diagram of a process with coprime factor uncertainty and a controller

We will now investigate how large the perturbations can be without violating the stability condition. For the system in Figure 4.14, we have

$$z = \frac{D^{-1}}{1 + GK}w_1 - \frac{D^{-1}K}{1 + GK}w_2 = D^{-1}\left(\frac{1}{1 + GK} \quad -\frac{K}{1 + GK}\right)\begin{pmatrix}w_1 \\ w_2\end{pmatrix}$$

and

$$\begin{pmatrix}w_1 \\ w_2\end{pmatrix} = \begin{pmatrix}\Delta D \\ \Delta N\end{pmatrix}z$$

The system can thus be represented with the block diagram in Figure 4.15. We can then invoke the small gain theorem and conclude that the perturbed system will be stable if the loop gain is less than 1 [49]. Hence

$$\left\|\begin{pmatrix}\Delta N \\ \Delta D\end{pmatrix}\right\|_\infty \left\|D^{-1}\left(\frac{1}{1 + GK} \quad -\frac{K}{1 + GK}\right)\right\|_\infty < 1 \tag{4.53}$$

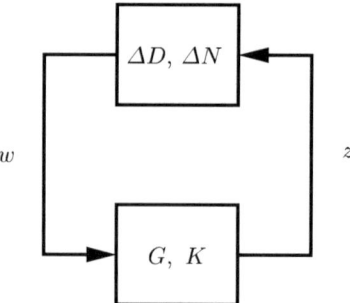

Fig. 4.15. Simplification of the block diagram in Figure 4.14

This condition can be simplified if we use the fact that the pair (N, D) is a normalised coprime factorisation. This gives

$$\left\| D^{-1} \begin{pmatrix} \dfrac{1}{1+GK} & -\dfrac{K}{1+GK} \end{pmatrix} \right\|_\infty = \left\| \begin{pmatrix} D \\ N \end{pmatrix} D^{-1} \begin{pmatrix} \dfrac{1}{1+GK} & -\dfrac{K}{1+GK} \end{pmatrix} \right\|_\infty$$

$$= \left\| \begin{pmatrix} I \\ G \end{pmatrix} \begin{pmatrix} \dfrac{1}{1+GK} & -\dfrac{K}{1+GK} \end{pmatrix} \right\|_\infty = \|F(G,K)\|_\infty$$

where $F(G, K)$ denotes the system matrix

$$F(G, K) = \begin{pmatrix} \dfrac{1}{1+GK} & -\dfrac{K}{1+GK} \\ \dfrac{K}{1+GK} & -\dfrac{GK}{1+GK} \end{pmatrix} \tag{4.54}$$

Introducing

$$\gamma(G, K) = \sup_\omega \|F(G(i\omega), K(i\omega))\|_\infty \tag{4.55}$$

we thus find that the closed-loop system is stable for all normalised coprime perturbations ΔD and ΔN such that

$$\left\| \begin{pmatrix} \Delta N \\ \Delta D \end{pmatrix} \right\|_\infty < \dfrac{1}{\gamma} \tag{4.56}$$

This equation is a natural generalisation of Equation (4.8) in classical control theory. Notice that since the systems have one input and one output, we have

$$|F(G,K)| = \bar\sigma F(G, K) = \dfrac{\sqrt{(1+|K|^2)(1+|G|^2)}}{|1+GK|} \tag{4.57}$$

and Equation (4.55) can thus be written

$$\gamma(G, K) = \sup_\omega \dfrac{\sqrt{(1+|K(i\omega)|^2)(1+|G(i\omega)|^2)}}{|1+G(i\omega)K(i\omega)|} \tag{4.58}$$

4.5.2 $\mathcal{H}_\infty$ loop shaping

The goal of $\mathcal{H}_\infty$ techniques is to design control systems that are insensitive to model uncertainty. It follows from Equations (4.55) and (4.56) that this can be accomplished by finding a controller K that gives a stable closed-loop system and minimises the $\mathcal{H}_\infty$-norm of the transfer function $F(G, K)$ given by Equation 4.54. It follows from Equation (4.56) that such a design permits the largest deviation of the normalised coprime deviations. It is interesting to observe that the transfer function G also describes the signal transmission from the disturbances d and n to v and x in Figure 4.2. A robust controller obtained in this way will also attenuate the disturbances very well. A state-space solution to the $\mathcal{H}_\infty$ control problem was given in [52]. A loop shaping design procedure was developed in [111, 114].

Frequency weighting

In the design procedure presented in [114], it is also possible to introduce a frequency weighting W as a design parameter. The $\mathcal{H}_\infty$ control problem for the process $G' = GW$ is then solved to give the controller K'. The controller for the process G is then $K = K'W$. In this way, it is possible to obtain controllers that have high gain at specified frequency ranges and high-frequency roll-off.

Generalised stability margin

A generalization of the classical stability margin was also introduced in [111]. For a closed-loop system consisting of the process G and the controller K as in Figures 4.2 and 4.14, we define the generalised stability margin as

$$b(G, K) = \begin{cases} \dfrac{1}{\gamma} & \text{if } (G, K) \text{ is stable,} \\ 0 & \text{otherwise} \end{cases} \qquad (4.59)$$

Notice that the generalised stability margin takes values between 0 and 1. The margin is 0 if the system is unstable. A value close to 1 indicates a good margin of stability. Reasonable practical values of the margin are in the range of 1/3 to 1/5. The $\mathcal{H}_\infty$ loop shaping in [111] gives a controller that maximises the stability margin, giving

$$b_{opt} = \sup_{K} b(G, K) \qquad (4.60)$$

4.5.3 Vinnicombe's theorems

A number of interesting theorems that relate model uncertainty to robustness have been derived in [154, 155]. These results, which can be seen as the natural conclusion of the work that began in [158], give very nice relations between robust control and model uncertainty. Vinnicombe has proven the following results.

Proposition 4.1. *Consider a nominal process G and a controller K and a parameter β. Then, the controller K stabilises all plants G_1 such that $\delta_v(P, G_1) \leq \beta$, if and only if $b(G, K) > \beta$.*

Proposition 4.2. *Consider a nominal process G, a perturbed process G_1 and a number $\beta < b_{opt}(G, K)$. Then, (G_1, C) is stable for all compensators K, such that $b(G_1, C) > \beta$ if and only if $\delta(P, G_1) \leq \beta$.*

The first proposition tells us that a controller K designed for process G with a generalised stability margin greater than β will stabilise all processes G_1 in a δ_v environment of G, provided that $\delta_v(P, G_1) < \beta$. Proofs of these theorems are given in [155]. Vinnicombe has actually given sharper results, which only requires the inequalities to hold pointwise for each frequency.

Connections with the classical control theory

The $\mathcal{H}_\infty$ loop shaping cannot be directly related to the classical robustness criteria. The classical robustness criteria such as M_t and M_s depend only on the loop transfer function $L = GK$. But the generalised stability margin $b(G, K)$ and the generalised sensitivity $\gamma(G, K)$ depend on both G and K. The generalised stability margin will therefore change if the process transfer function is multiplied by a constant and the controller transfer function is divided by the same number. One reason for this is that the criterion (4.57) implicitly assumes that the disturbances d and n have equal weight. This is a reasonable assumption if sufficient information about the disturbances are available, but very often we do not have this information. One possibility in formulating the design problem in this case is to choose the most favourable disturbance relation. This can be done by introducing a weighting of the disturbances. Let $G' = PW$ be the weighted process and let KW^{-1} be the weighted controller KW^{-1}. We have

$$L' = P'C' = GK = L$$

The loop transfer function is invariant to W. The generalised sensitivity γ becomes

$$\gamma(P', C') = \sup_\omega \frac{\sqrt{(1 + |K(i\omega)W^{-1}(i\omega)|^2)(1 + |G(i\omega)W(i\omega)|^2)}}{|1 + G(i\omega)K(i\omega)|}$$

A straightforward calculation shows that γ' is minimised for the weight

$$W = \sqrt{|K|/|G|}$$

see [122]. This weight gives the following expression for the weighted sensitivity

$$\gamma^* = \sup_{\omega}(|S(i\omega)| + |T(i\omega)|) \tag{4.61}$$

Notice that γ^* only depends on the loop transfer function $L = GK$. Using the weighted sensitivity function, we thus obtain an interesting connection between $\mathcal{H}_\infty$ loop shaping and classical robustness theory. The number γ defined by Equation (4.55) and Equation (4.58) is thus a natural generalisation of the maxima M_s, M_t of the sensitivity function and the complementary sensitivity. It is useful to introduce a combined sensitivity M by requiring that both M_s and M_t should at most be equal to M. The combined sensitivity implies that the Nyquist curve of the loop transfer function is outside a circle with diameter on

$$\left(-\frac{M+1}{M-1}, -\frac{M-1}{M+1}\right)$$

We have the following inequalities

$$2M - 1 < \gamma < 2M$$

$$\frac{\gamma}{2} < M < \frac{\gamma+1}{2}$$

see [122] where sharper results are also presented.

The generalised stability margin b is also a natural generalisation of the classical stability margin A_m. There are, however, some scale changes. The normal stability margin takes values between 1 and ∞ while the generalised stability margin takes values between 0 and 1. To get compatibility, the classical gain margin should be replaced by a stability margin defined by

$$A_m^* = 1 - \frac{1}{A_m}$$

where the number A_m^* will be between 0 and 1 for stable systems.

4.6 Summary

In this chapter we have discussed the role of model uncertainty and robustness, starting from the classical results of Bode and terminating with recent results on loop shaping. The presentation has essentially been done for single-input single-output systems. The key ideas extend to multivariable systems, which are discussed in the references. In the presentation, we naturally touch subjects such as Bode's ideal loop transfer function and fractional systems. The connections between classical based robust designs like QFT and $\mathcal{H}_\infty$ and loop shaping have also been established. Other important problems discussed include fundamental limitations and the question when two systems are close.

5 Open-loop Iterative Learning Control

Antonio Sala and Pedro Albertos

Abstract. In this chapter, the issue of introducing iterative improvement strategies in two-degrees-of-freedom structures will be addressed. In particular, for repetitive tasks, reference tracking and repetitive deterministic disturbance rejection can be carried out either in open-loop or in closed-loop. If some open-loop improvement mechanisms are introduced in the control strategy, then the burden on the feedback loop can be reduced so lower gain loops can yield improved robustness.

5.1 Introduction

In various chapters of this book, an iterative improvement of closed-loop performance is sought, using different techniques. All of them rely upon linear plant models, identified using different variants of least squares methods, over different frequency ranges.

Closed-loop control is the only way to achieve disturbance rejection if these are non-deterministic. Notwithstanding, in the case of deterministic disturbances, an open-loop strategy could be used, because sensors are not needed as the disturbance values can be calculated (sensors could be needed to determine initial conditions, *etc.*). A particular set-up is the "repetitive task" case, in which a particular reference tracking or disturbance rejection task is repeatedly executed from the same starting state. At any run, the same sequence of control actions should be applied to the plant. If the plant is invertible and time-invariant, the disturbances are known in advance and the system is output controllable, it is possible to control the system without any error. In that case, a feedforward approach could generate the set of control actions to carry out the task. To complement this control, the feedback controller would need only to undertake corrections based on non-deterministic disturbances and modelling errors. This kind of separation is called *control with two degrees of freedom*. If the plant is non-linear, inverting the non-linear model could generate the feedforward actions and it could be the case that a sufficiently robust linear regulator could give an accurate enough tracking.

Iterative learning control is a well-established methodology [37, 115] to derive the sequence of control actions to be applied to a process working under repetitive conditions, either in tracking a reference or in rejecting a deterministic disturbance. This approach has been well used in the robotic area, when a manipulator has to repeat a given trajectory many times. One of the disadvantages of the approach is the dependence on the reference trajectory, as it results in an open-loop control with pre-computed actions. The approach is a black box one, in the sense that no prior knowledge of the physics of the plant is used to synthesise the control actions.

The plant evolution is assumed to start always at the same initial conditions and the control action refining is a batch process. That is, the control updating is done out of the control performing. This is a main difference with respect to the older concept of repetitive control[1] [157]. In this way, in the iterative learning control approach, there are no stability problems (in fact, the control system acts in an open-loop way). Nevertheless, the control convergence to a steady control signal is not guaranteed and some cautions should be taken, as analysed later on.

Also, this setting assumes perfect measurement conditions. The errors in the measurements, as well as the changes in the plant or disturbance behaviours, are left to the feedback controller.

For linear plants, the sequence of control actions can be scaled with respect to the magnitude of the reference or the disturbance, if it has been computed for unitary signals. In this sense, some kind of generalisation is possible, but, in general, the control sequence is only valid for the trained conditions.

The aim of this chapter is to show how control structures with some open-loop components can improve the reference tracking and deterministic disturbance rejection and simultaneously decrease the feedback effort needed, hence improving the robustness of the overall control system. Feedback will then tackle only the task of compensating random disturbances and model mismatch and variations, if any.

Another possibility to be presented (in the repetitive task case) is to learn the sequence of control actions needed to follow a prescribed trajectory. Although that sequence could be derived from the same linear model used in the feedback control design, incorporating the learning mechanism could enable the system to compensate some unmodelled non-linearities in open-loop, maybe avoiding the need of non-linear identification and non-linear controllers. Non-linear controllers in the feedback loop would be only needed in the cases that significant non-deterministic disturbances arise (assuming that linear identification gives useless models) and/or strong operating condition changes (plant variations) occur.

For example, a typical case for applying these techniques would be a robot arm executing a trajectory-following task requiring quick and precise movements. The linearised models are time-varying so that achieving robustness against unmodelled dynamics needs low feedback gains. If the task includes grabbing and leaving an object, then the parameters also vary. Furthermore, there might be unmodelled irregularities in a bearing producing friction spikes always at the same arm position, *etc.* Iterative learning control could generate feedforward control actions so that the feedback regulator (either linear

---

[1] In the repetitive control framework, the reference is periodically time-varying and the control updating takes place in closed-loop. Thus, only a task is considered though its behaviour is repetitive or periodic.

or using some sort of linearisation) could have reduced gains and hence be less sensitive to inaccurate modelling of inertias, friction, *etc*.

In this case, the main advantages of this approach rely on both the use of a lower feedback gain and in getting a faster response. The high feedback gain to counteract the error presence increases the sensitivity to noise and easily leads to actuator saturations. On the other hand, the feedforward control avoids the plant's delays and allows for a faster response and a reduction in the magnitude error.

The main drawback is that the generated control action is only valid for the learned operating conditions.

5.2 Control with Two Degrees of Freedom

In this section, some of the ideas in section 4.2.7 will be reviewed to set up a framework motivating the iterative learning control laws discussed in later sections.

Let us have a process control loop such as the one depicted in Figure 5.1.

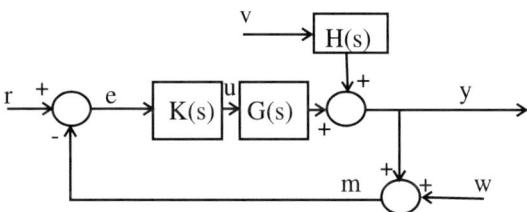

Fig. 5.1. Closed-loop control with set-point changes, process and measurement noise

In that set-up, the different transfer functions are

$$\frac{y(s)}{v(s)} = H\frac{1}{1+KG} \stackrel{\text{def}}{=} T_v \qquad \frac{y(s)}{-w(s)} = \frac{y(s)}{r(s)} = \frac{KG}{1+KG} \stackrel{\text{def}}{=} T_r \qquad (5.1)$$

On the other hand, in open-loop, the plant equation is:

$$y(s) = Gu(s) + Hv(s) \qquad (5.2)$$

so measurement noise does not affect the output. The classical conclusions for those equations is that high-gain regulators produce $T_v \approx 0$ (process disturbance rejection) and $T_r \approx 1$ reference tracking. As high gain cannot be achieved all over the frequency range due to stability and robustness constraints (see Chapter 4 or [140, 142]), the loop designer usually shapes one of the two closed-loop functions (T_v or T_r) and the other one is left with whatever the result of that shaping is. In many cases where combined tracking and disturbance rejection is sought, the open-loop transfer function KG is shaped to have the highest possible gain while not unstabilising the closed-loop (with reasonable robustness margins).

The referred shaping of closed-loop transfer functions can be made using pole placement or cancellation controllers (defining either T_r or T_v so that some internal stability requirements are met and then obtaining K from the appropriate expression in (5.1)), or via $\mathcal{H}_\infty$ techniques such as stacked sensitivity S/KS, etc. [114, 142, 164]. If one of the functions is shaped, then the other one must verify:

$$T_v = (1 - T_r)H \tag{5.3}$$

In many cases, an inaccurate model hinders gain increase so tracking performance is unacceptable, but disturbance following may be satisfactory if they are small or concentrated in a frequency range where robust stability constraints are not stringent.

5.2.1 Feedforward Control

To achieve some prefixed behaviour in both T_r and T_v, it is necessary to include a so-called two-degrees-of-freedom structure (2DOF) in the control loop such as the ones depicted in Figure 5.2 or any other obtained from them with valid block diagram transformations. Other than the inner feedback loop already mentioned, the input is fed forward to the control variable without waiting for an output error to occur. In both of these, the internal regulator K is designed to take into account non-deterministic disturbance rejection and robustness, given a certain model quality. Of course, many of the techniques in this book could be used to design appropriated experiments so as to improve the model quality and the closed-loop performance.

The remaining blocks are designed as follows, in each of the two approaches:

Set-point pre-filtering

The block diagram is depicted in Figure 5.2(a). If the output (let us call it $y_1(s)$) with a starting default pre-filter $Q_0 = 1$ is unsatisfactory, then a pre-filter $Q(s)$ is designed so that $Q(s)y_1(s)$ exhibits good behaviour. Although the pre-filter would involve, in a general case, inversion of the achieved closed-loop transfer functions, in many cases, the pre-filter is manually tuned, by trial and error, once the closed loop design has been finished. The manual tuning usually is based on a pre-filter such as:

$$Q(s) = K_e \frac{\alpha s + 1}{\beta s + 1} \tag{5.4}$$

or an equivalent discretised version.

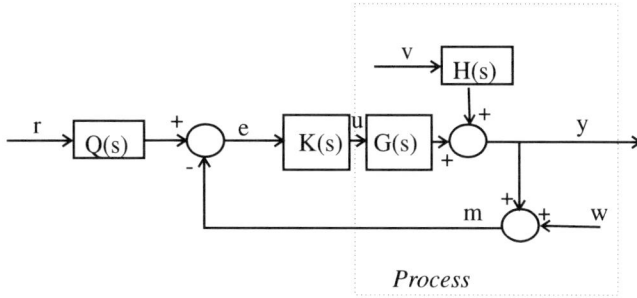

(a) Set-point pre-filtering

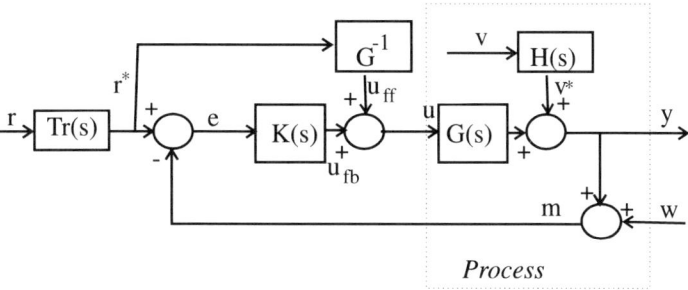

(b) Feedforward open-loop

Fig. 5.2. Two-Degrees-of-freedom configurations

- K_e is used to compensate the position error e_p of the closed-loop regulator, i.e., $K_e = 1/(1 - e_p)$;
- $\alpha > \beta$ is used to "accelerate" the output rise phase, usually in non-oscillating transients;
- $\alpha < \beta$ delays the rise phase in the output and alleviates some overshoot effects. In many practical applications, non-linearities advise against the use of high amplitude inputs and, in order to achieve that, rate-limited set-point changes are usually enforced, with a similar effect to this last pre-filtering. This type of pre-filtering is implemented into many commercial PID regulators.

Feedforward open-loop

The scheme can be analysed in Figure 5.2(b). In this case, it is easy to check that the transfer function from r to y is $T_r(s)$ whatever the internal regulator R is. On the other hand, the T_r block does not modify the stability or performance of the closed-loop regarding v. In this way, the reference tracking design is fully decoupled from the disturbance rejection and stabilisation ones. The tracking design is only based on open-loop calculations[2]. In this way, tracking performance and stability robustness can be balanced in a design. This will be the basic configuration for the iterative learning approach to be presented in the next sections, and it is similar to the set-up depicted in Figures 4.8 and 4.9.

This schema can be also applied to disturbance rejections, if they are deterministic and known in advance. In this case, $T_r(s)$ is replaced by $G_v(s)$ in the forward path from $v(s)$.

So, to summarise, 2-DOF configurations allow for separate design of, on one hand, disturbance and modelling error rejection and closed-loop stability robustness (limited by model quality) and, on the other hand, reference tracking behaviour (limited by actuator power[3] and bandwidth).

Iterative improvement can be carried out by identifying better and better models, so a better estimate of G will lead to a better u_{ff} and u_{fb} calculation in Figure 5.2(b). Notwithstanding, as pointed out in other chapters, maybe some reduced order models could be useful to design feedback controllers even if their open-loop step response is very different to that of the plant. In that case, a different model of G should be used for the design of K and for the design of the inverse plant G^{-1} in the referred figure.

In fact, the point in this chapter is to realise that maybe the "model" needed for disturbance rejection in closed-loop control could be simpler, and the regulator more cautious if the tracking of references were carried out in open-loop and learnt (identified) separately: the closed-loop model needs accuracy ONLY at stability-risk frequencies. There might be no need of accuracy over a wide range of frequencies to follow abrupt, quick reference changes.

In the case of repetitive tasks, the control actions u_{ff} can be learnt from repetition of experiments and progressively improve tracking performance. Of course, if the reference changes arbitrarily in an unpredictable way, then a general model G is needed. The first case (repetitive task) will be presented in more detail in the next sections. For the general case, the reader is referred to the open-loop and closed-loop identification techniques described in Chapter 2 and [108].

---

[2] Perfect tracking is not possible in non-minimum-phase plants. Notwithstanding, the 2-DOF approach is applicable to unstable plants as long as the feedback controller internally stabilises the loop.

[3] Actuator power is also an issue in stabilising unstable plants, where unrecoverable states can be reached [85].

5.3 The Iterative Learning Paradigm

As usual implementations of complex control laws are coded into a computer, we will consider the discrete time case, in where a sequence of control actions has to be learnt.

In the sequel, an analysis of a particular set-up for repetitive tasks will be presented. However, in a general case, direct use of the most updated system model in the schemes in Figure 5.2 can also be considered. That model can be obtained from previous identification experiments. An important point is that the model in the feedforward path can be more "aggressive" and complex in the sense that its modelling errors will not compromise closed-loop stability, but only tracking performance, so less caution is required.

In the following, a direct learning of control actions will be undertaken, without resorting to a model. That will ease convergence, but at the expense of not being a "general" plant inverse but one specialised for the particular repetitive task under execution. Although the learnt action is applied in open-loop, the learning algorithm obviously needs feedback from sensors to evaluate its performance, but the loop is only closed once the repetitive task has ended, and the dynamics will be simpler, as shown in later sections.

The assumptions

The iterative learning methodology can be understood as follows.

Let us have a repetitive system

$$y_k(t) = f_H(u_k(t), t) \quad t = t_0, t_0 + 1, \ldots, t_f \qquad (5.5)$$

where $f_H : U \rightarrow Y$ is a non-linear operator and $u_k(t)$ is the input sequence applied to the system in the repetition (iteration) $k = 0, 1, , \ldots$ at time t. It will be assumed that

- every trial lasts a fixed time T;
- for the desired output $y^d(t)$, a reasonable control action sequence $u^d(t)$ exists so that it achieves the desired output. Reasonable in this case means with acceptable bounds in amplitude and frequency. To accomplish this requirement, the desired signal shouldn't exhibit abrupt changes, and restrictions on the smoothness and minimum "gain" of f_H also apply. The desired output y^d is kept the same during all the iterations (repetitive task);
- the initial state of the plant is the same in all iterations. In 2DOF configurations, the initial condition error will exhibit the closed-loop dynamics;
- the system is time-invariant through the iterations. That means: f_H does not depend on k. Time variance inside each iteration is allowed (for example, grabbing or dropping a certain amount of mass, leading to a change in mechanical parameters) as long as it occurs at the same time instant in each iteration.

Note that minimum-phase systems can be "inverted" to obtain u_d if the duration of the requested task is short enough, or if the desired trajectory does not excite the inverse unstable dynamics (although generating the desired output in that case for long-lasting tasks is impossible in practice because of errors in estimation of the zero dynamics).

The objectives

The objective of the iterative learning is to track references and repetitive disturbances, with progressively lower tracking error by improving its open-loop control action estimates.

A learning operator f_L will provide the input $u_{k+1}(t)$ to be applied to the plant in the next iteration, taking into account the current input-output data pair, $(u_k(t), y_k(t))$, and the desired output, $y^d(t)$

$$u_{k+1}(t) = f_L(u_k(t), f_H(u_k(t), t), y^d(t), t) \tag{5.6}$$

This setting is represented in Figure 5.3.

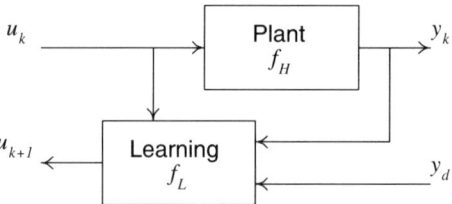

Fig. 5.3. Iterative learning

Thus, the iterative learning problem is to design an iterative algorithm (5.6) that converges to an input $u^*(t)$ that minimises the system output tracking error $y^d(t) - f_H(u^*(t), t)$ in some norm $\forall t \in [t_0, t_f]$ [115].

5.4 The Learning Algorithms

In order to avoid the closed-loop interaction between the current control action and tracking error, the following strategies are usually adopted:

- use next error (in current iteration) to update the control action in current time for the next iteration

$$u_{k+1}^{ff}(t) = u_k^{ff}(t) + \eta e_k(t+1) \tag{5.7}$$

where η is a learning factor. In this setting, the mismatch between the current control action and the one required to obtain the desired output is assumed to produce a tracking error in the next time instant. If it is assumed that the plant presents a pure time delay d, the learning algorithm would be:

$$u_{k+1}^{ff}(t) = u_k^{ff}(t) + \eta e_k(t+1+d)$$

- add the previous iteration feedback control actions to the current feedforward one:

$$u_{k+1}^{ff}(t) = u_k^{ff}(t) + \eta u_k^{fb}(t+1) \tag{5.8}$$

- the previous approaches are a particular case of the use of a general filtered measure of error in the last iteration to update the current control actions. In the next iteration, all time instants are known so the filter can be "non-causal" if so wished.

5.4.1 Convergence (Linear Case)

In this section, the convergence of some learning strategies will be analysed. The feedback loop is not considered and so the control actions refer to the feedforward part.

SISO case

For simplicity, let us firstly consider a single-input single-output (SISO) linear plant.

Denote by

$$\{h_k\} = \{0, \ldots, 0, h_m, h_{m+1}, \ldots\} \tag{5.9}$$

the pulse response of a sampled linear time-invariant system, where m is the relative degree of the system (if $m > 1$ that means that a pure delay is present). The system output is given by the usual convolution formula

$$y(i) = \sum_{j=0}^{i-m} h_{i-j} u(j) \tag{5.10}$$

If a finite length response is analysed ($i = m, \cdots, N$), then (5.10) can be written in matrix form. By defining vectors $u, y \in R^{N-m+1}$

$$u = \begin{bmatrix} u(0) \ u(1) \ \cdots \ u(N-m) \end{bmatrix}^T \tag{5.11}$$

$$y = \begin{bmatrix} y(m) \ y(m+1) \ \cdots \ y(N) \end{bmatrix}^T$$

$$\tag{5.12}$$

then

$$y = Hu \tag{5.13}$$

$$H = \begin{bmatrix} h_m & 0 & \cdots & 0 \\ h_{m+1} & h_m & \cdots & 0 \\ \vdots & \vdots & & \vdots \\ h_N & h_{N-1} & \cdots & h_m \end{bmatrix} \tag{5.14}$$

The inverse system can be also represented in finite time by a matrix relation $u = My$, where $M = H^{-1}$, and another convolution:

$$u(i-m) = \sum_{j=m}^{i} a_j y(m+i-j) \tag{5.15}$$

The structure of matrix M is also lower triangular:

$$M = \begin{bmatrix} a_m & 0 & \cdots & 0 \\ a_{m+1} & a_m & \cdots & 0 \\ \vdots & \vdots & & \vdots \\ a_N & a_{N-1} & \cdots & a_m \end{bmatrix} \tag{5.16}$$

so from any desired finite time output trajectory sequence $y = y^d$, the input sequence $u^*(t)$ can be derived given the feedforward matrix M. The output sequence must satisfy trivial conditions regarding causality (delay). For unstable systems, although the reasoning so far is valid, the H and M matrices can be badly conditioned so it is recommended to learn the control actions in a reference pre-filtering approach (Figure 5.2(a)), after the feedback loop has stabilised the system instead of feedforward open-loop injection (Figure 5.2(b)). For example, the condition number of H for a system with impulse response $h_k = 0.4e^{3.5*(0.2k)}$ for a task duration of 11 samples is 4.35×10^3. The problem of iterative learning in reference pre-filtering is that if the feedback controller is changed then the learned actions are no longer valid so the process must start again, whereas the second approach totally decouples tracking and disturbance rejection, as previously presented.

If a system model (state-space or transfer function) is known, M can be easily obtained. If a model is not available, the feedforward matrix elements a_i ($i = m, \cdots, N$) can be easily estimated from any previously recorded experimental input-output pairs (u, y), as well as an estimation of the total delay m, but the approach is very sensitive to model uncertainty, noise and disturbances. Also, it does not apply to non-linear systems. The iterative learning approach will compute the control actions directly, without resorting to calculation of M, even for non-linear systems where it cannot be constructed. The delay estimation is still needed, and a wrong one can hinder the convergence.

Let us consider y_k and u_k, the vectors containing the input and output samples in repetition (iteration) k, and $e_k = y^d - y_k$, the vector of errors. The size of the vectors is $N - m + 1$.

Let us apply an update of control action given by:

$$u_{k+1} = u_k + \Lambda e_k \qquad (5.17)$$

where Λ is a learning matrix of dimensions $(N - m + 1) \times (N - m + 1)$.

Then,

$$e_{k+1} = y^d - y_{k+1} = y^d - H(u_k + \Lambda e_k) = (I - H\Lambda)e_k \qquad (5.18)$$

so Λ must be designed so that the eigenvalues of $(I - H\Lambda)$ have a modulus of less than 1. As $eig(I - H\Lambda) = 1 - eig(H\Lambda)$, that condition can be shown to be equivalent to matrix $H\Lambda$ having eigenvalues in the unit circle centered at $+1$, so if oscillatory complex eigenvalues are ruled out, the desired values are real between 0 and 2. For full oscillation avoidance, values in $[0, 1]$ are advised.

Due to the triangular structure of H, if Λ is a diagonal matrix with all diagonal elements η such that

$$0 < \eta < 2/h_m \qquad (5.19)$$

the stability of the error dynamics in the iterations time axis is guaranteed. This applies to the SISO case, and corresponds to the learning rule

$$u_{k+1}(t) = u_k(t) + \eta e_k(t + m) \qquad (5.20)$$

From this perspective, the "optimal" learning factor would be

$$\eta = 1/h_m \qquad (5.21)$$

assigning the eigenvalues of $(I - H\Lambda)$ to the origin. Of course, depending on the approximate plant model, a non-diagonal Λ (meaning some kind of filtering on the error measures) can also yield stable error dynamics with less sensitivity to random disturbances (see later). That would be the philosophy under learning rules such as (5.8) or "non-causal" filterings previously mentioned.

If an approximate model is known, then the bound on the learning rate (5.19) can be used to design the iterative learning set-up. If the plant is unknown, an estimate of it could be obtained via impulse or step experiments. Note that the learning rate in (5.19) is limited by the maximum possible value of h_m in case that an uncertain model description is available. The uncertainty in models will be addressed later on.

Oscillations in the final control u^* may appear as the poles in the ideal control action cancel the zeros of the plant, unless those effects are accounted for in generating y^d. Usually, as learning control is applied to poorly known plants, it is not possible to *a priori* know the location of the plant zeros.

MIMO case

In a multiple-input-multiple-output (MIMO) case, input and output vectors at each sampling instant $u_k(t)$, $y_k(t)$ should be concatenated to form bigger vectors u_k, e_k. In that case, H is block lower triangular but its determinant and eigenvalues are still given by those of the block diagonal. Let us assume that the plant incorporates m actuators and m outputs (square plant) so that H is square and there exists a unique u^* achieving y^d.

In this case, the impulse response h_m is a matrix of dimensions $m \times m$, so the requisite for the learning rate matrix η (with dimensions $m \times m$, being η the building block of a block-diagonal Λ matrix) would be that the eigenvalues of $h_m \eta$ are between 0 and 2. That can be solved with straightforward pole placement techniques. The "optimal" learning rate matrix is $\eta = h_m^{-1}$, in the sense that poles are placed at the origin.

Note that, however, the design of learning rates requires at least an approximate knowledge of h_m and its directionality effects, which in the MIMO case can be harder to obtain than in a single variable experiment. If an approximate $\hat{h}_m$ is known such that the relative error

$$\Delta = (h_m - \hat{h}_m)\hat{h}_m^{-1} = h_m \hat{h}_m^{-1} - I \tag{5.22}$$

has a known bound in maximum singular value $\bar{\sigma}(\Delta) < \sigma_m$ strictly less than 1, then using a learning rate matrix $\eta = \hat{h}_m^{-1}(I - F)$, as the closed loop dynamics is determined by the eigenvalues of

$$I - h_m \eta = I - h_m \hat{h}_m^{-1}(I - F)$$
$$= I - (I + (h_m \hat{h}_m - I))(I - F) = F - \Delta(I - F) \tag{5.23}$$

that the maximum eigenvalue of $F + \Delta(I - F)$ must be less than one, being F a target dynamics that would be achieved in the case of exact matching between model and real plant. For the particular case of symmetric F (so its singular values are the absolute value of the eigenvalues λ_i), it can be shown that robust stability will hold if:

$$\max_i |\lambda_i| + \sigma_m \max_i (1 - \lambda_i) < 1$$

and, after some operations:

$$\sigma_m < \frac{\min_i(1 - |\lambda_i|)}{\max_j(1 - \lambda_j)} \tag{5.24}$$

For example, to safely use learning rates near $2\hat{h}_m^{-1}$ (i.e., $F = -I$) the modelling error must be very small. For positive definite F, the result is $\sigma_m < 1$. As most singular value stability results, the uncertainty bound is conservative.

Deterministic perturbations

When subject to deterministic repetitive disturbances, the output of the system is $Y = HU + D$, where D is a vector of disturbances invariant under iterations. In that case, the error equation still converges to zero, because $e_k = y^d - (Hu_k + D)$ so:

$$e_{k+1} = y^d - y_{k+1} = y^d - (H(u_k + \Lambda e_k) + D) = (I - H\Lambda)e_k \tag{5.25}$$

Note that the error converges to zero irrespective of D being directly measurable or known, as long as it is repetitive.

Non-deterministic perturbations

In the presence of non-deterministic perturbations, D is not constant over iterations so it cannot be cancelled as in the previous formula. Hence, the dynamics are:

$$\begin{aligned} e_{k+1} = y^d - y_{k+1} &= y^d - H(u_k + \Lambda e_k) - D_{k+1} = \\ &(e_k + D_k) - D_{k+1} - H\Lambda e_k = (I - H\Lambda)e_k + (D_k - D_{k+1}) \end{aligned} \tag{5.26}$$

If the disturbance D_k is stationary, the ILC law converges to zero-mean error, as $E(D_k) = E(D_{k+1})$ where $E()$ denotes mathematical expectation. Thus, the algorithm converges successfully for slowly-varying disturbances ($D_k - D_{k+1}$ small) but general stochastic disturbances produce variance effects on e_k, in many cases amplifying its effect, with amplification values depending on the learning rate (note that the ILC law feeds back in the next iteration the disturbances from the current one, adding in this way a second source of them). That is why the iterative learning control is advised in problems where the tracking or deterministic disturbance component is predominant over the stochastic one (although a valid approach is the 2DOF framework where a feedback regulator diminishes the effect of random disturbances to an acceptable level).

To evaluate the effect of disturbances, in the case that D_k is independent of D_{k+1}[4] the stationary error covariance matrix, after convergence to zero mean, would be given by the solution Ξ_e to the discrete Lyapunov equation:

$$\Xi_e = (I - H\Lambda)\Xi_e(I - H\Lambda)^T + 2\Xi_D \tag{5.27}$$

where Ξ_D is the covariance matrix of D_k. For instance, a white noise disturbance would have a diagonal Ξ_D.

---

[4] Note that coloured noise on the system output would qualify for that independent disturbance if the time between repetitions (iterations) is high enough to fade away significant correlations.

Delay estimation errors

If the delay m is underestimated, then the diagonal of H contains zeros, so the learning eigenvalues are always 1, i.e., the control actions increase forever and the learning behaves in an unstable way (ramp-like). In the case that it is overestimated, the learning algorithm can converge and, in some cases, be even better than in the original case but the calculation of the admissible range of learning rates is not straightforward.

Let us consider, as an example, a system with impulse response given by:

$$\{h_k\} = \{0, 0.01, 0.06, 0.5, 0.7, 0.8, 0.8, \ldots\}$$

and a task length of 11 samples. In that case, h_1 and h_2 are almost nil. It appears that, although a pure delay of $d = 2$ could be considered, the measurement noise masks it. Assuming a delay $m = 1$, the optimal learning rate (5.21) is 100, but in that case, the non-deterministic perturbations such as measurement noise are amplified by that gain, disturbing the algorithm in such a way that even if it will converge in average, the performance could be very poor due to that effect. If a more sensible choice $\bar{m} = 3$ is made, then the matrix H is no longer diagonal, but with a choice of $\eta = 1.51$ the eigenvalues of $I - H\Lambda$ have a maximum modulus of 0.66, so the convergence is reasonably quick and the amplification of the non-deterministic components is much less than in the previous case. Note that in this case, adopting the "optimal" learning rate from (5.21), i.e., $\eta = 1/0.5 = 2$ would have led to an unstable iteration scheme.

5.4.2 Non-linear Case

In a non-linear plant, the finite time length response has an expression $y = f_H(u)$, y and u being the output and input sequences as a vector, respectively. With the plant being causal, the operator f_H is also triangular.

The non-linear convergence for small learning rates can be proved in a local sense, as follows. These kind of iterations are, in a certain sense, a reformulation of gradient methods to solve $y^d = f_H(u^*)$ for u^*.

If the increments of control actions between iterations are small, the output approximately verifies:

$$y(u_0 + \Delta u) \approx f_H(u_0) + J(u_0)\Delta u$$

where J is the Jacobian matrix of first derivatives of H (local linearised model for u_k):

$$J = \frac{\partial (f_H)_i}{\partial u_j}$$

The Jacobian depends as well on the initial state, but it is assumed to be fixed. Hence, assuming the non-linear model is smooth enough so that $J(u_0)$

is continuous and bounded, then if the learning rule (5.17) is applied with a small $\|\Lambda\|$ (maximum singular value norm), it follows

$$e_{k+1} \approx (I - J\Lambda)e_k \tag{5.28}$$

hence, for a small η in (5.20), the linearised model's validity and a condition similar to (5.19) can be both fulfilled, if the diagonal elements in J do not change sign (monotonic instantaneous gain). Note that the non-linear case J is lower triangular, but all the diagonal elements need not be identical. At least, they should not change sign in any operating point.

However, J depends on u_k so in the next iteration, the Jacobian J will be at a different point, hence dynamics in (5.28) will be time-varying. Hence, convergence is unclear and, anyway, the idea could only work for infinitesimal learning rates. These expressions, however, provide some hints for more powerful results.

In the following, a proof of global convergence in the non-linear case will be outlined by induction, assuming the learning rate is below a certain bound, for the SISO case.

Let us divide the operator $H(u)$ into an off-diagonal $H_o()$ and a diagonal one $H_d()$, i.e., each output y_{ki} (reflecting iteration k, time $i+m$) will be expressed as

$$y_{ki} = H_{oi}(u_{k0}, u_{k1}, \ldots, u_{k(i-1)}) + H_{di}(u_{ki})$$

where the second subscript represents actual time in input and output variables.

We will assume that some knowledge is available on the diagonal instantaneous gain-like term H_{di}. In particular, we assume that H_{di} is positive and verifies, for any two feasible inputs (at time i) a and b, the Lipschitz condition:

$$\gamma_l(b-a) \leq H_{di}(b) - H_{di}(a) \leq \gamma_h(b-a)$$

where $0 < \gamma_l \leq \gamma_h$ and γ_h is approximately known. That will be the case, for example, if it has a derivative with respect to u_{ki} bounded by $0 < \gamma_l \leq \frac{\partial H_{di}}{\partial u_i} \leq \gamma_h$, but in fact it is not even needed for H_o or H_d to be differentiable functions for the learning algorithm to work.

Let us assume that the learning rate η is less than $1/\gamma_h$, and let us consider each element of e_{k+1}.

Starting with $e_{(k+1)1}$,

$$e_{(k+1)1} = y_{d1} - H_{d1}(u_{k1} + \eta e_{k1}) \leq y_{d1} - H_{d1}(u_{k1}) - \gamma_l \eta e_{k1} = (1 - \gamma_l \eta)e_{k1}$$

and

$$e_{(k+1)1} = y_{d1} - H_{d1}(u_{k1} + \eta e_{k1}) \geq y_{d1} - H_{d1}(u_{k1}) - \gamma_h \eta e_{k1} = (1 - \gamma_h \eta)e_{k1}$$

hence:

$$0 < (1 - \gamma_h \eta)e_{k1} \leq e_{(k+1)1} \leq (1 - \gamma_l \eta)e_{k1}$$

so, if $\eta < 2/\gamma_h$, e_{k1} converges as it is bounded from below and above by stable first-order linear dynamics.

By considering $e_{(k+1)j}$, assuming all e_{ki} converge to zero for $i < j$, we have, when enough iterations have elapsed that, defining

$$\psi_j = H_{oj}(u_{k1} + \eta e_{k1}, \ldots, u_{k(j-1)} + \eta e_{k(j-1)}) - H_{oj}(u_{k1}, \ldots, u_{k(j-1)})$$

it follows that

$$\begin{aligned}e_{(k+1)j} &= y_{dj} - H_{oj}(u_{k1} + \eta e_{k1}, \ldots, u_{k(j-1)} + \eta e_{k(j-1)}) - H_{dj}(u_{kj} + \eta e_{kj})\\ &< y_{dj} - H_{oj}(u_{k1}, \ldots, u_{k(j-1)}) - H_{dj}(u_k) - \gamma_l \eta e_{kj} - \psi_j = (1 - \gamma_l \eta)e_{kj} - \psi_j\end{aligned} \quad (5.29)$$

and

$$\begin{aligned}e_{(k+1)j} &= y_{dj} - H_{oj}(u_{k1} + \eta e_{k1}, \ldots, u_{k(j-1)} + \eta e_{k(j-1)}) - H_{dj}(u_{kj} + \eta e_{kj})\\ &> y_{dj} - H_{oj}(u_{k1}, \ldots, u_{k(j-1)}) - H_{dj}(u_k) - \gamma_u \eta e_{kj} - \psi_j = (1 - \gamma_u \eta)e_{kj} + \psi_j\end{aligned} \quad (5.30)$$

where ψ_j tends to zero when k is big if H_o is assumed to be continuous. Hence, the error is bounded by two convergent linear dynamics with an input approaching zero, thus e_{kj} converges also to zero.

In that way, by induction, as e_{k1} converges, then $e_{k2}, e_{k3}, \ldots$ also converge in the long run.

A similar proof would show convergence in the non-linear case with presence of deterministic repetitive perturbations, as in (5.25).

Non-linear MIMO case

Convergence for the MIMO non-linear case needs additional considerations. Based on previous considerations, an accurate enough estimate of the Jacobian must be available, in the sense of (5.22). With that estimate, convergence will be achieved by using its inverse.

Let us outline the full generalisation of the approach, defining $d_i(b, u)$ such that:

$$H_{di}(u_i + b) = H_{di}(u_i) + d_i(b, u_i)$$

so that if $u_{k+1} = u_k + b$, then

$$e_{(k+1)i} = e_{ki} - d_i(b, u_{ki}) + H_{oi}(\ldots) \quad (5.31)$$

If and invertible approximator $\hat{d}$ of d (linear or non-linear) is known (i.e., $m = \hat{d}(n, u)$ can be solved for n with known m and u, denoting the solution as $\hat{d}^{-1}(m, u)$) such that the uncertainty in the approximation, defined as:

$$\Delta(e, u) = d(\hat{d}^{-1}(e, u), u) - e \tag{5.32}$$

verifies, for any desired increment of output e and input u

$$\|\Delta(e, u)\|_2 < \sigma_m \|e\|_2 \quad \sigma_m < 1 \tag{5.33}$$

In that case, it can be shown that incrementing the control actions by

$$b_{ki} = \hat{d}_i^{-1}(\beta e_{ki}, u_{ki}) \tag{5.34}$$

will yield, for the first control action

$$\begin{aligned} e_{(k+1)1} &= e_{k1} - d_1(\hat{d}_1^{-1}(\beta e_{k1}, u_{k1}), u_{k1}) \\ &= e_{k1} - \beta e_{k1} - \Delta(\beta e_{k1}, u_{k1}) = (1 - \beta) e_{k1} - \Delta(\beta e_{k1}, u_{k1}) \end{aligned} \tag{5.35}$$

so

$$\|e_{(k+1)1}\| \leq |(1 - \beta)| \cdot \|e_{k1}\| + \sigma_m \beta \|e_{k1}\| \tag{5.36}$$

hence, for $0 < \beta < 1$, e_k converges. For reduced uncertainty levels, the convergence bound for β approaches 2, and letting $\beta = (I - F)$ a similar expression to (5.24) can be obtained. An induction argument similar to the SISO case can be put forward for the convergence of the full scheme.

An alternative approach to convergence proofs for iterative laws based on non-linear state-space models can be found in [37].

5.4.3 Reparameterisation

The previous approach directly learns the control actions that are applied at each stage of the repetitive task.

There is the option of generating the control actions from a parameterised function approximator. Of course, if the number of parameters is small, only an approximation to the control time history could be achieved, but note that in a 2DOF structure, the feedback part would handle those approximator errors. Learning rates may need to be further reduced, even more if the proposed approximator is non-linear in parameters.

The possible advantages of these parameterisations are two:

- for some references (such as steps), the complexity of the control actions reduces with time (all are equal to the same constant near the settling time);

- maybe the task to be carried out needs to be executed for different values of a certain measured variable (a mass, a reference scaling, *etc.*). In that case, including that variable in the approximator inputs could lead to useful interpolations for non-linear systems, in a sense similar to gain scheduling on the feedforward blocks, learning them from experience.

For linear plants, taking the model reference control structure in mind, the parameterised function approximator could be implemented, in some cases, as a filter. The desired output is defined as the signal generated by a model under some excitation, and the sequence of control actions is generated as the output of a filter with the same excitation. In this setting, the adaptation to changes in magnitude of the desired output is naturally achieved.

The heuristic procedure will be:

- define the desired output, $y^d(t)$ as the signal generated by the reference model $G^d(s)$ under the input $\bar{u}$;
- consider a suitable filter $G(s, \alpha)$, where α is a vector of parameters to be learned. This filter should have, at least, the poles in $G^d(s)$. Let us call $u_\alpha(t)$ the filter output for the input $\bar{u}$;
- apply a learning algorithm to the parameters α, considering as error the difference between $y^d(t)$ and the output of the plant excited by $u_\alpha(t)$;
- if the error does not converge, try a new filter $G(s, \alpha)$, with a different structure/set of parameters.

Of course, this heuristic approach does not have any convergence proof but, in some cases, can lead to good results, as illustrated later on in some examples.

The main advantages are the reduced number of parameters, the easy implementation and the implicit generalisation for different reference magnitudes. Although some generalisation in the excitation input $\bar{u}$ can be expected, the results are only valid for the trained kind of inputs.

5.5 Examples

Example 5.1. In this example, the set of control actions to control a linear plant will be iteratively learnt. The plant will be:

$$G(s) = \frac{1.8}{(s+1)(s+2)(s+9.5)}$$

and discretised at 1.5 s sampling time. The instantaneous gain h_1 of the discretised model is 0.0535 so the "minimum-time" learning rate is $\eta = 1/0.0535 = 18.7$. A feedback PD regulator is in charge of trying to achieve the reference behaviour when the iterative approach for the repetitive task hasn't yet learned the feedforward actions.

The closed-loop equations in Figure 5.2(b) are:

$$y = \frac{G}{1+GK}u^{ff} + \frac{GK}{1+GK}(r^*) + \frac{1}{1+GK}v^* \qquad (5.37)$$

$$u = \frac{1}{1+GK}u_{ff} + \frac{K}{1+GK}(r^* - v^*) \qquad (5.38)$$

$$u^{fb} = \frac{-GK}{1+GK}u_{ff} + \frac{K}{1+GK}(r^* - v^*) \qquad (5.39)$$

The learning rate used in the simulation was 9.2, half of the previously calculated optimal one. Note that in practice, that optimal gain may be a very rough estimate so it's better to play on the safe side, especially when significant non-linearities could be present. The MATLAB® code that carries out the simulation is:

```
ng=1.8;dg=poly([-1 -2 -9.5]); T=1.5;
qe=9.2;uff=0*range'; [ngz,dgz]=c2dm(ng,dg,T,'zoh'); nkz=[9.5500
-3.5500]; dkz =[1.0000 0.0000]; nsamples=60;range=1:(nsamples);
r=sin(0.16*(range'-1)*T)+square(0.2*(range'-1)*T);r(1)=0;
v=0.4*sin(0.96*(range'-1)*T); %deterministic perturbation
%simulate loop:
[cng,cdg]=feedback(ngz,dgz,nkz,dkz,-1); %g/(1+gk)
[t1,t2]=series(nkz,dkz,ngz,dgz); %gk
[cngk,cdgk]=cloop(t1,t2,-1); %gk/(1+gk)
[cnk,cdk]=feedback(nkz,dkz,ngz,dgz,-1); %k/(1+gk)
[cn1,cd1]=feedback(1,1,t1,t2,-1); %1/(1+gk)
for iters=1:20
%calculate output:
[yy1,x]=dlsim(cng,cdg,uff); [yy2,x]=dlsim(cngk,cdgk,r);
[yy3,x]=dlsim(cn1,cd1,v); y=yy1+yy2+yy3;
%calculate control action (feedback):
[yy1,x]=dlsim(cngk,cdgk,uff); [yy2,x]=dlsim(cnk,cdk,r-v);
ufb=-yy1+yy2; e=r-y;
%prepare next iteration:
uff=uff+qe*[e(2:nsamples); 0];
end
```

and the results are plotted in Figure 5.4.

When the trials end, all the reference tracking and disturbance cancellation are handled by the feedforward control action, and the feedback controller receives a zero input. Of course, if would still be acting in the case of random disturbances. In that case, a reduction of the learning rate would also be advised to avoid excessive variation of u_{ff} between trials due to those random process disturbances.

Example 5.2. A second example, over the same system, will approximate the desired control action to track a constant desired trajectory given by $r(t) = 1 - e^{-.3t}$.

The approximator will use a control action given by:

$$u(t) = \alpha_1 + \alpha_2 e^{-.3t} + \alpha_3 t e^{-.6t}$$

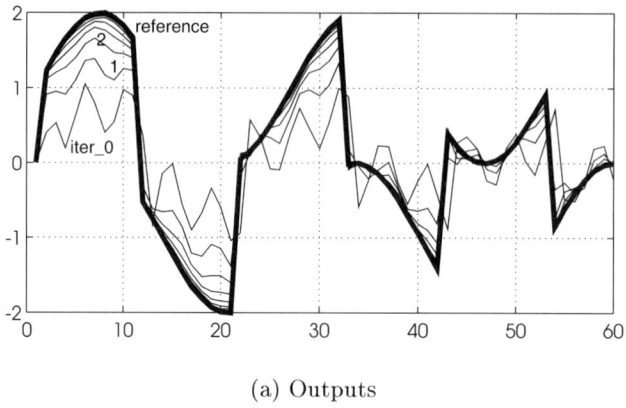

(a) Outputs

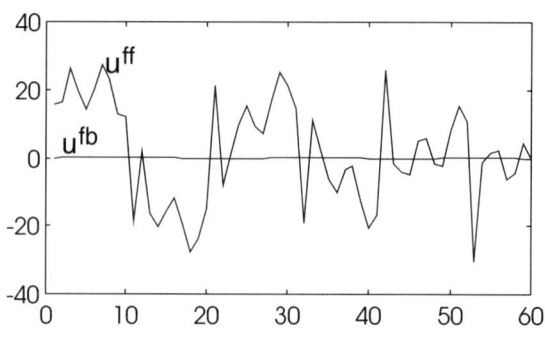

(b) Final control actions u^{ff} and u^{fb} ($u^{fb} \approx 0$)

Fig. 5.4. Example results

everything being discretised at $T = 1.5$ s. The adjustment calculates the parameter increments from the control increments via least squares fitting. The MATLAB® code is the same as in the previous example, except changes such as:

```
time=(range'-1)*T; r=1-exp(-0.3*time);
v=0; %deterministic perturbation
psi2=exp(-.3*time);psi3=ones(nsamples,1);
psi4=time.*exp(-.6*time); psi=[psi2 psi3 psi4];
theta=zeros(3,1);
```

in the initialisation code, and

```
 uffdes=uff+qe*[e(2:nsamples);0];
theta=inv(psi'*psi)*psi'*uffdes; uff=psi*theta;
```

in the iterations loop. Figure 5.5 shows the output in several iterations. The final quadratic error is 0.004. If the $te^{-.6t}$ regressor was not used, the error would have been 0.02. Note that very acceptable tracking is achieved with just two or three parameters. If other deterministic disturbances were present, additional regressors should be included, acting at the time points where the referred disturbances take place.

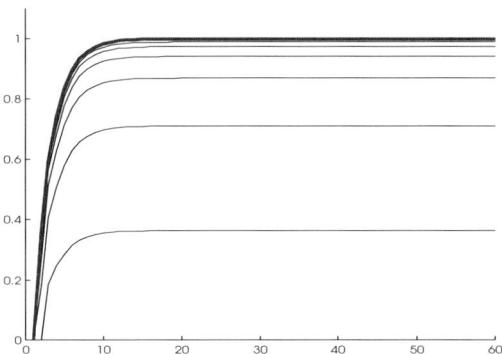

Fig. 5.5. Reparameterisation of open-loop actions: Plant output

5.6 Conclusions

In this chapter, the possibilities of the use of two-degrees-of-freedom structures in the context of iterative performance improvement are analysed. One particular possibility, iterative learning for repetitive tasks, has been analysed. Iterative learning control is able to achieve good tracking performance even for some non-linear processes with non-smooth non-linearities, so that at least in the repetitive case, there is no need to increase feedback gains (thus reducing robustness) to compensate for non-linearities.

Of course, apart from the approach previously presented, direct use of the most updated system model (from standard open- or closed-loop estimation techniques) can also be thought of for feedforward control blocks.

Acknowledgements

This work benefited from discussions with M. Olivares from Universidad Santa María, Valparaíso, Chile.
The authors thank Generalitat Valenciana and UPV for its support grant GV-00-100-14.

6 Model-based Iterative Control Design

Pedro Albertos

Abstract. In this chapter, the main ideas about the iterative identification and control design are illustrated by means of a number of academic examples. The main purpose is to emphasise the advantages and drawbacks of the approach. Thus, no attention is paid to the power or adequacy of the control design, identification or model parameter estimation algorithms. The common feature of the plants used in the designs is their apparent frequency decomposition due to either the presence of resonant modes or high-frequency components.

6.1 Introduction

Two main approaches can be followed to design a control system: to use a model of the process to be controlled and design a model-based control system, or to use the knowledge and the experience about how to control the process and implement a controller performing in a similar way. The first approach is applied here but taking advantage of the experience obtained by controlling the system, allowing for a progressive enhancement of the model. Thus, it is clear that a process model is needed. It has been previously discussed that there are many ways of process modelling and there is no *perfect* model of a process. Any model represents a partial behaviour of the system.

In a complex process with different timescale modes, high-frequency modes can be masked by measurement noise, being difficult to locate under relatively low-gain inputs. Also, fast modes require strong or high-frequency actions to be excited and only appear if fast response of the controlled system is required. On the other hand, to achieve a non-stringent control, simple control efforts are applied and a very simple, low frequency model of the process could be valid.

Most control design approaches look for a fast controlled system response to either input changes or as a reaction to external disturbances. That is, to reduce as much as possible the control errors, under some control constraints. If the control design is based on a simple model of the plant, undesirable controlled system features could appear if hidden oscillating modes are excited enough.

For instance, to tune a PID controller with low performances, very little information from the process is needed. The experimental Ziegler-Nichols rules [165] for tuning such a controller only require a few data from the open-loop step response of the process or the oscillating conditions under proportional control. Of course, a PID optimised controller [87] can achieve much better performances if a parametric process model is available or on-line tuning of the controller parameters is performed.

This is always the case; the total controller design effort is split into the modelling, the controller computation and the final controller tuning phases. Another option is to perform these phases in an iterative framework, as already discussed in the previous chapters as well as [1, 61].

The problem of a control system design for a plant for which the initial model is not well known, and where the main goal is the maximisation of a kind of performance index, can be approached following an iterative schema, such as:

Assume an initial model of the plant, design the most performing controller, and iteratively repeat:

- estimate a more accurate model, until required;
- design a suitable controller improving the performances, until required.

Specifications may be also updated for a given model, as the iterations proceed, if there is no constraint violation in the actual controlled plant. The idea is to get a basic model, design a controller and apply it. Check for the constraint's fulfilment. If possible, improve the specifications and redesign the controller. Do it iteratively until no further improvement is possible. Under the current operating conditions, that is, on the limit of the constraints, get new data from the process and estimate a more accurate model. Closed-loop identification algorithms should be applied.

This idea, decoupling of the identification and control design activities, which is the main goal of this book, has been also studied in different frameworks, [9] and different approaches are described in the following chapters. These two steps, in the final control design, could be fully decoupled, see [103, 124, 160].

The approach is rather general and different identification, analysis, design, and iteration algorithms may be applied. In this first method, we focus our attention on a simple set of such algorithms, in order to explore the main issues of this approach. In [3], the interaction between the two steps has been expressed in the frequency domain. The framework in this chapter is also related to the features of the frequency domain response of the plant. Given an initial low-frequency range model of a plant, the target consists of iteratively designing a controller that stabilises the plant while increasing the closed-loop bandwidth. The model parameter estimation is very simple and based on frequency analysis [2].

Very simple models are used for the plant and, wherever it is possible, closed-loop identification settings are avoided, just looking for the appearance of new modes of oscillation in the control output. The ultimate goal is the achievement of a controlled system, as fast as possible, fulfilling some operating constraints.

For discrete time control, the influence of the sampling period is considered and it is shown that its selection is also critical. A progressive reduction can be achieved, as far as the improvement on the process model allows for stronger control actions. All these issues are analysed in the following sections.

In the next section, the general approach is outlined and an already presented example [3] is revisited, the main goal being to increase the system bandwidth under the restriction of a maximum step response overshoot. The whole closed-loop system is identified and a cascade of controllers is suggested.

Then, two new situations are considered that attempt to simplify the structure of the controller and reduce the control system to a single loop. First, a direct application of frequency scale decomposition is shown. Signal power spectrum analysis is used to estimate the additional modes. Then, a generalised predictive controller, (GPC), where the weight in the control is iteratively decreased to get a faster response, is designed. The issue of the sampling period selection is then introduced and discussed.

6.2 Iterative Control Design

The approach is as follows: given a process, some control goals and constraints, design by an iterative approach the most performing control system without violating the constraints.

At stage j of the design, the process to be controlled, G, is approximately represented by a model $\hat{G}_j$, belonging to $\bar{G}$, a pre-defined set of models ($\hat{G}_j \in \bar{G}$). As control goal, a desired controlled system behaviour, such as that given by the reference model $\bar{M}_{j,i}$ or the optimal one with respect to the cost index $J_{j,i}$, where the index i stands for a series of more performing controlled systems, using the same process model, is assumed. That is, $\bar{M}_{j,i+1}$ is better (faster) than $\bar{M}_{j,i}$, and/or the errors in $J_{j,i+1}$ are more weighted with respect to the control action u than they are in $J_{j,i}$.

By means of a suitable control design methodology, a controller $K_{j,i}$ is computed and implemented in the control system. The controlled plant behaviour will differ from that expected at the design stage due to the differences between the model used in the computation and the actual plant. Thus, the requirements can be raised and the controller redesigned, based on the same model. This procedure can be repeated until a controller K_{j,l_j}, denoted as the *edge controller*, is obtained without any possible improvement in the performances. Any attempt to go further will result in the constraints' violation.

The deviation from the designed behaviour can be evaluated by the degrading of some required performances, the violation of control, state or error constraints, or by means of a cost index $Q_{j,i}$.

The process will be driven by the controller actions. The more performing the control is, the more exciting the control actions are. So, new data from the input/output process will allow for the selection of a better process model.

Under the process controlled by the limiting controller K_{j,l_j}, a new model $\hat{G}_{j+1}$ can be estimated and the procedure above repeated.

As the final goal of the design is to improve the control, the identification processes should be defined in order to get a model good for these purposes, not requiring, in general, a global accuracy. In the applications below, a basic spectral analysis of the process signals, under the action of the edge controller, will provide enough information.

The iterations will end if no further improvement is achieved by refining the model, or if the performances are satisfactory.

In the first application, the closed-loop controlled system is identified, and the controller is designed in such a way that the actual closed-loop system meets some performance criterion, as expressed by a reference model. This leads to a cancellation controller. This identification scheme simplifies the experimental set-up but involves some risks, such as over-dimensionality of the model, modes mismatching, or conservative designs. As the control is based on the cancellation of the model, the controlled system could present the well-known problems of hidden oscillations or instability. All these issues are discussed in the following example.

6.2.1 Example Revisited

In [3], the following set-up is followed:

- the plant is available and, by means of an open-loop test signal, a simple model transfer function, $\hat{G}_0$, of the plant is obtained;
- at the j-model iteration, the desired controlled system behaviour is given by the reference model $\bar{M}_{j,i}$,

$$\bar{M}_{j,i} = \left(\frac{1-\alpha_{j,i}}{z-\alpha_{j,i}}\right)^{n_j}$$

where i is the performance improvement iteration. It is desired to keep α and n as small as possible, maximising the bandwidth;
- the control constraint is expressed as the maximum admissible step response overshoot δ_0.

For each model $\hat{G}_j$, the design, $K_{j,i}$, is carried out for $i = 1, 2, \ldots, l_j$, where the closed-loop system behaviour is gradually improved until δ, in the actual controlled plant, reaches the maximum allowed value δ_0. The trespassing of this threshold value shows the relevance of unmodelled modes. At this step, the best controller, K_{j,l_j}, is applied.

Now, in order to improve the design, on real time operation, a new plant model $\hat{G}_{j+1}$ of the current controlled system, $\frac{K_{j,l_j}G}{1+K_{j,l_j}G}$, is estimated and a new iteration is performed.

As a result of the controller design approach, the closed-loop system will show a higher frequency behaviour, reflected in the estimated model.

The procedure stops if no further improvement in control performance is achieved without trespassing the overshoot constraint and a higher order model is considered to be too complex. If the designed controller is applied to the previous closed-loop system, the global controller will have a multiloop structure.

In order to clarify the ideas exposed above, an example where the controller design is based on pole assignment and cancellation is shown.

The actual plant is represented by a third order process with transfer function

$$G(z) = \frac{0.019z + 0.017}{z^3 - 2.28z^2 + 1.71z - 0.41}$$

which comes from the sampling of a continuous time system, with sampling period $T = 0.1\ s$. Given its step response (or equivalently, the frequency one), a reduced order (first order) model is estimated, such as

$$\hat{G}_0(z) = \frac{0.36}{z - 0.8}$$

with a rather weak approximation. A controller is designed, by cancellation, to get the closed-loop transfer function

$$\bar{M}_{0,i}(z) = \frac{1 - \alpha_{0,i}}{z - \alpha_{0,i}} \rightarrow K_{0,i}(z) = \frac{1 - \alpha_{0,i}}{0.36} \frac{z - 0.8}{z - 1}$$

A sequence of $\alpha_{0,i} = 0.95, 0.9, \ldots$ is taken and the overshoot is checked. It appears that for $\alpha_{0,4} = 0.8$, the overshoot $\delta > 0.1$. That is, the constraint is violated. Thus the first "best" controller is

$$K_{0,3}(z) = \frac{1 - 0.85}{0.36} \frac{z - 0.8}{z - 1}$$

A new first order model, $\hat{G}_1$, is then tuned for the closed-loop system

$$M_{0,3} = \frac{K_{0,3} G}{1 + K_{0,3} G}$$

and the procedure above is repeated, now with a sequence $\alpha_{1,i} = 0.95, 0.93, \ldots$ As little improvement is achieved, it means that a first order model is no further a good approximation. So, a third order model $\hat{G}_2$ is now estimated.

The reference model for the controller design would be, in this case,

$$\bar{M}_{2,i}(z) = \left[\frac{1 - \alpha_{2,i}}{z - \alpha_{2,i}} \right]^3$$

and the whole process is repeated again. The new controller will feedback again the output, $y(z)$, of the previously controlled plant leading to a multi-loop control. At the end, denoting by $K_j = K_{j,l_j}$, if this process is repeated m times, the control law for a reference signal $r(z)$, will be:

$$u(z) = (K_0.K_1.\ldots .K_m)r(z) - (K_0.(1 + K_1.(1 + \ldots .K_m)))y(z)$$

showing the complexity of the resultant control.

A number of conclusions can be outlined from this example.

1. This approach is closely related to the human experimental learning. The plant is cautiously controlled and its response is observed. Then, some extra features are added to the basic control already implemented, until the model is not good anymore and a new identification experiment is carried out.
2. Too-conservative requirements at the earlier stages of the design will imply further stronger control actions to get the same final performances. "A controlled plant loses some of its possible open-loop wilderness or vivacity, becoming louder." This is the general concept that controlling a system is limiting its freedom. It is counter-posed to the general (false) feeling that a controller permits achievement of something that is not possible without control.
3. The approach can be applied using different cost indices, design criteria or estimation methods.
4. The closed-loop modelling and the control cascade introduces unnecessary complexity in the control.

The main reason for this example is to illustrate that all the design steps are very simple and easy to perform. In order to show some alternatives, the control of resonant systems is illustrated following two different controller design approaches.

6.3 Application 1: Single Control Loop

An approach similar to the previous one is now applied to control a plant with high-frequency resonant modes[1], but a single control loop is implemented. This requires estimation of the plant open-loop transfer function at each identification step. Simple identification approaches are used: step response for the stable initial plant, and a signal power spectrum analysis to estimate the resonant modes, as soon as they are detectable with respect to the noise.

Assume a given, and unknown plant, $G(s)$, such as

$$G(s) = \frac{10^6 s + 10^9}{s^6 + 0.402 s^5 + 1.003 \cdot 10^6 s^4 + 4.008 \cdot 10^5 s^3 + 2.5 \cdot 10^9 s^2 + 7.5 \cdot 10^8 s + 5 \cdot 10^7}$$

---

[1] Example worked out by Alicia Esparza.

In gain-zero-pole representation, it is

$$G(s) = \frac{10^6(s+1000)}{(s^2+0.002+1000^2)(s^2+0.1+50^2)(s+0.1)(s+0.2)}$$

There is a measurement white noise, $n(t)$, with $\sigma = 10^{-2}$.

The control goal is to achieve a stable and fast step reference tracking, without oscillations.

The open-loop step response of the process is represented in Figure 6.1, (solid line). Following the Strej approximation to estimate the model of the

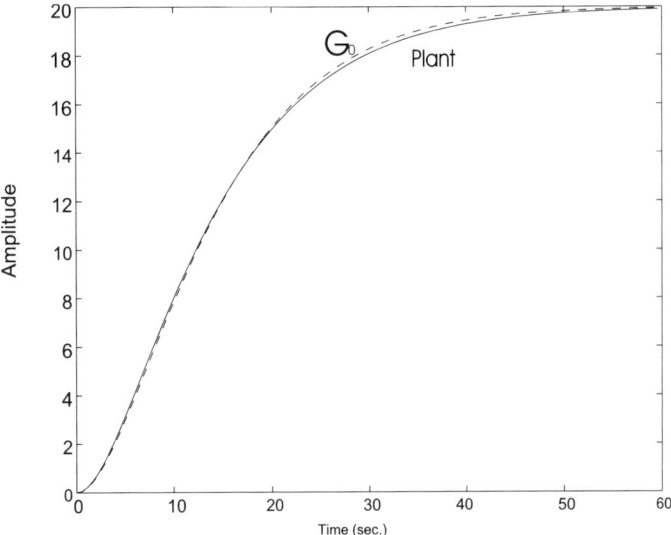

Fig. 6.1. Plant and second order model step response.

process, a double 7.4 s time constant is derived. The initial second order transfer function

$$\hat{G}_0(s) = \frac{20}{(1+7.4s)^2}$$

can be assumed to be an initial process model. In fact, the model step response is also represented (dashed line) in Figure 6.1, and the matching is, apparently, very good.

If a frequency response identification is performed, matching the lower frequency ranges (until 3 rad/sc), a similar second order model could be obtained. Higher frequencies, with very low gains (lower than –80 dB), are difficult to excite and the noise masks the output signal. In Figure 6.2, the frequency responses of the actual plant (solid line) and the model, $\hat{G}_0(s)$, (dotted line) are compared. The two peaks (marked on the figure) at the resonant frequencies are missing, but the approximation is pretty good at low frequencies.

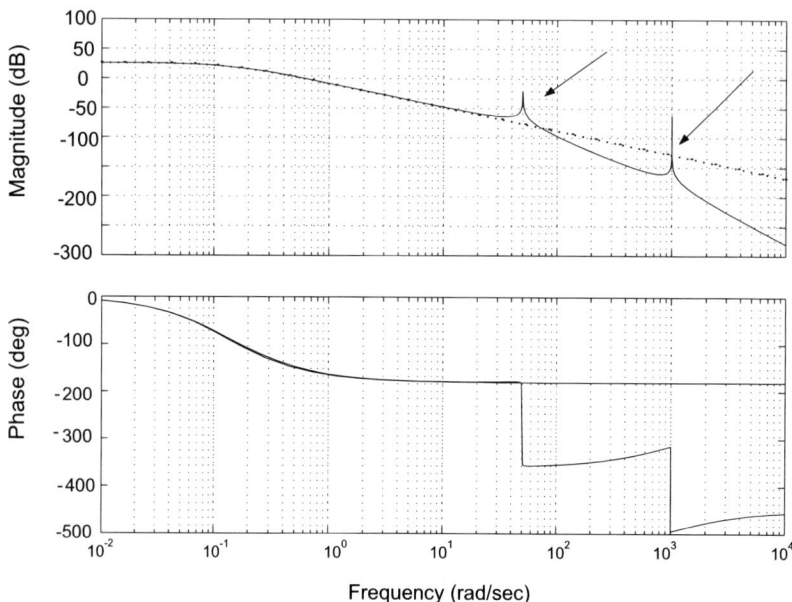

Fig. 6.2. Frequency responses of the plant and the model, $\hat{G}_0$

6.3.1 Controller Design

First iteration. Assume the plant model is $\hat{G}_0$. In the reference model, the time constants cannot be very low. The design is conservative. For $\tau = 2$ s, the reference model is

$$\bar{M}_{01} = \frac{0.5^2}{(s+0.5)^2}$$

and the cancellation controller is

$$K_{01} = \frac{\bar{M}_{01}}{\hat{G}_0(1-\bar{M}_{01})} = \frac{0.68s^2 + 0.185s + 0.0125}{s^2 + s}$$

The actual and designed step responses are plotted in Figure 6.3. The controlled response is good enough, so the requirements can be more exigent.

New iterations. Assume a double time constant $\tau_i = 2, 1, \ldots$ and repeat the controller design. For $\tau_i = 0.2$ s, the matching in the transient section seems to be also good, as shown in Figure 6.4(a). The model reference is

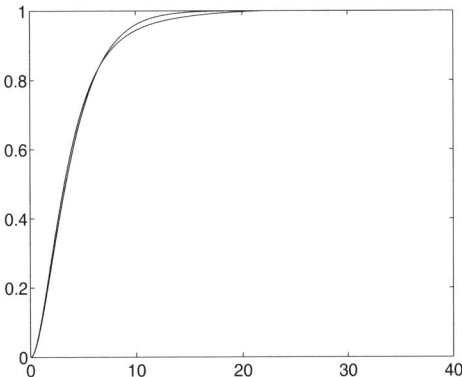

Fig. 6.3. Controlled closed-loop step response, for $\tau = 2$ s: actual plant and plant model

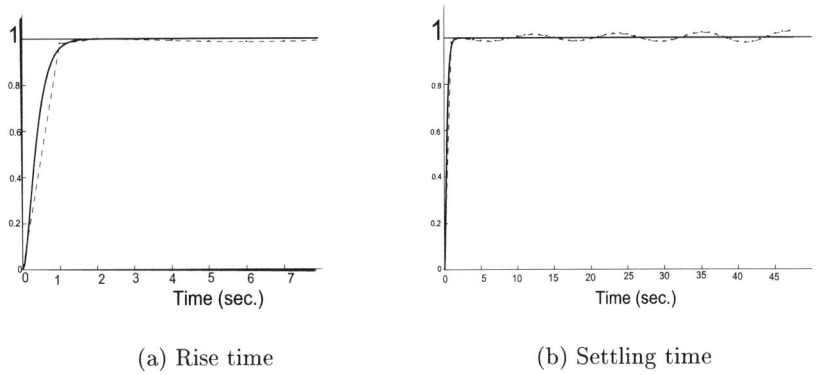

(a) Rise time   (b) Settling time

Fig. 6.4. Controlled closed-loop step response, for $\tau = 0.2s$: actual plant (dotted), plant model (solid)

$\bar{M}_{03} = \frac{5^2}{(s+5)^2}$, and the current controller is

$$K_{03} = \frac{\bar{M}_{03}}{\hat{G}_0(1 - \bar{M}_{03})} = \frac{68.45s^2 + 18.5s + 1.25}{s^2 + 10s}$$

Nevertheless, as time is going on, the actual plant response starts to oscillate, Figure 6.4(b).

The frequency responses are plotted in Figure 6.5. As shown in this figure, the bandwidth has been enlarged in such a way that the first resonance peak of the plant shows a higher gain and a sudden jump in the bandwidth appears.

If the gain or the nominal performances are further increased, the actual plant will become unstable. This evolution is smooth, and the resonance frequency

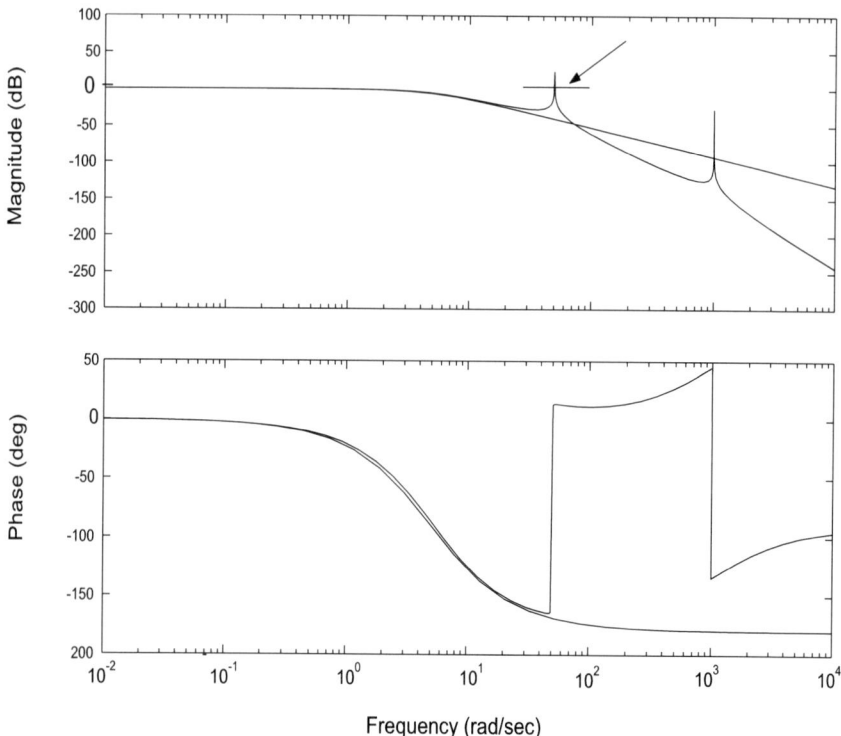

Fig. 6.5. Frequency response of the designed (solid) and actual (dotted) controlled processes

can be estimated. For that purpose, the output signal is analysed, as shown in Figure 6.6. From the spectral analysis, an approximated resonance frequency of 50 rad/s is suggested.

6.3.2 Model Improvement

The new model could be chosen such that

$$\hat{G}_1(s) = \hat{G}_0(s) \frac{\sigma_h^2 + \omega_h^2}{(s + \sigma_h + \omega_h \cdot j)(s + \sigma_h - \omega_h \cdot j)}$$

As these new modes are not dominant, it is clear that:

$\sigma_h \ll \min(\{\sigma_1, ..., \sigma_n\})$
$\omega_h \gg \max(\{\omega_1, ..., \omega_n\})$

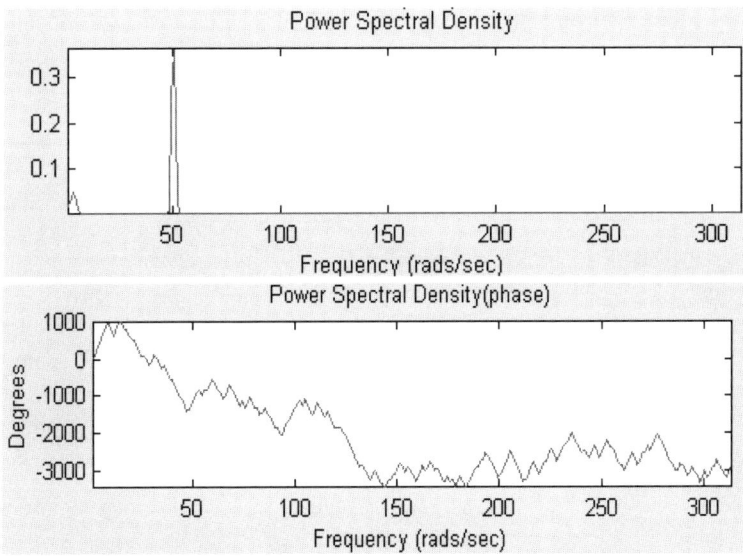

Fig. 6.6. Spectral density analysis of the controlled system output

where:

σ_h, ω_h: real and imaginary part, respectively of the hidden mode.

σ_i, ω_i ($i = 1,...,n$): real and imaginary part of the stable dominant poles of the previous model.

Thus, a good approximation would be to consider ω_h to be equal to the oscillation frequency, whereas σ_h is chosen to be small enough to keep the other poles dominant.

So, two additional poles with imaginary part 50 are added to the old model. The real part, being very small and not so relevant, is assumed to be 0.01. Based on the previous result, the following approximated model is considered:

$$\hat{G}_1(s) = \hat{G}_0(s) \frac{0.01^2 + 50^2}{(s + 0.01 + 50i)(s + 0.01 - 50i)}$$

$$\hat{G}_1(s) = \frac{5 \cdot 10^4}{54.76 s^4 + 15.9 s^3 + 1.369 \cdot 10^5 s^2 + 3.7 \cdot 10^4 s + 2500}$$

The matching between the step input responses of both the plant and the model are almost indistinguishable. The frequency responses are shown in Figure 6.7.

The frequency domain matching enlarges the range until 200 rad/s, the first resonant peak being captured by the additional poles. Thus, better results are expected in the newly-designed controllers.

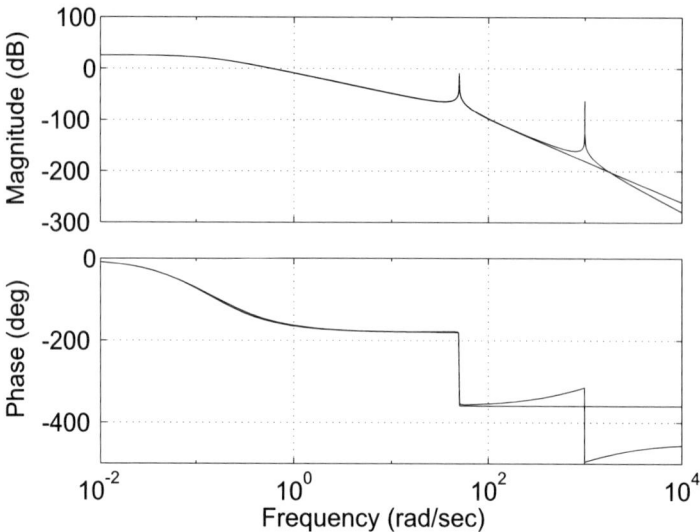

Fig. 6.7. Plant and model frequency responses in the second iteration

6.3.3 Control Design

The reference model should be now a fourth order system. A multiple $\tau = 0.2$ s time constant is assumed,

$$\bar{M}_{11} = \frac{5^4}{(s+5)^4}$$

given a cancellation controller:

$$K_{11} = \frac{\hat{M}_{11}}{\hat{G}_1(1-\hat{M}_{11})} = \frac{34225 s^4 + 9935 s^3 + 8.556 \cdot 10^7 s^2 + 2.313 \cdot 10^7 s + 1.563 \cdot 10^6}{5 \cdot 10^4 s^4 + 1 \cdot 10^6 s^3 + 7.5 \cdot 10^6 s^2 + 2.5 \cdot 10^7 s}$$

The controlled plant step response is now stable, and remains so even if the reference model is improved by assigning a much shorter time constant. In particular, for $\tau = 1/300$ s, the controlled plant step response starts to oscillate but it is still stable.

The edge controller is

$$\hat{M}_{16} = \frac{300^4}{(s+300)^4} \qquad K_{16} = \frac{\hat{M}_{16}}{\hat{G}_1(1-\hat{M}_{16})}$$

$$K_{16} = \frac{4.436 \cdot 10^{11} s^4 + 1.288 \cdot 10^{11} s^3 + 1.109 \cdot 10^{15} s^2 + 2.997 \cdot 10^{14} s + 2.025 \cdot 10^{13}}{5 \cdot 10^4 s^4 + 6 \cdot 10^7 s^3 + 2.7 \cdot 10^{10} s^2 + 5.4 \cdot 10^{12} s}$$

6.3.4 Model Improvement

Just to evaluate the new resonance peak, the required controller's performances are increased until the oscillation appears. In particular, for $\tau = 1/625$ s, the reference model is given by $\bar{M}_{18} = \frac{625^4}{(s+625)^4}$, leading to an ideal cancellation controller such as

$$K_{18} = \frac{\bar{M}_{18}}{\hat{G}_1(1 - \bar{M}_{18})}$$

$$= \frac{8.356 \cdot 10^{12} s^4 + 2.425 \cdot 10^{12} s^3 + 2.089 \cdot 10^{16} s^2 + 5.646 \cdot 10^{15} s + 3.815 \cdot 10^{14}}{5 \cdot 10^4 s^4 + 1.25 \cdot 10^8 s^3 + 1.172 \cdot 10^{11} s^2 + 4.883 \cdot 10^{13} s}$$

Again, the differences in the time behaviour, as shown in Figure 6.8, are due to additional modes. Similar differences appear in the frequency plots.

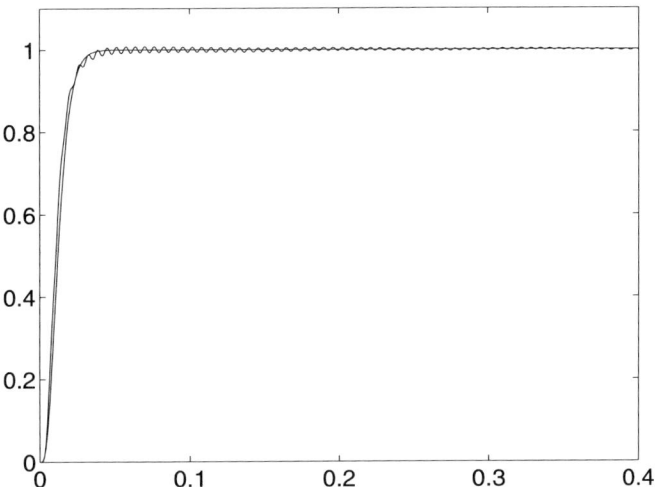

Fig. 6.8. New iteration, step response

In order to perform a new spectral analysis of the controlled process output, the design is forced to oscillate, as depicted in Figure 6.10. The new PSA shows up a resonance frequency around 950 rad/s. Thus, the plant model is enlarged by this new pair of poles with the real part, as before, kept low. The new model will be

$$\hat{G}_2(s) = \hat{G}_1 \frac{0.001^2 + 950^2}{(s + 0.001 + 950i)(s + 0.001 - 950i)}$$

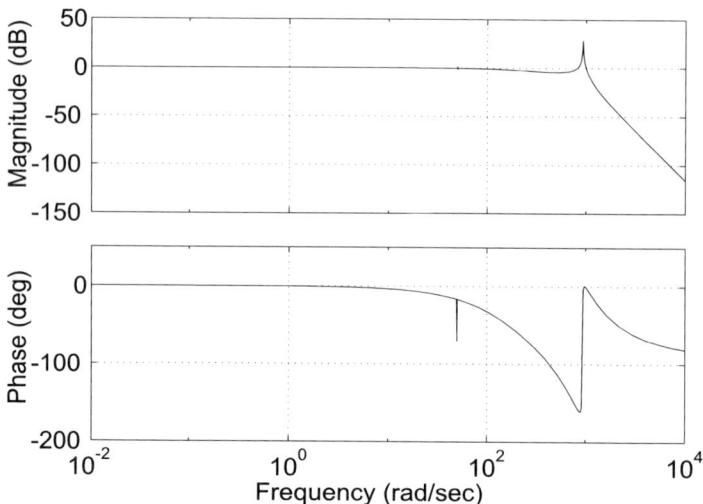

Fig. 6.9. New iteration, Bode diagram

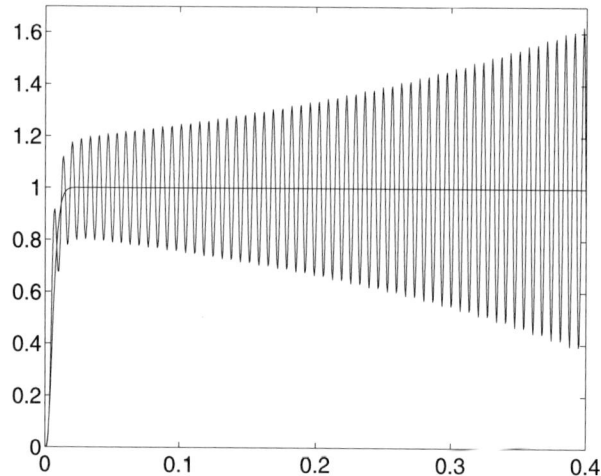

Fig. 6.10. Unstable response

and, altogether:

$$\hat{G}_2(s) = \frac{4.513 \cdot 10^{10}}{54.76s^6 + 16s^5 + 4.956 \cdot 10^7 s^4 + 1.438 \cdot 10^7 s^3 + 1.236 \cdot 10^{11} s^2 + 3.339 \cdot 10^{10} s + 2.256 \cdot 10^9}$$

Of course, the new model matches almost perfectly the given plant, mainly in the frequency response, as shown in Figure 6.11.

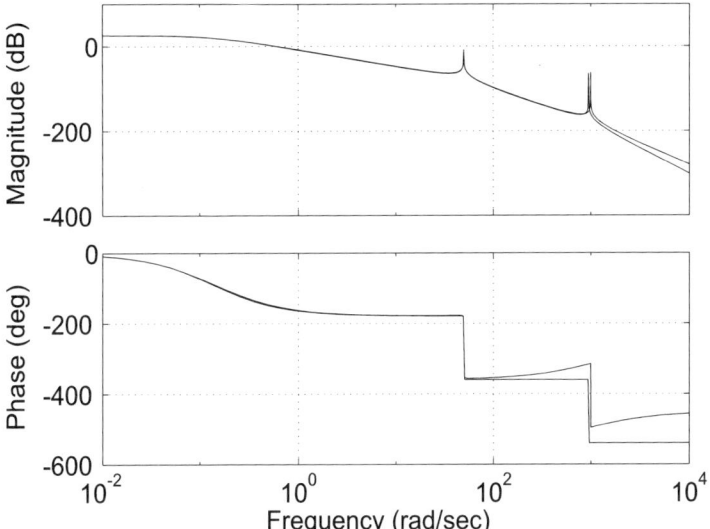

Fig. 6.11. Next improvement, frequency response

This almost perfect model will allow, in theory, for a much larger process bandwidth.

The improvement in the time settling or the bandwidth, as the design proceeds can be easily realised comparing the open-loop responses and those of the controlled systems.

Of course, this design is unfeasible due to the lack of robustness, but the same approach can be followed if the control design procedure is not just a simple cancellation controller but a more robust one, as illustrated in the next application.

It is interesting to realise that the rise time of the iterative designs is being continuously decreased, as can be seen in the closed-loop step responseplots.

6.4 Application 2: GPC Control Approach

In this application[2], the identification approach is as before, leading to very simple and approximated models. Also, the plant is assumed to be showing different timescale behaviours, which involves the appearance of different resonance frequency peaks. Two main differences are considered. First, the control design methodology is based on the GPC [38]. This results in a three-term controller. The second main difference is that the control is digital, and the selection of the sampling period for the discrete time (DT) controller will be discussed.

Again, the worked out example is a simple one, trying to illustrate the main options, without claiming a "real" process control problem.

6.4.1 Iterative Design of the GPC

The generalised predictive controller is a model-based optimal control strategy where uncertainties and disturbances are handled by just applying the first part of the computed control sequence and recomputing again the optimal control over a finite optimisation horizon.

There are many different settings for this approach. The basic cost index is assumed to be:

$$J = \sum_{i=N_1}^{N_2} \alpha_i \left[y(k+i|k) - w(k+i)\right]^2 + \sum_{j=N_1}^{Nu} \lambda_j \left[\Delta u(k+j-1)\right]^2$$

where:

$y(k+i|k)$: estimated output at $k+i$ based on data at k
$w(k+i)$: reference at $k+i$
Δu: increment in the control action
N_1: minimum prediction horizon
N_2: maximum prediction horizon
N_u: control horizon
α_i: error weighting factor
λ_j: control action weighting factor

The last parameter is directly connected to the control effort and system response "vivacity": the smaller the parameter λ_j is, the stronger the controller actions are and the system response is faster. That is, the smaller this parameter is, the greater the bandwidth of the controlled system.

But, as for any model-based control design technique, the GPC resulting controller is only good for a given range of processes, not far from the model used for the computation. If strong control actions are applied, non-expected high-frequency process modes can be excited and the behaviour greatly differs from the one designed.

The design procedure will be as follows:

---
[2] Example worked out by Julio Romero.

- given a CT approximated process model $\hat{G}_j(s)$, a DT step response equivalent $\hat{G}_j(z)$ is obtained;
- for this DT model, a "passive" GPC controller is designed and the controlled system response is evaluated. By passive GPC, it is denoted a controller where the weight of the control actions in the cost index, $\lambda_{j,i}$, is high.
  The resulting controller is applied, as shown in Figure 6.12;
- evaluate the controlled system response. For that purpose, an index based on the system error,

$$J_e = \sum_{i=1}^{N} (y(k+i|k) - w(k+i))^2$$

$$J_{\lambda_1} > J_{\lambda_2} > ... > J_{\lambda_n} \quad \lambda_1 > \lambda_2 > ... > \lambda_n$$

the bandwidth or any other related performance can be established;
- design, for the same model, a more "active" controller, that is, reducing $\lambda_{j,i+1} < \lambda_{j,i}$;
- if the control becomes bang-bang like, consider the option of reducing the sampling period, for the same CT process model. Otherwise, proceed to a new identification procedure;
- repeat the whole iterations, until the control is satisfactory.

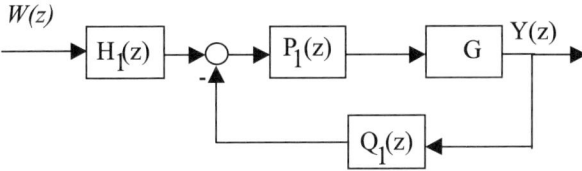

Fig. 6.12. GPC Controller

6.4.2 Numerical Example

Based on the time step response of the given plant, a second order model seems to be appropriated, with two time constants of 2 and 10 s (see Figure 6.13):

$$\hat{G}_0(s) = \frac{0.4}{s^2 + 0.6s + 0.05} = \frac{0.4}{(s+0.1)(s+0.5)}$$

In principle, although the idea is to design a controller to be as fast as possible, a conservative sampling period of $T_s = 1$ s is selected.

The DT process model is

$$\hat{G}_0(z) = \frac{0.1547z + 0.1349}{z^2 + 1.511z + 0.5488}$$

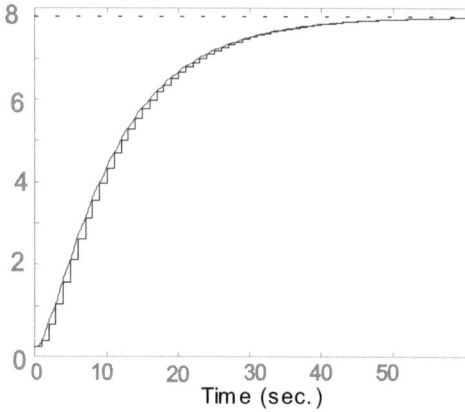

Fig. 6.13. Step response, plant and DT model

The step response of the DT model is also shown in Figure 6.13. The frequency responses are in Figure 6.14, where the hard line represents the actual plant.

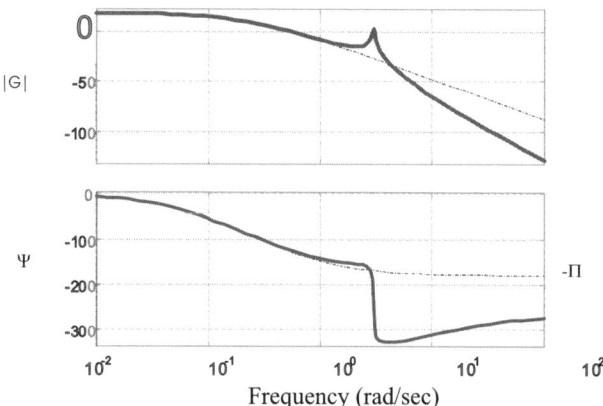

Fig. 6.14. Plant frequency response

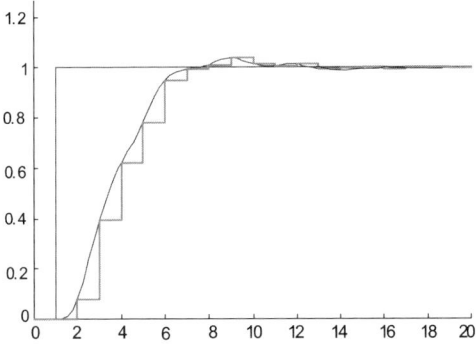

Fig. 6.15. Sampled and CT output from the actual plant, GPC, $\lambda = 1$

Initially, the design should be very conservative, as no good knowledge of the process is obtained from just a step response. Once the error index is evaluated and the design is validated, that is, there is an acceptable match between the designed and the actual closed-loop responses, lower values are assigned to the weighting factor $\lambda_{j,i+1}$. This coefficient can be reduced until an increment in the index J_e is detected. This is in contradiction of the design criterion, where the errors are becoming more weighted than the control actions. This should be related to the appearance of oscillations.

The GPC design parameters are taken as: $\lambda = 1$, $N_1 = 1$, $N_2 = 40$, $N_u = 4$, $\alpha = 1$, with polynomial T= 1.

For this design, the behaviour is plotted in Figure 6.15, showing an adequate matching.

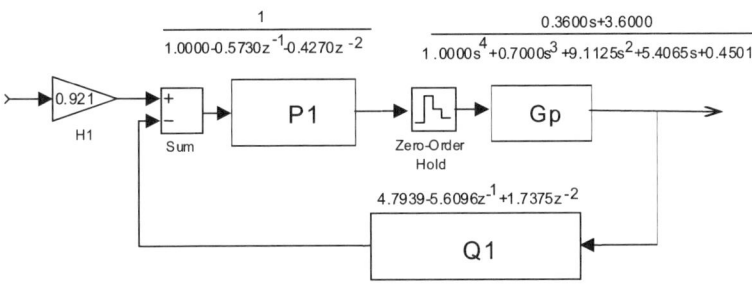

Fig. 6.16. GPC controlled plant, $\lambda = 0.3$

Based on this good result, a new controller is designed, taking a lower weight for the control actions. For $\lambda = 0.3$, the following simulation schema is built, Figure 6.16.

The new system response is depicted on the left of Figure 6.17. The designed plant response is on the right of Figure 6.17.

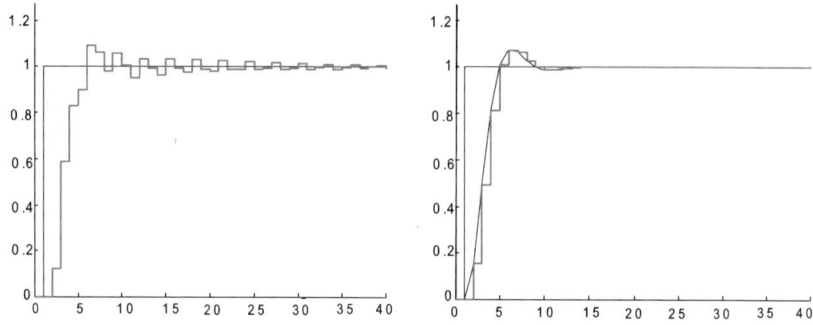

Fig. 6.17. DT and CT plant step response: (a) actual plant, (b) modelled plant

As expected, high-frequency oscillations appear. Applying the same technique as before, *i. e.*, looking for the resonance frequency, a good estimation is $\omega_h = 2.5$ rad/s. A low real part, such as $\sigma_h = 0.01$, is assumed, leading to the model:

$$\hat{G}_1(s) = \hat{G}_0(s) \frac{6.25}{(s + 0.01 + 2.5j)(s + 0.01 - 2.5j)}$$

The Bode diagrams for the new model and the actual plant are depicted in Figure 6.18.

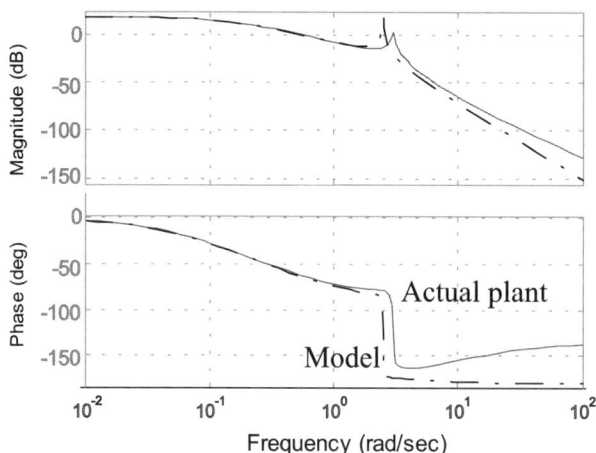

Fig. 6.18. Frequency response comparison

The model, in this case, covers most of the frequency range, although the resonance frequency is not perfectly estimated.

Now, the new model is discretised with sampling period $T_s = 0.2$ s. The GPC is redesigned with the following parameters:
$$N_1 = 1, N_2 = 80, N_u = 4, T(z^{-1}) = 1, \alpha = 1, \lambda = 2$$
and the following controller is obtained

$$H = 0.944$$
$$P(z) = \frac{1}{1 - 0.478z^{-1} - 0.225z^{-2} - 0.272z^{-3} - 0.025z^{-4}}$$
$$Q(z) = 197.2 - 688.37z^{-1} + 944.6z^{-2} - 600z^{-3} + 146.95z^{-4}$$

The controlled system step responseis drawn in Figure 6.19.

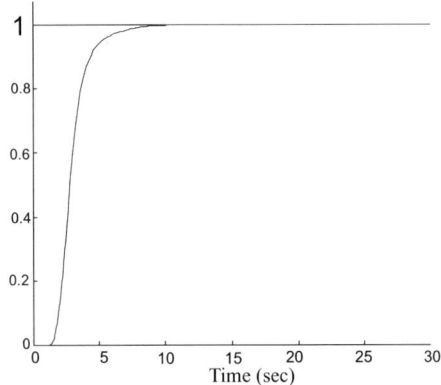

Fig. 6.19. Step response. Second model GPC design

A lot of improvement is obtained. The settling time is reduced and the oscillations do not appear.

It could be argued that, for the initial CT model, better results could be obtained if a shorter sampling period is used. But as previously mentioned, this could be dangerous, as the hidden modes could be excited. As an example of this issue, for the initial model, a DT model for the same sampling period, $T_s = 0.2$ s, is used. The GPC design, with the following parameters

$$\lambda = 1, \alpha = 1, N_1 = 1, N_u = 4, N_2 = 150, T = 1$$

leads to the unstable response in Figure 6.20.

That is, a too-short sampling period applied to a low-frequency adequate model becomes inadequate to design a highly-performing controller. In this case it was a GPC controller, but similar results were obtained using other control design techniques. This is particularly important if the designed controller is a DT controller and the original model is a CT one.

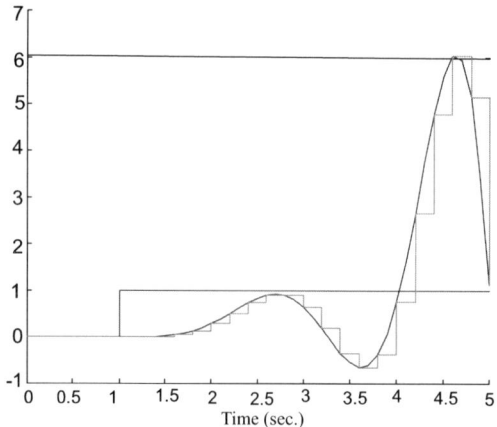

Fig. 6.20. Inappropriate sampling period selection

6.5 Conclusions

In this contribution, some different approaches for model-based iterative control design have been outlined.

The basic conclusions are:

- an approximated model should be used in its region of validity;
- this applies both for frequency ranges and sampling periods;
- as far as a control designed for a plant is applied, new data from the actual plant can be used to improve the model;
- approximated models are usually good enough for control design purposes, mainly if there is a final on-site tuning of the controller;
- the sampling period should be chosen according to the range of frequencies the model is representing.

This very simple methodology has been recently applied [4] to the design of controllers for flexible robot manipulators where the dynamics can be characterised by one single or double integrator plus additional resonant modes, which are very difficult to estimate under low excitation.

Acknowledgements

The author acknowledges the valuable inputs to this work of Alicia Esparza and Julio Romero from Universidad Politécnica de Valencia (DISA).

7 Windsurfing Approach to Iterative Control Design

Brian D.O. Anderson

Abstract. Iterative identification and control relies on a sequence of controllers, deemed satisfactory for a sequence of models, being connected to a real plant, and thereby at the very least not introducing an unstable closed loop. Given that the models used for control design are not identical to the real plant, a technique is needed to assure stability of the model-derived controllers with the real plant. This is provided by the windsurfer approach to iterative control design, which seeks to model and control over successively larger bandwidths, with any one bandwidth enlargement step being comparatively modest.

7.1 Introduction

What on earth can the "windsurfing approach" to iterative control design be, and how can windsurfing ever have anything at all to do with automatic control? While the reader will have to wait to read some of the details of the answer, let us record here that iterative control is a technology that enables a controller to learn something about the plant being controlled. This is just what a human must do when he or she learns windsurfing, and if one thinks about how the human acquires the necessary knowledge, this can suggest an approach to iterative control.

There were two streams of ideas that provided the genesis of this work. The first arose when the author, working at the time with R. Kosut on adaptive control problems being addressed by "conventional" adaptive control theory, pondered how a human learnt windsurfing, realising that conventional control theory shed no light on the phenomenon. The natural follow-on question is then whether the human learning process can inspire some new style of adaptive control theory. The second stream arose when the author, often working in collaboration with others including especially M. Gevers and R. Kosut, concluded that much adaptive control theory, being advanced, failed to address some very significant practical issues that could even invalidate or make dangerous application of the theory. The follow-on question of how to address these issues also stimulated many ideas.

Adaptive control is technically not the same as iterative control, or iterative control and identification. Nevertheless, they are first cousins at least. Both are concerned with controlling in the presence of plant uncertainty, and seek to learn something about that uncertainty in order to design an appropriate controller. This is in contrast to, say, the approach of robust control, where one attempts to design round the uncertainty, *i.e.*, to provide a fixed control design that will deliver adequate performance no matter what the values of the uncertain parameters are. The difference between adaptive control on the one hand and iterative control and identification on the other essentially rests in one fact. In adaptive control, processes of identification and controller design are normally fully interleaved, *i.e.*, each new measurement is used to improve the parameter estimates, and at each new time instant adjustment of the controller parameters is permitted. In contrast, in iterative control and identification, one identifies over an interval in which the controller stays fixed, then one adjusts the controller, then one identifies again, and so on. The conceptual difference is not great.

Hence in this chapter, although we are technically describing an iterative control and identification method, we see it within a broader framework, and in particular see it as a valid replacement for some adaptive control methodology, which under some circumstances may be flawed.

Let us sketch an outline of this chapter. In Section 7.2, we set out three conceptual difficulties with conventional adaptive control theory. No theory is perfect, least of all that of this chapter. We do not mean to suggest that all adaptive control theory is flawed, or nearly always flawed. But we do wish to make some very cautionary remarks, and to suggest that the approach of this chapter can sometimes circumvent these difficulties.

In Section 7.3, we try to capture some aspects of what the human is doing when learning to windsurf, and then we set up our underlying assumptions for the iterative control and identification algorithm we are going to present. Of course, any set of assumptions can be criticised; there will always be some situations in which they are plainly not fulfilled. The real issue is whether or not there is a substantial set of situations in which they are fulfilled. The assumptions also differ from those in many adaptive control algorithms, especially in relation to fluidity over orders of plants and models of those plants. However, we restrict attention to linear systems and, as it turns out, largely restrict attention to stable plants, or possibly plants with no more instability than a pole at the origin. This is a tighter restriction than applies in many adaptive control contexts, though it is not an absolute one for us.

Section 7.3 also discusses the identification-control design cycle, and the details are fleshed out in Section 7.4. Some of the key questions that arise are:

- how can one design to get a series of closed-loop designs in which variation of one parameter yields a variation of the closed-loop bandwidth?
- how can one identify a possibly unstable plant in a stable closed loop, in the presence of noise that is fed back around the loop?
- how can one build a low-order model that is a good approximation of a high-order model, in the sense of giving close closed-loop transfer function matching with a prescribed controller?

Section 7.5 contains examples, and Section 7.6 discusses the promised extension to open-loop unstable plants. We explain in Section 7.7 just how the conceptual problems of many adaptive control algorithms are avoided by the approach. Section 7.8 gives a flavour of what the next steps might be.

7.2 Fundamental Problems in Adaptive Control

In this section, we will recall certain aspects of conventional adaptive control algorithms, and then isolate three differing problems with these algorithms, that raise fundamental questions about their practical utility.

7.2.1 Typical Adaptive Control Algorithms

For the best part of two decades, many of the ideas regarding adaptive control design have been linked by a common thread, perhaps initiated in book form first by reference [65]. Broadly speaking, at any instant of time, one has a controller attached to a plant. One also has a design objective in mind for the closed-loop, which might be a reference trajectory following, pole positioning, or minimising a linear quadratic index. Measurements are collected on the closed-loop comprising the plant and the controller, and an identification algorithm is run on-line. This identification algorithm may seek to estimate the plant directly or, perhaps using some not particularly obvious parameterisation, seek to estimate the parameters directly of a controller that will achieve a particular design objective. Either way, there is an implicit or explicit plant identification procedure going on.

For the sake of simplicity, let us assume that the identification algorithm focuses directly on estimating plant parameters. Using the current estimate of the plant parameters, the control design algorithm computes the controller that would be appropriate or even optimum for the situation, in the event that the true plant was exactly modelled by the plant parameters. With this controller connected, more measurements are then collected. As more measurements are collected, the plant estimate is updated, the controller is updated using the plant estimate, and so on.

A typical theorem in adaptive control starts with a set of premises concerning the plant. For example, it is very often assumed that the plant is linear and time-invariant; perhaps it has a transfer function of a certain degree or relative degree, perhaps it is minimum-phase, perhaps one knows the sign of the DC gain, and so on. The theorem then postulates that a certain algorithm or combination of algorithms is used to estimate the plant parameters, and to compute the controller from these estimates. The theorem then usually goes on to say that as time tends to infinity, the plant parameter estimates converge to limiting values, the controller parameters converge to limiting values, in the steady state correct behaviour is achieved, and in the process all the signals remain bounded. Of course, not every adaptive control theorem is exactly like this, but many are.

In the later subsections of this section, we indicate some problems that arise with this paradigm. The ideas set out in the later sections in the main avoid these problems.

7.2.2 The Problem of Changing Experimental Conditions, Given Accurate but Inexact Models

Most, but not all, people know that one can find two plants whose Nyquist diagrams or impulse responses are almost indistinguishable, but whose responses become extremely different when the plants are placed in a closed-loop control loop, and an identical controller and input used in both loops.

Indeed, in [141] one reads: "Modelling principle 1: arbitrarily small modelling errors can lead to arbitrarily bad closed-loop performance".

The issue here is associated with change of experimental conditions. Models can only have their quality (as an approximation of whatever it is they are modelling) evaluated for a particular set of experimental conditions, and changing from open-loop operation to closed-loop operation with a specified controller is a particular change in experimental conditions. So, in fact, any change of controller is a change of experimental conditions.

It follows that, with a particular controller in the loop, one could have a good model of a plant secured through identification, where good in this sense means that the closed-loop performances match. But when one changes the controller, perhaps as part of the adaptive control process that calls for a controller change in the light of the current identified model, the performance of a new controller with the identified model may be very different to the performance of the new controller with the true plant. *The model that one has at any instant of time, which may be good with the current controller, is not necessarily guaranteed to be good with a changed controller.* To the extent that in adaptive control or even its first cousin, iterative control and identification, where one changes the controller from time to time, one must reckon with the possibility that with the change of controller, the model will become ineffective. One could even contemplate instability.

7.2.3 Impractical Control Objectives

Suppose one has a state variable model of a plant that is completely controllable and completely observable. There is a well-developed body of theory, the LQG design theory, [8] for example, which sets out how one can design an LQG optimal controller. Such a controller can always be constructed to be stabilising, given the minimality of the state variable model of the plant. However, a celebrated paper [54] establishes that in general for such problems, there are no guaranteed gain or phase margins. The reference, in fact, shows with an example that one can formulate an LQG problem that results in a phase margin of ε degrees, where ε is an arbitrarily small positive number.

Now, consider what would happen if one happened to formulate an adaptive LQG problem where the true but unknown plant coincided with the plant that gave the unacceptably low phase margin of the example. In running an adaptive control problem, one would expect very large transient signal values, one would have great difficulty learning the controller and, in fact, in practical terms one would never expect the algorithm to converge. Signals would just be impossibly large.

Underlying this behaviour is the fact that the control objective posed is an impractical one. Hard enough when the plant is known, the objective becomes effectively impossible to meet in an adaptive situation. This is serious enough.

But what makes this problem even more serious for "conventional" adaptive control paradigms is that it is intrinsic in most adaptive control problems that the plant at the outset is unknown. The unknownness of course is only partial, *but it may logically prevent the assessment of whether or not the control objective is feasible.* So, before the adaptive control algorithm has even run any distance, one may not only be posing an impossible control problem, *but one may lack the knowledge to understand that it is impossible.*

The style of adaptive control algorithms that are the subject of the various theorems mentioned in Section 7.2.1 is such that no hints at all are given for dealing with this sort of difficulty.

7.2.4 Transient Instability Problem

The prototypical adaptive control theorems mentioned in Section 7.2.1 normally include the assertion that all signals in the closed loop remain bounded for all time. Left unmentioned is the issue of how big this bound may be. They allow values of control input 10^6 or 10^9 times the value of the steady-state input appearing in response to a unit reference step because instantaneously unstable closed loops are permitted, *i.e.*, closed loops which, if parameters were frozen, would be unstable. The theoretical assurance is only about $t \to \infty$. *Without a quantitative indication of what the signal bound might be, the theorems provide insufficient assurance* to a prospective user of an adaptive controller wanting to rely on the algorithm that is the subject of the theorem.

7.3 High-level Overview of the Windsurfer Approach to Adaptive Control

In this section, we shall seek to capture some aspects of the way a human learns windsurfing, and translate these into a high-level statement of what an adaptive control algorithm may look like.

7.3.1 Learning Windsurfing

How does a human learn windsurfing? It is quite clear that one thing they do not do is identify the coefficients in a transfer function linear model of a windsurfer and set the parameters of the controller (and the human is the controller) to provide good closed-loop control. Some other mechanisms are operating. Any human learns windsurfing by initially coping with very gentle wave conditions, light winds, and non-variable winds. Crudely speaking, the human's first successes on a windsurfer involve low gain, low bandwidth, and low disturbance signals. Increasing experience of the windsurfer progressively expands the bandwidth over which the human can control (thus he or she learns to cope with fast shifts in wind direction, choppy seas), and the human also learns to provide a higher gain loop so that he or she can move faster than a novice, under the same wind and sea conditions. The path from low-gain, low-bandwidth control to high-gain, high-bandwidth control is a progressive one, *i.e.*, as the human learns, there are incremental changes in bandwidth and gain that he or she learns to apply. In effect, the human mentally builds some kind of a model that is initially a very primitive one, and which becomes more and more accurate over a wider and wider bandwidth as experience is acquired. Increasing accuracy of the model allows the higher gain controller.

7.3.2 The Starting Point for Windsurfer Adaptive Control

Let us first remark that the words adaptive control in the term "windsurfer adaptive control" are somewhat of a misnomer. "Iterative identification and control" is more accurate. In adaptive control, one typically contemplates identification and control redesign essentially occurring simultaneously. In discrete time, as each new measurement is collected, there is the opportunity to update the plant estimate, and to update the controller parameters. In iterative identification and control design, on the other hand, one contemplates a period in which the controller stays fixed and identification occurs; then the controller is updated, and a new identification process starts with the new controller. The controller remains fixed during this new identification process. This is how we contemplate the windsurfer approach to adaptive controllers working.

As with any adaptive control problem, one makes some *a priori* assumptions concerning the plant. Different adaptive control problems make different *a priori* assumptions. The ones we shall make here are as follows:

Assumption 7.1. The plant is linear, time invariant, and with a transfer function that has no poles in the right half plane nor on the imaginary axis with the possible exception of the origin.

Assumption 7.2. An initial controller is known, almost certainly of low bandwidth and low authority, which stabilises the plant.

If the plant is open-loop stable, any constant controller with small enough gain will not destabilise it. If the plant has a simple pole at the origin, knowledge of the sign of the residue is enough to enable definition of the controller. If the plant has a multiple pole at the origin, somewhat more information is needed.

What we do not need to make any assumption about is the order of the plant, the relative degree of the plant, and whether it is minimum-phase.

One more assumption is needed:

Assumption 7.3. A rational transfer function model of the plant is available, with no poles in the right half plane or on the imaginary axis, except possibly at the origin. The controller that stabilises the plant also stabilises the model.

Note that the model of the plant could be highly inaccurate. There is no requirement, for example, that the model has the same order of the plant, let alone that the model has a collection of parameters which, if adjusted, would give exact modelling of the plant.

7.3.3 The Identification/Control Design Cycle

Beginning at the initial time, and thereafter at all stages in the algorithm, one has available the true plant (as a physical entity, not in terms of a mathematical description), a model of the plant (which is a mathematical description) and a controller (which is available also as a mathematical description). One postulates also that one has the controller connected to the plant, and that the closed loop is excited, with the signals in it observable. At the same time, a copy of the controller is assumed to be available to be connected to the model, with this latter closed loop excited by the same external signals, apart from possible disturbance signals, as those that excite the true plant–controller combination. The outputs of the two closed loops will, in general, be different as a result of modelling errors, and as a result of disturbance signals associated with the true plant. Both outputs are assumed to be available for measurement.

The various steps of the process are as follows:

Step 1. Using the initial model and the initial controller, and with identical input excitation of the two loops, compare the outputs and assess whether or not the current model is a good model of the plant in the presence of the particular controller.

Step 2. If it is a good model, proceed to Step 3. If not, reidentify the model, using some form of closed-loop identification algorithm (discussion of a procedure appears later).

Step 3. Using the model, design the controller to achieve a somewhat wider closed-loop bandwidth (subsequently, a particular controller design algorithm for doing this will be described).

Step 4. Examine the outputs of the two closed-loop arrangements resulting from the new controller with the true plant and the existing model. If, with this controller, the true model is a good model of the plant, return to the previous Step 3. If not, reidentify.

Step 5. Redesign the controller for the new model, and return to Step 3.

In the above set of steps, we have not set out how the process ends. In fact, it will end when

- either the signal-to-noise ratio is such that effective identification is simply not possible; or
- the controller objectives are being satisfactorily met; or
- with the latest model, the control objectives are manifestly in practice unobtainable.

7.4 More Detailed Description of the Algorithm

The two crucial steps in the algorithm are apparently the controller design step, and the closed-loop identification step. We shall discuss these in Sections 7.4.1 and 7.4.2 below. However, as we will review further below in Section 7.4.3, there is a possible requirement for model reduction. Also, it is desirable to understand the conditions that cause the algorithm to stop in some more detail and this is done in Section 7.4.4.

7.4.1 Internal Model Control Design

The internal model control (IMC) method, set out in, for example, [116] is a control design method that is particularly suited to the task of obtaining a prescribed closed-loop bandwidth when the open-loop plant is stable. In particular, there is a way of parameterising the controller effectively directly in terms of this closed-loop bandwidth.

Let us describe the algorithm. Suppose at some step in the process, one has a model of the plant given by a transfer function G_i. Factor this transfer function as $[G_i]_a [G_i]_m$, where $[G_i]_a$ is the all-pass factor associated with G_i, and $[G_i]_m$ is the minimum-phase factor associated with G_i. This factor may have a zero at infinity; we assume it has no other imaginary axis zero. Then, the controller to achieve a notional bandwidth of λ_i is given by the formula

$$K_i = \frac{Q_i}{1 - G_i Q_i} \tag{7.1}$$

Here, Q_i is defined as

$$Q_i = [G_i]_m^{-1} F_i \tag{7.2}$$

where F_i involves the bandwidth λ_i and an integer n:

$$F_i = \left(\frac{\lambda_i}{s+\lambda_i}\right)^n \qquad (7.3)$$

The integer n simply needs to be chosen so that Q_i is proper.

The resulting designed closed-loop transfer function $\bar{T}_i$ is given by

$$\bar{T}_i = \frac{G_i K_i}{1 + G_i K_i} \qquad (7.4)$$

If one introduces expressions (7.1) to (7.3) into Equation (7.4) for the closed-loop transfer function, one in fact finds that

$$\bar{T}_i = F_i [G_i]_a \qquad (7.5)$$

What this equation reveals is that *the magnitude response of the designed closed loop is given directly by the magnitude response of F_i*. As one would expect, the unstable zeros of the plant appear in the closed-loop transfer function, but in such a way as to not disturb the magnitude response. The attenuation of the closed-loop response is $20n$ dB/decade above the designed closed-loop bandwidth λ_i.

7.4.2 The Identification Step

Consider Figure 7.1, which depicts the plant with unknown transfer function G, a stabilising controller with known transfer function K, and output measurement noise or disturbance obtained by passing white noise through a filter of transfer function H. Suppose one wants to identify G. The signals r, u and y are available for measurement, and the signals r and d are assumed to be independent. While it is true that

$$y = Gu + d \qquad (7.6)$$

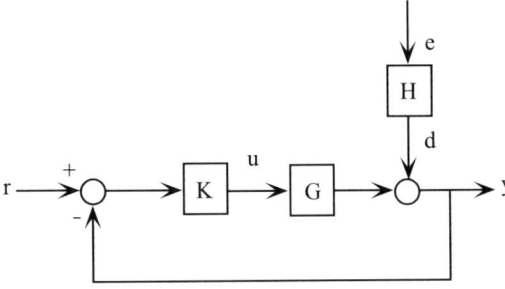

Fig. 7.1. Set-up for closed-loop identification

it is not true that u and d are independent. This is because d is fed round to u by the feedback loop and K. Accordingly, many conventional methods of (open-loop) identification cannot be used for identifying G based on measurements of u and y. Additionally, should G be open-loop unstable, some of those methods could also give problems, in the sense that they may presuppose the open-loop stability of G. These few remarks indicate that closed-loop identification is a more difficult task than open-loop identification.

A major step forward in handling closed-loop identification was made by [72]. This work showed how to convert a closed-loop identification problem into a conventional open-loop identification problem, and we shall outline the technique now. Let us suppose that we have available a model G_i which is stabilised by K. Let us also represent G_i as a ratio of stable proper transfer functions, thus

$$G_i = \frac{N_i}{D_i} \tag{7.7}$$

It is also possible to represent K as a ratio of stable proper transfer functions, thus

$$K = \frac{X}{Y} \tag{7.8}$$

Because K is stabilising, it is also possible to require that X, Y satisfy a Bezout identity:

$$N_i X + D_i Y = 1 \tag{7.9}$$

A result originally due to Youla and Kucera, (set out, for example, in [152]) establishes that the set of all plants stabilised by the controller K is precisely the set of transfer functions

$$\mathcal{G} = \left\{ \frac{N_i + RY}{D_i - RX} : R \text{ proper and stable} \right\} \tag{7.10}$$

Evidently, if K stabilises the plant with unknown transfer function G as well as the model with transfer function G_i, the unknown transfer function G must be an element of the set $\mathcal{G}$ of (7.10). Put another way, there must be a proper stable transfer function R such that the following equation holds:

$$G = \frac{N_i + RY}{D_i - RX} \tag{7.11}$$

However, the contribution of [72] was to recognise not only that the task of identifying G was equivalent to the task of identifying R, but that the task of identifying R was one that could be simply cast as an open-loop identification problem.

Consider the arrangement of Figure 7.2, which is in effect a redrawing of Figure 7.1, using the form of G given by (7.11) and introducing the noise d at a slightly different point.

7 Windsurfing Approach to Iterative Control Design

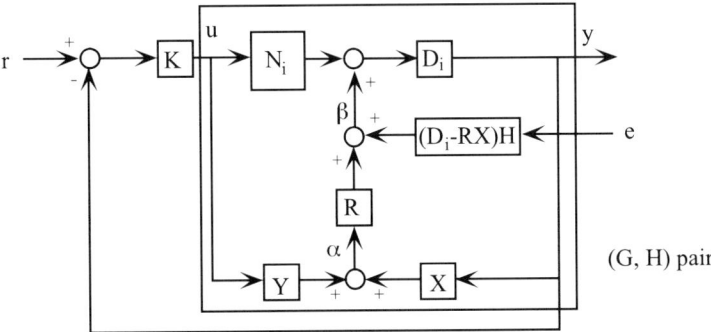

Fig. 7.2. Redrawing of Figure 7.1 using Youla-Kucera parametrisation of plant G

It is not hard to check that the signals α and β can be expressed using known quantities derived from the measurable signals r, u and y:

$$\alpha = Xr \tag{7.12}$$
$$\beta = D_i y - N_i u$$

But more than this is true. As suggested by Figure 7.2, there holds:

$$\beta = R\alpha + (D_i - RX)He \tag{7.13}$$

Notice that because e and r are independent, so must e and α be independent. Also, R must be stable. The task of identifying R is therefore a standard open-loop identification problem, given the availability (from measurements) of α and β, the presumed stability of R and the independence of e and α. Thus, at least in principle, we can see how to solve the closed-loop identification problem.

In open-loop identification, it is typical to contemplate issues of signal-to-noise ratio, use of filters and the like. In this chapter, we will make simple observations on these matters. The first observation can be made in terms of a filtered output error. The unfiltered output error is $\beta - R\alpha$, but we shall look at the filtered output error

$$\xi = Y(\beta - R\alpha) \tag{7.14}$$

[Recall that Y was introduced in (7.8).] Some manipulations will show that:

$$\xi = \left(\frac{GK}{1+GK} - \frac{G_i K}{1+G_i K} \right) r + \frac{1}{1+GK} H e \tag{7.15}$$

This equation reveals several things:

- modulo filtering by Y, the approximation error between the behaviours of G and G_i is not the absolute error, but rather the error associated with the two closed-loop transfer functions. Thus, G_i will be a good model of G with the power norm of ξ (or some other measure of the size of ξ) as the approximation criterion exactly when the two closed-loop transfer functions are close;

- suppose the closed-loop system formed by the unknown plant and the controller has a similar transfer function to the closed-loop system formed by the model and the controller. Then, the signal-to-noise ratio within (7.13) will be small. Accordingly, it will be difficult to identify R. Turning this round, it only makes sense to run an identification step when one knows that there is a genuine closed-loop mismatch;
- as one extends out the bandwidth over which one controls, it is likely to be the case that there is good closed-loop matching over lower bandwidths, and at some stage poorer closed-loop matching near the edge of the new pass band. This suggests that a parameterisation for R should be one that captures well the behaviour near the edge of the current pass band. Thus, R can be of a comparatively low order, even if G_i is of high order, so long as it is able to adequately represent a transfer function that is essentially band-pass in character.

7.4.3 Model Reduction

Recall that in the course of operating the iterative algorithm, one would expect to identify the closed-loop plant (with the aid of an open-loop Youla-Kucera parameter) at least several times. If precautions are not taken, this can result in a dimensionality explosion. To be more precise, suppose that prior to a re-identification of a plant, one is using a model G_i of degree n. Suppose the associated controller also has degree n. It is, in general, possible to find the factorisation representations in Equations (7.7) through (7.9) so that N_i, D_i, X and Y all have degree about n. Suppose that R only has degree 2 (and this may well be an underestimate). Then, as Equation (7.11) shows, the new estimate of the plant is likely to have degree approximately $2n + 2$. Thus, even if a very uncomplicated R is chosen, in an attempt to capture the true plant's contribution to a small part of the pass-band of a closed-loop system, one is faced with a large increase in the degree of the new plant model. It makes sense then to allow for the possibility of a model reduction step at this point. There are many ways to carry out model reduction; suffice to say here that model reduction taking into account the presence of a controller producing a stable closed loop is almost certainly more effective than open-loop model reduction. For a recent treatment of many model reduction ideas, see [119].

7.4.4 Stopping Conditions for the Algorithm

One reason to stop the algorithm would be that the desired closed-loop design has been achieved. However, there are at least two circumstances that might cause premature termination. First, it might be that one is unable to identify accurately, the signal-to-noise ratio being too low. Second, it might be that the most recently identified model and the most recently obtained controller together constitute a closed-loop system of such high sensitivity that further adjustment of the controller in an attempt to produce even wider band performance would be ruled out on the grounds of unacceptably low phase and gain margins, unacceptably high sensitivity and so on.

The paper in [104] analyses those situations that result in difficult or, in practical terms, impossible identification problems. The signal-to-noise ratio associated with the closed-loop output error can be poor because of unstable zeros of the plant that lie inside the pass band, and poor gain or phase margins of the closed-loop. The first phenomenon can only arise when the designed closed-loop bandwidth has been sufficiently pushed out. The second, of course, may well indicate difficulty or inability to expand the bandwidth of the closed-loop system, even given the availability of a good model, perhaps because the sought bandwidth starts to substantially exceed the plant's open-loop bandwidth.

7.5 Examples

Example 7.1. Our first example is drawn from [104]. The plant is a flexible-link robot arm with transfer function

$$G(s) = 0.5196 \frac{\prod_{i=1}^{5}(s-z_i)}{\prod_{i=1}^{6}(s-p_i)}$$

The poles are

$$p_1, p_2 = -0.0996 \pm j\, 3.0017$$
$$p_3, p_4 = -0.3339 \pm j\, 12.131$$
$$p_5, p_6 = -1.845 \pm j\, 31.481$$

and the zeros are

$$z_1 = -13.162$$
$$z_2, z_3 = -10.646 \pm j\, 12.27$$
$$z_4, z_5 = 7.169 \pm j\, 11.54$$

Evidently, the structure is resonant, with three pole pairs fairly well separated and all lightly damped. The three left half plane zeros are likely to present little problem. The two right half plane zeros will serve to limit the closed-loop bandwidth that can be attained.

The initial model is one that, not unreasonably, captures to a degree the lowest resonance, the roll-off behaviour at high frequencies (20 dB/decade)

– through the insertion of a zero well into the left half plane, and the low frequency gain. Thus

$$G_0(s) = 0.5188 \frac{s + 13.31}{(s + 0.0903 + j\,3.0027)(s + 0.0903 - j\,3.0027)}$$

The frequency response of G and G_0 are shown in Figure 7.3.

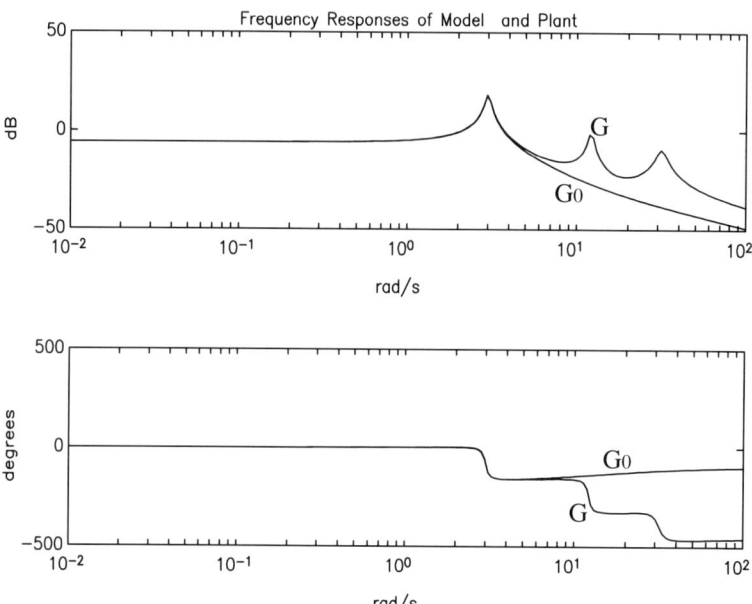

Fig. 7.3. Frequency response of model G_0

A conventional adaptive control algorithm, unless it had built in a sufficiently high degree for estimating the plant, could have great difficulty in coping with unmodelled resonances. One of the features of the windsurfer approach is that *it allows adjustment of the degree of the model transfer function*.

Following the IMC design method, the controller is chosen to secure a closed-loop transfer function $\lambda(s + \lambda)^{-1}$ where λ is variable corresponding to the design closed-loop bandwidth. A first order closed-loop transfer function is acceptable because $G_0(s)$ has no non-minimum-phase zeros, and relative degree one. The controller is second order, with an integrator. As it turns out here, the constant λ defining the bandwidth simply scales the controller transfer function, which is otherwise independent of λ.

For the purposes of identification, the (reference) input consists of just four periods of a zero-mean square wave of amplitude 1 and period π/λ. The

7 Windsurfing Approach to Iterative Control Design

plant output was corrupted by zero-mean white noise with standard deviation 0.05. For a validation step (discussed below), the external input can be turned off as appropriate.

Since we have G_0 to begin with, we need to test the closed-loop designs with G_0 and G to validate G_0.

One procedure can be based on looking at power spectra. Recall from (7.13) that we identify R using measurements α, β with

$$\beta = R\alpha + (D_i - RX)He$$

and here e is the perturbing noise. Recall also (7.14) and (7.15), which relate the error in (7.13) to the way the two closed-loop systems (G, K) and (G_0, K) process r:

$$\xi = Y(\beta - R\alpha) = \left(\frac{GK}{1+GK} - \frac{G_0 K}{1+G_0 K}\right) r + \frac{1}{1+GK} He \ .$$

Evidently, the quantity $\left[GK(1+GK)^{-1} - G_0 K(1+G_0 K)^{-1}\right] r$ carries the useful information about the existing modelling error, i.e., about R, and the term $(1+GK)^{-1} He$ is the noise component obstructing the determination of R. Call these terms v_0 and w_0 respectively. Suppose r is a stationary random signal. We can obtain the power spectrum $\Phi_{w_0}(\omega)$ by turning r off. We can also determine the power spectrum of ξ:

$$\Phi_{\xi_0}(\omega) = \Phi_{\vartheta_0}(\omega) + \Phi_{w_0}(\omega) \ .$$

If this spectrum is significantly larger than $\Phi_{w_0}(\omega)$ in some frequency band (especially around the cut-off frequency), then the model G_0 is invalidated in this band.

A second procedure for validation uses correlation techniques, see [108].

Roughly speaking, the cross-correlation method involves plotting (time-lagged) values of the cross-correlation of an input signal and an output error signal. If identification is perfect, the true cross-correlation will be zero. In practice, the sample cross-correlation will not equal the true cross-correlation, and the identification is not perfect. So, non-zero values will be obtained. Should these be within appropriate confidence limits, then that part of the system excited by the known input can be assumed to be well-modelled. No information is gleaned about the noise model; on the other hand, this means that one does not need to try to identify the noise model. Another attractive feature is that the confidence limits are easy to calculate, and are independent of the lag. For details, see [108, pp 511–516].

In terms of our previous notation, this means that we can examine the correlation between α and $\beta - R\alpha$ (when checking the validity of a model of R), or filtered versions of these quantities. Equivalently, one can look at the correlation between r and ξ (or filtered versions of these quantities), and then we are checking more directly the adequacy of G_0 as a model of G, in the presence of a controller K.

When the designed closed-loop bandwidth reaches 1.5 rad/s, the power spectrum validation method reveals no significant difference between $\Phi_{\xi_0}(\omega)$ and $\Phi_{w_0}(\omega)$. On the other hand, the method of correlations suggests that G_0 is ceasing to be a good model of G. (Note that at this frequency, one is well short of including the unmodelled resonances in the closed-loop bandwidth. However, the primary resonance, which is in the pass-band, certainly has some modelling error, and this may be responsible.)

The model G_0, however, is retained, and the closed-loop bandwidth pushed out slowly. By the time the bandwidth reaches 3 rad/s, both validation methods indicate poor modelling. The power spectrum method gives a huge discrepancy between $\Phi_{\xi_0}(\omega)$ and $\Phi_{w_0}(\omega)$ at 12 rad/s. This is obviously a consequence of G_0 failing to model the resonances in G corresponding to the poles $-0.3339 \pm j\, 12.131$, coupled with the fact that this resonance is close enough to the closed-loop cut-off frequency that it shows up.

The fact that $\Phi_{\xi_0}(\omega)$ and $\Phi_{w_0}(\omega)$ are so different means that there is adequate signal-to-noise ratio to identify the Youla-Kucera parameter R. Having assumed R is second order and then having identified it, one can validate it by the method of correlations before calculating the new G_1. It passes the test, and so G_1 can be constructed. The G_1 that is found is

$$G_1 = 0.5189 \frac{\prod_{i=1}^{5}(s - z_{i1})}{\prod_{i=1}^{6}(s - p_{i1})}$$

with poles

$$p_{11}, p_{21} = -0.0895 \pm j\, 3.0026$$
$$p_{31}, p_{41} = -0.4834 \pm j\, 12.03$$
$$p_{51}, p_{61} = -2.475 \pm j\, 31.502$$

and zeros

$$z_{11} = -12.967$$
$$z_{21}, z_{31} = -7.336 \pm j\, 11.05$$
$$z_{41}, z_{51} = 9.098 \pm j\, 12.07$$

As expected, G_1 has higher order than G_0. Frequency responses of G and G_1 are shown in Figure 7.4. This new model G_1 captures the imaginary parts of the zeros and poles remarkably well.

The closed-loop bandwidth can now be progressively pushed out. Notice that the defined closed-loop transfer function now is the product of an all-pass transfer function and $\lambda(s+\lambda)^{-1}$, because of the non-minimum-phase zeros in $G_1(s)$. When the bandwidth reaches 12 rad/s, both the validation methods suggest that G_1 is inadequate. When re-identification is performed and the controller readjusted for the new model, *performance cannot be improved*. Why? Notice that G and G_1 have unstable zeros with a bandwidth of about 12 rad/s. This will set a "fundamental limit" (see Section 4.4 and [140]) on the

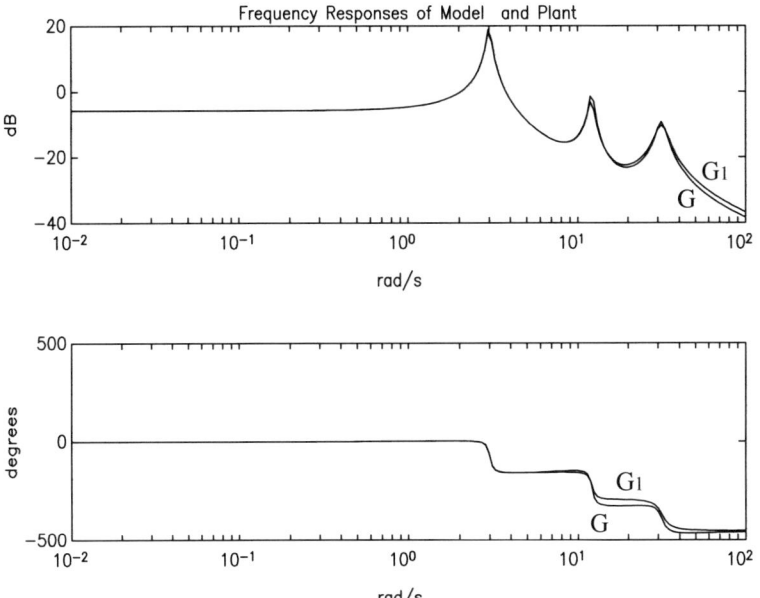

Fig. 7.4. Frequency response of model G_1

practically-achievable closed-loop bandwidth, and even very small modelling errors will not really be able to be properly compensated in a sound design.

The "fundamental limit" concept manifests itself in different ways; one is that, when one is designing at the fundamental limits, the true plant has to be known extremely accurately in order to get the designed-for performance. In our situation, the accuracy requirement for design cannot be matched by the accuracy offered by the identification algorithm.

Example 7.2. Our second example is drawn from [102]. The true plant is

$$G(s) = \frac{9}{(s+1)(s^2 + 0.06s + 9)}$$

with $y = Gu + e$, and disturbance e being zero mean noise with 100 Hz bandwidth and flat spectrum from 0 to 100 Hz of height 0.0025.

The initial model $G_0(s)$ mismodels the DC gain and fails to capture the resonance:

$$G_0(s) = \frac{0.8}{s + 1.2}$$

Any adaptive control algorithm that forces the plant model to remain first

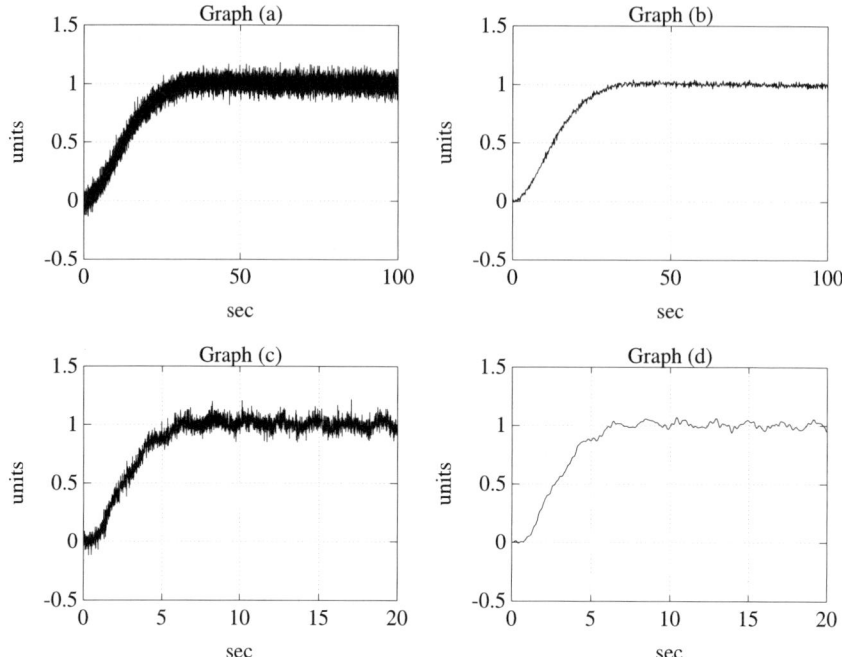

Fig. 7.5. Simulation results 1

order will be very challenged by this example. The design closed-loop transfer function is $\lambda^2(s+\lambda)^{-2}$ with λ variable. Figure 7.5 shows in (a) and (b) the actual closed-loop system unit step responses with $\lambda = 0.1\,\text{rad/s}$ (in (b), low-pass filtering has been applied). In (c) and (d), the bandwidth λ is $0.5\,\text{rad/s}$. Bearing in mind that the ideal response is that from $\lambda^2(s+\lambda)^{-2}$ excited by a unit step, we can conclude that by $\lambda = 0.5\,\text{rad/s}$, the model is inadequate.

In order to improve the identification, low-amplitude sinusoidal signals near $0.5\,\text{rad/s}$ were superimposed on the unit step excitation. A second order R was also assumed. This resulted in a new model, of higher degree than $G_0(s)$:

$$G_1(s) = \frac{0.0625s^2 - 0.34s + 10.28}{s^3 + 1.28s^2 + 9.12s + 10.32}$$

Redesign of the controller using $G_1(s)$ and with $\lambda = 0.5\,\text{rad/s}$ yields closed-loop performance with the true $G(s)$ shown in Figure 7.6. We can increase λ to $2\,\text{rad/s}$ without significant degradation.

It turns out that G_1 is a good model of G over this bandwidth, as seen in Figure 7.7.

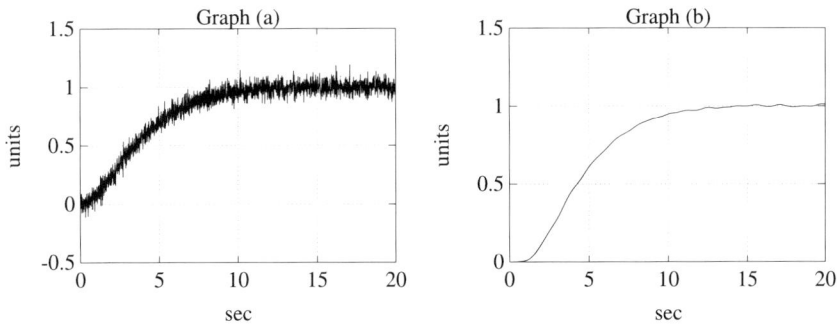

Fig. 7.6. Simulation results 2

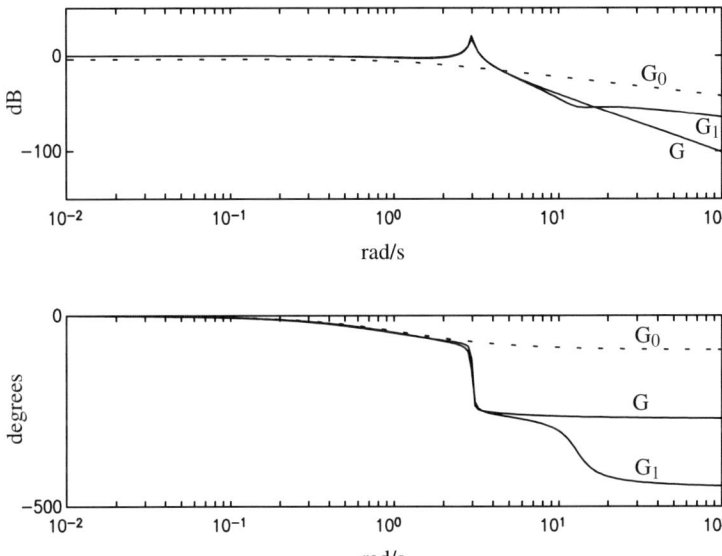

Fig. 7.7. Frequency responses of models and plant

7.6 Unstable Plants

The methods we have outlined up to now depend on the open-loop plant being stable. This is not because of any constraints in the identification process but rather because the IMC design method is far more efficacious for stable plants than for unstable plants [36, 116]. These techniques can often result in the controller having a non-minimum phase-zero, which may lie in the closed-loop pass band. Also, the techniques may involve more than a single parameter, and any parameters that they do involve are not as simply related to the closed-loop bandwidth as is the single parameter in the IMC design procedure for a stable plant. For these reasons, and as explained in more detail in the paper [105], a different structure is used, as depicted in Figure 7.8.

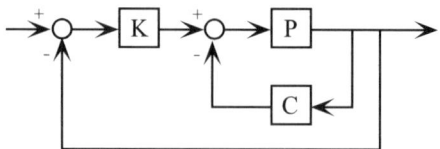

Fig. 7.8. Modification to IMC method for unstable plant P, with C stabilising the inner loop

The inner-loop controller C is chosen in order to stabilise the plant P. The result is an internal closed loop with transfer function $G = P/(1+PC)$. This inner closed-loop transfer function G is stable, and then IMC methods can be applied to the design of K.

The paper [105] sets out a number of guidelines for the choice of C. For example, should P have an unstable closed loop pole at $+a$, it is desirable for the closed-loop to have a pole at $-a$. Of course, control of an unstable plant is always harder than controlling a stable plant, and there are fundamental limitations expressible with Bode integral theorems, waterbed theorems and the like [51, 140]. Another example of such a difficulty arises when the plant P cannot itself be stabilised by an open-loop-stable C (it is easy to find examples of unstable plants for which there is no controller that is both stable and stabilising). In this case, the transfer function G necessarily acquires a right half plane zero, at any point where C has a right half plane pole. The controller K must then be found to stabilise a stable plant with an unstable zero, and the usual sorts of restrictions start to come into play when the closed-loop bandwidth starts to include that zero.

As these brief remarks should make clear, plant instability can be a major complication for design. Add to this the fact that simple single parameter adjustment of the closed-loop bandwidth is no longer possible, and one can see the situation is a good deal more complex.

7.7 Coping with the "Fundamental Problems"

In Section 7.2, we recorded some fundamental or conceptual problems of much adaptive control theory and thinking. The particular problems flagged were:

- the problem of changing experimental conditions, given accurate but inexact models; the issue is that once you change the controller, the model employed for controller redesign may become invalid;

- the problem of impractical control objectives; the issue is that the possibility or impossibility in practical terms of the control objective may not be assessable at the outset, given the (possibly partial) lack of knowledge of the plant;
- the transient instability problem: saying that all signals are bounded in a theorem does not preclude 1 MA current in a 1 kW motor.

Let us consider how these problems are addressed. The method of addressing the first is somewhat *ad hoc*. When we attempt a controller redesign, we make an adjustment of a bandwidth. If this adjustment is very small, we know that unsuitability of the model will not be an issue. In fact, the quality of the model (as displayed in its ability to match closed-loop performance with the same controller connected to model and to plant) will only degrade smoothly as the bandwidth of the closed-loop system is pushed out, and the controller is changed to do this. Thus, the adaptive algorithm must ensure that the trial expansions of bandwidth are not too large. Nor, of course, should they be so small that the whole process takes too long. An expansion of 20 per cent is a good target to begin with; but note that this could be too big if it suddenly swept into the bandwidth an extremely oscillatory mode. Finally, note that we cannot attempt a re-identification until there is a genuine difference between the two closed-loop transfer functions – otherwise the signal-to-noise ratio will be too small to allow an accurate identification.

How is the problem of impractical control objectives addressed? Here, the approach of this chapter comes in to its own. Effectively, a flag is raised that indicates the impracticality. One has to stop the algorithm early, on account of an inability to identify due to adverse signal-to-noise ratio, or an inability to find a controller giving a satisfactory (including acceptable gain and phase margin) closed-loop design.

Likewise, the problem of transient instability is dealt with well. At each step of the algorithm, it is guaranteed that an unstable (frozen or instantaneous) closed loop cannot be encountered. Smooth expansion of the bandwidth as part of the algorithm produces in the first instance degradation of the quality of the model, and that degradation sets in well before the model–controller closed loop is stable while the real plant–controller one is unstable.

7.8 Some Further Quantitative Directions

In this section, we will discuss briefly a direction in which the ideas of this chapter are being taken further. This concerns the formalising of approximation measures.

In Section 7.2, we alluded to a problem inherent in many adaptive control methodologies, which we termed "the problem of changing experimental conditions, given accurate but inexact models". The idea is that with one controller in place, a model may be a good model of a plant, while with another controller in place, it may not be. The point we want to make here is that this conceptual, rather qualitatively stated, difficulty is capable of quantitative refinement, using the concept of the ν-gap metric [153]. The first application to adaptive control appears in [7].

The question we are going to discuss in quantitative terms is the following. Suppose that G_0 is a good model of G in the presence of a controller K_0. Suppose K_1 is another controller for G_0. To what extent will G_0 be a good model for G in the presence of K_1?

Quantitative treatment of this question requires some preliminary work, to establish certain distance measures.

We shall first explain a distance measure for two transfer functions. At each frequency, we can define a certain distance, termed the *chordal distance*, which is a generalisation of the distance between points associated with the same frequency ω on two Nyquist diagrams. Then we define a single distance, by looking at all frequencies. First, for transfer functions $K_0(j\omega)$ and $K_1(j\omega)$, we have the chordal distance

$$\kappa[K_0, K_1](j\omega) = \frac{|K_1(j\omega) - K_0(j\omega)|}{\sqrt{1 + |K_1(j\omega)|^2}\sqrt{1 + |K_0(j\omega)|^2}} \tag{7.16}$$

If $K_1(j\omega_0) = \infty$, we compute the chordal distance by letting $\omega \to \omega_0$. We also define the ν-gap distance by

$$\delta_\nu(K_0, K_1) = \sup_\omega \kappa[K_0, K_1](j\omega) \tag{7.17}$$

provided

$$1 + K_1(-s)K_0(s) \neq 0, \quad \text{for all} \quad s = j\omega \tag{7.18}$$
$$wno[1 + K_1(-s)K_0(s)] + \eta(K_0) - \bar{\eta}(K_1) = 0 \tag{7.19}$$

where $\eta(K_0)$ and $\bar{\eta}(K_1)$ denote the number of open right half plane poles of K_0 and the number of closed right half plane poles of K_1 respectively, while $wnoX(s)$ denotes the number of clockwise encirclements of the origin by $X(s)$ as s moves around the standard Nyquist contour (up the imaginary axis), indented into the right half plane around any $j\omega$-axis pole of $X(s)$. If (7.18) or (7.19) fail, then

$$\delta_\nu(K_0, K_1) = 1 \tag{7.20}$$

Next, we shall look at closed-loop behaviour. In relation to a stable closed-loop plant G and controller K, defined the *generalised sensitivity matrix*,

$$T(G, K) = \begin{bmatrix} G \\ 1 \end{bmatrix} (1 + KG)^{-1} \begin{bmatrix} K & 1 \end{bmatrix} \tag{7.21}$$

7 Windsurfing Approach to Iterative Control Design

Notice that the four entries of T include the conventional sensitivity and complementary sensitivity functions. It is not hard to check that the maximum singular value of T at any frequency is

$$\bar{\sigma}[T(G,K)](j\omega) = \frac{\sqrt{1+|G(j\omega)|^2}\sqrt{1+|K(j\omega)|^2}}{|1+G(j\omega)K(j\omega)|^2} \tag{7.22}$$

A large value of $\bar{\sigma}[T]$ at some frequency generally signals a poor design. In terms of the complementary sensitivity M, $\bar{\sigma}_2[T]$ is given by

$$\bar{\sigma}^2(T) = \left[1 + \frac{1}{|G|^2}\right]\left[|M|^2 + |G|^2\,|1-M|^2\right] \tag{7.23}$$

This reveals, for example, that achieving $|M| \simeq 1$ in frequency bands where $|G|$ is small yields a large $\bar{\sigma}(T)$; and achieving very small $|M|$ when $|G|$ is very large, *i.e.*, suppressing a pole close to the $j\omega$-axis so that it is outside the closed-loop bandwidth, also yields a large $\bar{\sigma}(T)$.

We can say that G_0 is a good model of G when K_0 is connected when for some small ε, there holds,

$$\|T(G_0,K_0) - T(G,K_0)\|_\infty \leq \varepsilon \tag{7.24}$$

or possibly

$$\|T(G_0,K_0) - T(G,K_0)\|_\infty \leq \varepsilon \|T(G_0,K_0)\|_\infty \tag{7.25}$$

or

$$\bar{\sigma}[T(G_0,K_0) - T(G,K_0)](j\omega) \leq \varepsilon\bar{\sigma}\left[T(G_0,K_0)\right](j\omega) \quad \text{for all } \omega \tag{7.26}$$

The question we want to consider now is as follows: suppose one of (7.24) through (7.26) holds, and K_1 is a second controller so that (G_0,K_1) is an acceptable closed loop. What are (sufficient) conditions for (G,K_1) to be, firstly, stable and secondly, a loop with similar behaviour to (G_0,K_1)?

Stability is easiest. According to [153], a sufficient condition is:

$$\delta(K_0,K_1) < 1 \tag{7.27}$$

and

$$\bar{\sigma}[T(G,K_0)](j\omega)\kappa(K_0,K)(j\omega) < 1 \quad \text{for all } \omega \tag{7.28}$$

One has knowledge of $T(G_0,K_0)$; and so given one of (7.24), (7.25) and (7.26), one can check (7.28). Though we will not use the fact, we comment that

$$\bar{\sigma}\left[T(G,K_1)(j\omega)\right] \leq \frac{\bar{\sigma}[T(G,K_0)](j\omega)}{1 - \bar{\sigma}[T(G,K_0)](j\omega)\kappa(K_0,K_1)(j\omega)} \tag{7.29}$$

Next, the following inequality is known [7, 153],

$$\bar{\sigma}[T(G, K_1) - T(G, K_0)] \leq \frac{\bar{\sigma}[T(G, K_0)]^2 \kappa(K_0, K_1)}{1 - \bar{\sigma}[T(G, K_0)]\kappa(K_0, K_1)} \tag{7.30}$$

where we have suppressed dependence on ω.

Accordingly,

$$\begin{aligned}
\bar{\sigma}&[T(G, K_1) - T(G_0, K_1)] \\
&\leq \bar{\sigma}[T(G, K_1) - T(G, K_0) + T(G, K_0) \\
&\quad - T(G_0, K_0) + T(G_0, K_0) - T(G_0, K_1)] \\
&\leq \bar{\sigma}[T(G, K_1) - T(G, K_0)] + \bar{\sigma}[T(G, K_0) - T(G_0, K_0)] \\
&\quad + \bar{\sigma}[T(G_0, K_0) - T(G_0, K_1)]
\end{aligned} \tag{7.31}$$

For simplicity, neglect $\bar{\sigma}[T(G, K_0) - T(G_0, K_0)]$ on the grounds that G_0 is a good model of G with controller K_0 and take $\bar{\sigma}[T(G, K_0)] = \bar{\sigma}[T(G_0, K_0)]$. Then

$$\begin{aligned}
\bar{\sigma}&[T(G, K_1) - T(G_0, K_1)] \\
&\leq \frac{\bar{\sigma}[T(G_0, K_0)]^2 \kappa(K_0, K_1)}{1 - \bar{\sigma}[T(G_0, K_0)]\kappa(K_0, K_1)} + \bar{\sigma}[T(G_0, K_0) - T(G_0, K_1)]
\end{aligned} \tag{7.32}$$

A greater overbound again is possible, since it is known that [7, 153]

$$\kappa(K_0, K_1) \leq \bar{\sigma}[T(G_0, K_0) - T(G_0, K_1)]. \tag{7.33}$$

There results

$$\begin{aligned}
\bar{\sigma}&[T(G, K_1) - T(G_0, K_1)] \\
&\leq \left[\frac{\bar{\sigma}[T(G_0, K_0)]^2}{1 - \bar{\sigma}[T(G_0, K_0)]\kappa(K_0, K_1)} + 1 \right] \bar{\sigma}[T(G_0, K_0) - T(G_0, K_1)]
\end{aligned} \tag{7.34}$$

This formula suggests that changes in $T(G_0, K)$, when K varies from K_0 to K, are reflected proportionally in $T(G, K)$, but with a multiplier. The multiplier may be very large when the initial design $T(G_0, K_0) \simeq T(G, K_0)$ has large $\bar{\sigma}(T)$, i.e., is a poor, or non-robust design. The multiplier may also be large if the changes in K brings the closed-loop system near the stability boundary.

Acknowledgements

Support of the Office of Naval Research, Washington, is gratefully acknowledged. This work also benefited from discussions with R. Bitmead and M. Gevers.

8 Iterative Optimal Control Design

Robert R. Bitmead

Abstract. Interleaved stages of system identification using closed-loop data and model-based control design are studied in the framework of linear quadratic Gaussian optimal control (LQG) and least squares optimal identification. The tack is taken to explore methods in which the objective criterion for each phase is modified to reflect the overall control task at hand. This tutorial introduction examines the tools available for sculpting the respective control and identification criteria. The question of adding caution to the model and controller updates is also examined.

8.1 Introduction

We have previously seen earlier chapters on identification for control, closed-loop identification and robust control design. The content of each of these seeks to couple the *modelling for control* aspects of system identification with the *control design from approximate models* methods of robust control. In this chapter, we shall push this connection a little harder and further to develop an approach to joint modelling and control design in which related sequences of models and controllers are generated that attempt to reach some good model–controller combination, as measured by a performance criterion.

The presence of a criterion or objective function is a core issue. We distinguish between three such objectives; the overall control performance on the actual system, the model fitting criterion, and the design criterion for the controller based on the model with its attendant expected performance. One sees that, of these three measures, it is only the first that is of direct interest. Indeed, the other two measures are there only as vehicles to improve the first. Equally, it is clear that only the latter two criteria are within our scope to modify directly as part of our design. We shall consider how to manipulate these design criteria to develop controllers that perform well according to the main objective.

To expand a little further on this distinction, we distinguish between two feedback loops – *the achieved loop* and *the design loop* depicted in Figure 8.1. This achieved–designed distinction was introduced in Chapter 1. We also differentiate between *global* objectives defined on the real-world achieved loop and *local* objectives developed in our computer-based design loop.

Here, $G(z)$ is the real plant, $\hat{G}(z)$ is the model, $K(z)$ is the controller, $H(z)$ is the disturbance-shaping filter and $\hat{H}(z)$ the disturbance model. Signals are as follows: y_t, the actual process output; y_t^c, the model simulated closed-loop output; u_t, the plant input; u_t^c, the simulated closed-loop model input; v_t, the actual output disturbance process presumed to be generated by white

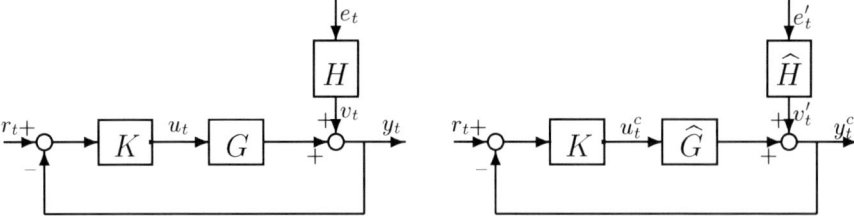

Fig. 8.1. Achieved and design loops

noise, e_t, fed through H; and v'_t, the modelled disturbance process generated (simulated) by a different white noise e'_t fed through $\hat{H}$. Signal r_t is a reference signal common to both loops, which may be used to assist the identification phase.

A helpful way to understand the connections between modelling and control is to consider the requirements demanded of a model by the current controller. Consider the robust stability condition,

$$\left| \left(G(e^{j\omega}) - \hat{G}(e^{j\omega}) \right) \frac{K(e^{j\omega})}{1 + K(e^{j\omega})\hat{G}(e^{j\omega})} \right| < 1, \quad \text{for all } \omega \in (-\pi, \pi] \quad (8.1)$$

In considering this inequality, which should hold over the entire frequency range, there will be some frequencies where the modelling error, $G - \hat{G}$, provides an active constraint so that the inequality is close to violation and thus limits the designed performance $\frac{K\hat{G}}{1+K\hat{G}}$, because of the need to ensure robust stability. There will be other frequencies where the requirement is overly satisfied or is an *inactive* constraint.

For a particular model $\hat{G}$, various controllers K will produce differing frequency regions of active constraints with the real plant. In this way, a specific controller indicates where it is limited due to the model quality. By understanding this, we are able to understand at which frequencies model quality improvement will remove an active constraint on the controller, opening the way for possibly improved performance with the subsequent controller. Because we consider a model change and then the subsequent controller designed on the basis of the new model, this fundamentally is an iterative approach in which small or cautious changes are made to the sequences of models and controllers.

In this chapter, we shall investigate the application of criterion-based methods for the successive identification and control stages. In particular, we shall explore mechanisms in which the local modelling and design criteria are both constructed to reflect the performance measured by the global objective. We commence by studying methods of controller refinement with a fixed model, known as *controller tuning*. This will give us some insights into the connections between local and global objectives and the role of experimental data in tuning. In Section 8.3 we move on to include model refinement as well as controller adjustment. This is done using optimal methods of identification and control, with manipulation of the criteria. Section 8.4 treats in brief a number of auxiliary issues and embellishments to the overall scheme of mixing optimal LQG control with (optimal) least squares identification. This includes the use of validation tests to ensure or guarantee the stabilising properties of the new controller before it is applied to the closed loop. There follows a brief discussion of successful applications and areas for further work.

8.2 Controller Tuning with Data

To understand the distinction between robust control design and iterative control design with data, it is important to stress the different information available to the designer. In robust control design, one accommodates known (or assumed) bounds on model error to constrain a model-based controller. A guarantee of stability is provided together with an expected performance, probably degraded to accommodate the plant uncertainty. In iterative controller design, one adds the extra information of exactly how well the designed controller performed in the achieved closed loop.

Not only is a performance measure derived from the experimental data of the new closed loop, but also it is possible to compute a stability margin for this new closed loop, which can measure the calibre of the stability guarantee coming from the robust control design. This information may actually be gleaned in different frequency bands, yielding a more detailed picture of the effects of model uncertainty. For stability evaluation, regions representing active constraints to the controller versus inactive constraints may be determined. From a performance perspective, this same information is available, thus indicating where the robust, model-based design has been overly conservative.

Considering the two loops of Figure 8.1, we see that the measured signals (y_t, u_t) and simulated signals (y_t^c, u_t^c) are available from operation with the reference signal $r_t = 0$. Denote the power spectral densities of these signals as $(\Phi_y(\omega), \Phi_u(\omega)$ and $(\Phi_{y^c}(\omega), \Phi_{u^c}(\omega))$, where the former pair is derived from measurements and the latter pair computed analytically. The measured pair indicates achieved performance and the computed pair the anticipated performance. The variation between the two indicates a mismatch between the assumed model error bounds and the actual, which may be used to refine the design without altering the current nominal model.

If the global performance objective is to minimise the quadratic criterion,

$$J_{global} = \lim_{N \to \infty} \frac{1}{N} \sum_{t=1}^{N} \left(y_t^2 + \lambda u_t^2 \right) \tag{8.2}$$

then we may reformulate this expression using Parseval's formula linking the signal spectra to time-domain sums.

$$J_{global} = \frac{1}{2\pi} \int_{-\pi}^{\pi} (\Phi_y(\omega) + \lambda \Phi_u(\omega)) \, d\omega \qquad (8.3)$$

The argument of this integral is able to be evaluated frequency-by-frequency to describe the achieved distribution of power in the closed loop of plant G with the current controller K.

By comparison, the designed performance also has such a representation,

$$J_{local} = \lim_{N \to \infty} \frac{1}{N} \sum_{t=1}^{N} \left((y_t^c)^2 + \lambda (u_t^c)^2 \right) \qquad (8.4)$$

$$= \frac{1}{2\pi} \int_{-\pi}^{\pi} (\Phi_{y^c}(\omega) + \lambda \Phi_{u^c}(\omega)) \, d\omega \qquad (8.5)$$

This describes the closed loop of K with the model $\hat{G}$. By comparing the spectra Φ_y and Φ_{y^c}, we are able to determine at which frequencies we have either over- or under-achieved in controlling the output power. Similarly, Φ_u and Φ_{u^c} describe the associated control power expended. The question that naturally arises is whether we might make use of this information to redesign or tune the controller.

8.2.1 Recasting the Local Control Criterion

Define the two stable and stably invertible filters, $F_1(z)$ and $F_2(z)$, such that

$$|F_1(z)|^2 = \left(\frac{\Phi_y(z)}{\Phi_{y^c}(z)} \right), \quad |F_2(z)|^2 = \left(\frac{\Phi_u(z)}{\Phi_{u^c}(z)} \right) \qquad (8.6)$$

Such filters are easily derived by, for example, fitting a low-order autoregressive model $A_1(z) y_t = e_t$ to the measured signal y_t and a similar model $A_1^c(z) y_t^c = e_t^c$ to y_t^c. (Here, A_1 and A_1^c are polynomials of chosen degree, which might be fitted using, say, the ar function of MATLAB®. These two polynomials are inherently stable.) We compute analogous filters for u_t and u_t^c. We may then compute

$$F_1(z) = \frac{A_1^c(z)}{A_1(z)}, \quad F_2(z) = \frac{A_2^c(z)}{A_2(z)} \qquad (8.7)$$

with the required stability properties and relationship (8.6) to the signal spectra.

From the achieved and design loops, we note that,

$$y_t = \frac{1}{1+KG} H e_t + \frac{KG}{1+KG} r_t \text{ and } y_t^c = \frac{1}{1+K\hat{G}} \hat{H} e'_t + \frac{K\hat{G}}{1+K\hat{G}} r_t$$

with related relationships holding for u_t and u_t^c. Note that, if the closed-loop signals are measured with the zero reference r_t, then automatically we have $F_1 = F_2$ and no extra computation is required.

After having conducted the above closed-loop experiment to permit collection of data, we may now recast the local control design objective function as follows.

$$J'_{design} = \lim_{N\to\infty} \frac{1}{N} \sum_{t=1}^{N} [(F_1(z)y_t^c)^2 + \lambda(F_2(z)u_t^c)^2] \tag{8.8}$$

$$= \frac{1}{2\pi} \int_{-\pi}^{\pi} \left(|F_1|^2 \Phi_{y^c}(\omega) + \lambda |F_2|^2 \Phi_{u^c}(\omega)\right) d\omega, \tag{8.9}$$

$$\simeq \frac{1}{2\pi} \int_{-\pi}^{\pi} \left(\Phi_y(\omega) + \lambda \Phi_u(\omega)\right) d\omega \tag{8.10}$$

$$= J_{global}, \text{ from (8.3)}.$$

This *trick* to recast the local design criterion to appear to be congruent to the globally achieved criterion raises several questions. How can one solve the recast optimal control problem with the criterion in (8.8)? Why is the relationship "$\simeq$" and not "$=$" in the third expression (8.10)? We shall deal with these questions in order.

8.2.2 Solution of the Recast LQG Problem

The frequency-weighted problem may be re-posed (but not *reposed*) by defining the signals $\bar{y}_t^c = F_1 y_t^c$, $\bar{u}_t^c = F_2 u_t^c$ and $\bar{v}_t' = F_1 v_t'$. The system equations may then be rewritten,

$$\bar{y}_t^c = F_1 y_t^c = F_1 \hat{G} F_2^{-1} F_2 u_t^c + F_1 \hat{H} e_t' = \left[F_1 \hat{G} F_2^{-1}\right] \bar{u}_t^c + \bar{v}_t'$$

$$\bar{u}_t^c = -F_2 u_t^c = -\bar{K} \bar{y}_t^c$$

Thus, to solve the frequency-weighted LQG problem, we solve an unweighted LQG problem with $(\hat{G}, \hat{H}, \lambda)$ replaced by $(F_1 \hat{G} F_2^{-1}, F_1 \hat{H}, \lambda)$ to produce a dynamic feedback controller $\bar{K}(z)$. The controller for the frequency-weighted problem is $K(z) = F_2^{-1}(z) \bar{K}(z) F_1(z)$.

We remark, however, that the use of frequency weightings above necessarily increases the order of the controller. In LQG design, the controller order is, in general, the same as that of the plant model plus that of the noise model. Here we see that both of these model orders are increased to compute $\bar{K}(z)$ and additionally the order of K is still further increased over that of $\bar{K}$.

8.2.3 The Need for Iteration

The "$\simeq$" in (8.10) above indicates an important feature of this controller tuning, which was alluded to before. The frequency weightings used to modify the control design relate to the previous controller and not that one which is being sought now. Thus, in the expression (8.8) for the revised local control design criterion, the weighting filters, F_1 and F_2, depend on the previous controller designed and not the current controller being sought. This is an inherent feature of iterative design that the closed-loop experimental data reflects information about the previous design and, necessarily, not about the subsequent one. If, however, we limit ourselves to small adjustments in the nature of tuning, then there will be a close correspondence. We shall return to these issues of caution and small adjustments later in this chapter and in Chapter 9.

At this stage, we should also note that these requirements of controller tuning and of iteration disappear with the use of an exact plant model. That is, if $\hat{G}(z) = G(z)$ precisely, then the need to distinguish between the achieved and designed loops evaporates. However, in practical applications, the prospect of an exact model is pathological. Since, as remarked above, the introduction of frequency weights increases the controller order, one may view the weights as accommodating model–plant mismatch in a different and underhanded way. This is, in part, correct and one should avoid the temptation to use frequency weights of too high an order. In general, we have found third order weights to capture most useful information.

8.3 Iterative Identification and Control Design

While controller tuning on the basis of achieved closed-loop data is an attractive idea for ameliorating performance degradation due to model–plant mismatch, there are limits to its capabilities if the initial model is too far divorced from the important aspects of the real plant. Similarly, we have already commented that this approach tends to increase the controller order dramatically as the frequency weightings cancel the nominal model. The logical approach to dealing with these limitations is to introduce a new model, which captures the most appropriate aspects of the plant for the current controller. That is, we conduct system identification of the plant to revise the model, rather than relying on implicit (and high-order) model adjustments via frequency weighting. We shall see, however, that many of the ideas from controller tuning remain pertinent, especially that of linking the local criteria to the global objective.

Considering the performance aspects of this iteration, we note from the figures that the achieved and designed output signals without excitation are given by

$$y_t = \frac{H}{1+GK} e_t, \text{ and } y_t^c = \frac{\hat{H}}{1+\hat{G}K} e_t'$$

where e_t and e'_t are white noise processes of unit variance. Similarly, the input signals are given by

$$u_t = \frac{-KH}{1+GK} e_t, \text{ and } u^c_t = \frac{-K\hat{H}}{1+\hat{G}K} e'_t$$

Global performance in rejecting the disturbance process is therefore measured by the closed-loop transfer functions $\frac{H}{1+GK}$, $\frac{KH}{1+GK}$, $\frac{\hat{H}}{1+\hat{G}K}$ and $\frac{K\hat{H}}{1+\hat{G}K}$. Specifically, Parseval's formula may be invoked to write the global objective as

$$J_{global} = \frac{1}{2\pi} \int_{-\pi}^{\pi} \frac{\left(1+\lambda|K(e^{j\omega})|^2\right)|H(e^{j\omega})|^2}{|1+G(e^{j\omega})K(e^{j\omega})|^2} d\omega$$

For the designed systems we have equivalently,

$$J_{local} = \frac{1}{2\pi} \int_{-\pi}^{\pi} \frac{\left(1+\lambda|K(e^{j\omega})|^2\right)|\hat{H}(e^{j\omega})|^2}{|1+\hat{G}(e^{j\omega})K(e^{j\omega})|^2} d\omega$$

Performance robustness might be defined as the closeness of the two transfer function magnitudes, weighted by the control criterion. Define the following two transfer functions.

$$W(z) = \left(1+\lambda K(z)K(z^{-1})\right)^{\frac{1}{2}} \tag{8.11}$$

$$\Delta^p = \frac{WH}{1+GK} - \frac{W\hat{H}}{1+\hat{G}K} \tag{8.12}$$

We have an immediate relationship, using the triangle inequality.

$$\left\|\frac{W\hat{H}}{1+\hat{G}K}\right| - |\Delta^p|\right| \leq \left|\frac{WH}{1+GK}\right| \leq \left|\frac{W\hat{H}}{1+\hat{G}K}\right| + |\Delta^p| \tag{8.13}$$

Thus, we might seek to keep Δ^p small in order that the performance of the modelled closed loop be close to the performance of the achieved closed loop. We note that this is *not* part of the controller tuning approach with a fixed model, where the designed performance was deliberately distorted to accommodate modelling inadequacy.

With the introduction of the identified model as part of the technique, we may focus on the minimisation of a variant of the performance error Δ^p as the local objective.

8.3.1 Local Identification Criterion

With the current controller K_i operating, an experiment is conducted on the achieved closed loop with external excitation signal r_t. This constitutes a closed-loop experiment with K_i. The data from this experiment may then be used to fit a plant model between the (achieved) measured signals u_t and y_t. We shall consider two approaches to this; the so-called "direct approach" in which the signals u_t and y_t are used as input and output in a standard system identification, and the "two-stage" method where an intermediate model is constructed before fitting the plant model.

Direct system identification

The criterion for plant and noise model fitting is to reduce the summed squares of prediction errors for filtered data signals. That is, we filter the data $\{u_t, y_t\}$ through a stable and stably invertible (data) filter $L(z)$ to produce signals $\{u_t^f, y_t^f\}$ and then fit a model between these signals using prediction error methods. We minimise a criterion

$$\sum_{t=1}^{N} \left(y_t^f - \hat{y}_{t|t-1}^f \right)$$

where the one-step-ahead prediction $\hat{y}_{t|t-1}^f$ is computed as

$$\hat{y}_{t|t-1}^f = \hat{H}^{-1}(z)\hat{G}(z)u_t^f + \left[1 - \hat{H}^{-1}\right] y_t^f$$

Using this closed-loop data in the criterion minimisation is equivalent to selecting $\hat{G}$ and $\hat{H}$ to minimise the following frequency domain quantity, as is demonstrated in earlier chapters of this book:

$$\int_{-\pi}^{\pi} \left[\left| \frac{\left[G(e^{j\omega}) - \hat{G}(e^{j\omega})\right] K_i(e^{j\omega})}{1 + G(e^{j\omega})K_i(e^{j\omega})} \right|^2 \Phi_r(\omega) \right. \tag{8.14}$$

$$\left. + \left| \frac{1 + \hat{G}(e^{j\omega})K_i(e^{j\omega})}{1 + G(e^{j\omega})K_i(e^{j\omega})} \right|^2 \Phi_v(\omega) \right] \frac{|L(e^{j\omega})|^2}{|\hat{H}(e^{j\omega})|^2} d\omega$$

where Φ_r is the power spectral density of the reference signal r_t and $\Phi_v = |H(e^{j\omega})|^2$ that of v_t.

This formula delivers a wonderful insight into the competing phenomena governing the frequency regions in which the identification algorithm will fit the model $\hat{G}$ to the plant G or even to the inverse of the controller $-K_i^{-1}$. The design variables of particular note are the filter L, which determines the relative importance of specific frequency bands, and the excitation spectrum Φ_r, which encourages fitting the plant rather than the controller. In this work, we do not address the other major design variable, which is the model structure of admissible $(\hat{G}, \hat{H})$ pairs.

Local criterion selection

Our aim is to reformulate the closed-loop identification criterion to reflect the global control objective. One important result in this direction relies very simply on the triangle inequality. We may rewrite the global performance objective as the 2-norm of the vector $(y_t \; u_t)^T$,

$$J_{global} = \lim_{N \to \infty} \sum_{t=1}^{N} (y_t^2 + \lambda u_t^2) = \left\| \begin{matrix} y_t \\ u_t \end{matrix} \right\|_2$$

Since this is a norm, we may immediately state that

$$\left| \left\| \begin{matrix} y_t^c \\ \lambda^{\frac{1}{2}} u_t^c \end{matrix} \right\|_2 - \left\| \begin{matrix} y_t - y_t^c \\ \lambda^{\frac{1}{2}}(u_t - u_t^c) \end{matrix} \right\|_2 \right| \leq \left\| \begin{matrix} y_t \\ \lambda^{\frac{1}{2}} u_t \end{matrix} \right\|_2 \leq \left\| \begin{matrix} y_t^c \\ \lambda^{\frac{1}{2}} u_t^c \end{matrix} \right\|_2 + \left\| \begin{matrix} y_t - y_t^c \\ \lambda^{\frac{1}{2}}(u_t - u_t^c) \end{matrix} \right\|_2 \tag{8.15}$$

This is the signal-domain equivalent of the performance robustness result (8.13).

Our local identification criterion is chosen to reflect this inequality.

$$J_{local}^I = \left\| \begin{matrix} y_t - y_t^c \\ \lambda^{\frac{1}{2}}(u_t - u_t^c) \end{matrix} \right\|_2 \tag{8.16}$$

This choice causes the identification stage of the iterative scheme to concentrate on the fitting of a model that maintains the controlled model signals (from the designed or simulated loop) close to those of the achieved loop with the actual plant and the same controller. Our next task is to connect (8.16) with the permissible variables in the identification problem.

To this end, we next consider the performance-based modelling error defined in (8.12)

$$\begin{aligned}
\Delta^p &= \frac{W_i H}{1 + GK_i} - \frac{W_i \hat{H}}{1 + \hat{G}K_i} \\
&= W_i \times \frac{H - \hat{H} + \hat{G}K_i H - GK_i \hat{H}}{(1 + GK_i)(1 + \hat{G}K_i)} \\
&= W_i \times \left[\frac{H - \hat{H}}{1 + GK_i} + \frac{(G - \hat{G})K_i \hat{H}}{(1 + GK_i)(1 + \hat{G}K_i)} \right]
\end{aligned} \tag{8.17}$$

This should be related to the frequency domain formulation of the identification criterion (8.14).

How might (8.14) and (8.17) be reconciled? We begin by making the following assumption.

Assumption 8.1 (modelling). The noise model $\hat{H}(z)$ coincides with the actual disturbance generation process $H(z)$.

While it is always unpleasant to introduce additional assumptions, this one is needed for our approach and is, in part, mitigated by real-world experience, as will be presented in a later chapter for a sugar-cane crushing mill control problem. Knowledge of the additive output disturbance is usually tied to the formal specification of the control objective, which is bound into the global criterion.

With the modelling assumption, $\hat{H} = H$, (8.17) becomes

$$\Delta^p = \frac{\left(G - \hat{G}\right) K_i \hat{H} W}{(1 + GK_i)\left(1 + \hat{G}K_i\right)}$$

The following selection of design variables is made for the local identification phase of the iterative design.

- $\hat{H}(z) = H(z)$, which is possible because of our assumption above;
- data filter

$$L(z) = \frac{\hat{H}(z)(1 + K_i(z)K_i(z^{-1})^{\frac{1}{2}}}{1 + \hat{G}_i K_i(z)} \tag{8.18}$$

- reference input spectrum selection

$$\Phi_r(\omega) = \gamma \Phi_v(\omega) = \gamma |H(e^{j\omega})|^2 \tag{8.19}$$

with $\Phi_r \gg \Phi_v$ for frequencies of control interest, should there be doubt about the validity of the assumption that $\hat{H} = H$ everywhere.

These design choices have the effect of forcing the leading term of (8.14) to resemble $\Delta^p \times \gamma$. The identification phase of iterative control design seeks to find the model $\hat{G}$ that best captures the closed-loop performance of the actual plant G with current controller K.

8.3.2 The Iterative Algorithm

A sequence of controllers and identified models is constructed as follows:

1. Commence with a stabilising controller K_0, an initial plant model $\hat{G}_0$ and a noise model $\hat{H}$. Begin loop $i = 0$.
2. Conduct a closed-loop experiment with K_i yielding data $\{y_t, u_t, y_t^c, u_t^c\}$.
3. Determine quality of current model by examining fit of closed-loop designed data $\{y_t^c, u_t^c\}$ to the achieved data set. Some tests are discussed in [102].
    (a) If satisfied with the model, adjust the controller by tuning with data to yield K_{i+1}. The model remains fixed $\hat{G}_{i+1} = \hat{G}_i$. This controller tuning phase uses the method of recasting the LQG design as in (8.8).

(b) If dissatisfied with the model, use the data to fit a new model $\hat{G}_{i+1}$ and design a new controller K_{i+1} using $\hat{G}_{i+1}$ and no extra data-weighting. That is, a new model is fitted with closed-loop data excited as recommended in (8.19) and the filtering in (8.18).

4. Stop if satisfied with performance, or improvement ceases or reverses. Else return to Step 2.

8.4 Introducing Caution into Iterative Design

One feature of iterative identification and control design is that it uses two *one-shot* methods that do not explicitly relate the previously identified model or computed controller to the current ones. Since the mappings from data to identified model or criterion to computed controller are largely opaque to the designer, it is problematic to ensure that new models or controllers are *close* in some sense to the previous ones. Since the underpinning philosophy of iteration is that we move in small steps improving the performance at each stage (*c.f.* the discussion of Section 8.2.3), it is important to ensure some continuity between successive stages. This is our task in this section. It will be taken up in more detail in the next chapter.

8.4.1 The Vinnicombe ν-metric

Before launching into a discussion of continuity, one should develop an appropriate measure of distance that relates the closed-loop stability and controlled performance of plant-model pairs to the separation between two plants or two controllers. We then need to look at how to manage that separation distance in the iterative updates. A suitable measure of distance between plants is given by Vinnicombe's ν-gap metric [153], which was introduced to us in Chapter 4.

Definition 8.1 (ν-metric). The ν-distance, $\delta_\nu(G_1, G_2)$, between two plants $G_1 = N_1 D_1^{-1}$ $G_2 = \tilde{D}_2^{-1} \tilde{N}_2$, which satisfy a winding number constraint [see (4.49) in page 91]

$$\frac{1}{2\pi} \Delta \arg_\Gamma (\tilde{N}_2 N_1^* + \tilde{D}_2 D_1^*) = 0 \qquad (8.20)$$

is given by

$$\delta_\nu(G_1, G_2) = \left\| (I + G_2 G_2^\star)^{-\frac{1}{2}} (G_2 - G_1)(I + G_1^\star G_1)^{-\frac{1}{2}} \right\|_\infty \qquad (8.21)$$

and by 1 if the winding number condition fails.

The generalised sensitivity function of the plant-controller feedback pair (G, K) is given by

$$T(G, K) = \begin{pmatrix} G(I + KG)^{-1} K & G(I + KG)^{-1} \\ (I + KG)^{-1} K & (I + KG)^{-1} \end{pmatrix} \tag{8.22}$$

The generalised stability margin of the plant–controller pair (G, K) is given by,

$$b_{G,K} = \begin{cases} (\|T\|_\infty)^{-1}, & \text{if } (G, K) \text{ is stable} \\ 0, & \text{else} \end{cases} \tag{8.23}$$

we have Vinnicombe's result:

Theorem 8.1 (Vinnicombe). *Consider a controller K and two plants G_1 and G_2, with G_1 stabilised by K. The following results hold:*

[I] *(G_2, K) is stable for all plants G_2 satisfying $\delta_\nu(G_1, G_2) \leq \beta$ if and only if $b_{G_1,K} > \beta$.*
[II] *If $\delta_\nu(G_2, G_1) < 1$ then*

$$\arcsin b_{G_2,K} \geq \arcsin b_{G_1,K} - \arcsin \delta_\nu(G_1, G_2) \tag{8.24}$$

and

$$\delta_\nu(G_1, G_2) \leq \|T(G_1, K) - T(G_2, K)\|_\infty \leq \frac{\delta_\nu(G_1, G_2)}{b_{G_1,K} b_{G_2,K}} \tag{8.25}$$

The ν-distance between two plants is a measure appropriate for the guarantee of stability with the same controller. The relationship in Theorem 8.1 for simultaneous stability is a sufficient condition only, and is therefore conservative. Plants stabilised by the same controller do not necessarily have ν-distance less than $b_{G_1,K}$, nor do they need to satisfy the winding number condition (8.20). Nevertheless, the ν-distance is focused on the core issue for iterative control design, stability.

The notion of a controller stabilising a plant or pair of plants, such as a real plant and its model, may be simply extended to that of a single plant stabilising (or being stabilised by) a pair of controllers. Thus, if (G, K_1) is stable with margin b_{G,K_1} and the next controller K_2 satisfies $\delta_\nu(K_1, K_2) < b_{G,K_1}$, then (G, K_2) is also stable. Further, equivalent variants of the inequalities (8.24) and (8.25) also hold. This variation of the theorem is important for allowing us to know how to adjust controllers with a guarantee of stability and performance flowing from the properties with the previous controller.

The generalised sensitivity matrix $T(G, K)$ provides a measure of performance. Usually, the closed-loop performance is specified in terms of a combination of the weighted norms of the transfer functions comprising T. Inequality (8.25) relates the guarantees of performance between two successive closed loops. The theorem also provides a connection between the stability margin and performance. As the stability margin drops, this is revealed through the peaking in the frequency response of T. This is a useful link that can be exploited in closed-loop experimental data analysis, since the value of $\|T\|_\infty$ can be estimated from data, using the $\mathcal{H}_\infty$-norm system identification described in earlier chapters. This will be revisited in Chapter 9.

8.4.2 Cautious Adjustment of Controllers and Models

Once the value of $b_{G,K} = \|T(G,K)\|_\infty$ is estimated from data, the question arises of how to move cautiously from one plant model or controller to a new one while respecting the ν-distance limits for guarantees of performance.

Consider first the control problem. Suppose we have a true plant G, a current plant model G_1, a current control K_1, that stabilises both G_1 and G with respective margins b_{G,K_1} estimated from data and b_{G_1,K_1} computed, and a new candidate control K_2, which by design stabilises G_1. The issue is to ensure that K_2 also will stabilise the true plant G.

Since K_1 and K_2 both stabilise the same plant G_1, we may write K_2 as a Youla-Kucera parameterised variation of K_1 as was done in Chapter 7. Express G_1 and K_1 as coprime factor descriptions over the ring of stable proper transfer functions $\mathcal{RH}_\infty$

$$G_1 = X_1 Y_1^{-1} = \tilde{Y}_1^{-1} \tilde{X}_1, \quad K_1 = N_1 D_1^{-1} = \tilde{D}_1^{-1} \tilde{N}_1$$

with these coprime factorisations related by the stabilisation condition $\tilde{D}_1 Y_1 + \tilde{N}_1 X_1 = I$ and $\tilde{Y}_1 D_1 + \tilde{X}_1 N_1 = I$. Then for some $Q \in \mathcal{RH}_\infty$, K_2 can be written as [66]

$$K_2 = (N_1 - Y_1 Q)(D_1 + X_1 Q)^{-1} \tag{8.26}$$
$$= \left(\tilde{D}_1 + Q\tilde{X}_1\right)^{-1} \left(\tilde{N}_1 - Q\tilde{Y}_1\right)$$

Further,

$$K_2 = \left(\tilde{D}_1 + Q\tilde{X}_1\right)^{-1} \left(\tilde{N}_1 - Q\tilde{Y}_1\right) = \tilde{D}_2^{-1} \tilde{N}_2, \text{ for the same } Q \in \mathcal{RH}_\infty$$

If (8.20) holds, then

$$\delta_\nu(K_1, K_2) = \left\| \left(\tilde{D}_2 \tilde{D}_2^\star + \tilde{N}_2 \tilde{N}_2^\star\right)^{-\frac{1}{2}} Q \left(D_1^\star D_1 + N_1^\star N_1\right)^{-\frac{1}{2}} \right\|_\infty, \tag{8.27}$$

otherwise, $\delta_\nu(K_1, K_2) = 1$.

The impact of (8.27) is that it relates the size of Q to $\delta_\nu(K_1, K_2)$. Note that Q appears in the numerator alone while it also forms part of N_2 and D_2. However, as $Q \to 0$, so does $\delta_\nu(K_1, K_2) \to 0$. Choosing parameter $\alpha \in (0, 1]$ we may define the set of controllers

$$K_\alpha = (N_1 - \alpha Y_1 Q)(D_1 + \alpha X_1 Q)^{-1} \tag{8.28}$$

This is a set of controllers, all of which stabilise the model G_1 and which varies continuously from K_1 at $\alpha = 0$ to K_2 at $\alpha = 1$. Given stability margin $b_{G,K_1} > 0$, there always exists an $\alpha_0 > 0$ such that

$$\delta_\nu(K_1, K_\alpha) < b_{G,K_1} \quad \forall \alpha \in (0, \alpha_0)$$

With this construction, it is possible to take an existing stabilising controller K_1 with margin b_{G,K_1} and a candidate controller K_2 (usually designed without any connection to K_1 in mind) and to produce a new composite controller K_α, which has guaranteed stability and performance properties. By exchanging the places of plant models and controllers above, a related approach may be generated that commences with current model G_1 stabilised by K_2 and candidate model G_2 also selected to be stabilised by K_2, so that it generates a class of plant models parameterised by a Youla-Kucera parameter R times scalar $\beta \in (0, 1]$. We might then choose G_β so that satisfies

$$\delta_\nu(G_1, G_\beta) < \gamma,$$

for some constant γ. In this way, the model movement may also be cautiously regulated. These constructions of close models and controllers have been discussed more extensively in [24].

8.4.3 Simultaneous Cautious Controller and Model Adjustment

Iterative identification and control design combines many ingredients; notably a true plant G, an initial controller K_0, and sequences of plant models $\{G_i\}$ and controllers $\{K_i\}$. The simultaneous stabilisation property exhibited by these sets is that each plant model, G_{i+1}, is stabilised by the pair of controllers K_i and K_{i+1}. Each controller, K_i, stabilises the pair of plant models G_i and G_{i+1}. This is illustrated in Figure 8.2. It should be pointed

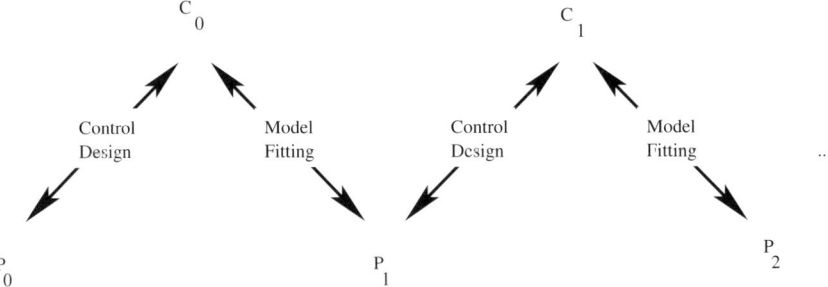

Fig. 8.2. Simultaneous stabilisation diagram for joint modelling and control design. Directed arrows indicate stabilisation.

out that underlying this simultaneous stabilisation of G_is and K_is is the stabilisation and closed-loop performance of G by the K_is. The question arises as how best to manage this stabilisation sequence so as to achieve stability and performance with G. For the following, we shall consider that $G_{\alpha,i+1}$ is given as a function of previous model G_i, Youla-Kucera parameter Q_i, scalar α_i and margin b_{G_i,K_i}. Similarly, $K_{\beta,i+1}$ is computed from K_i, R_i, β_i and γ_i.

Our interest is in achieving stability and performance of the $(G, K_{\alpha,i+1})$ loop, given measured properties of the (G, K_i) loop. Write $G = XY^{-1} = \tilde{Y}^{-1}\tilde{X}$, $G_i = X_i Y_i^{-1} = \tilde{Y}_i^{-1}\tilde{X}_i$, and $K_i = N_i D_i = \tilde{D}_i^{-1}\tilde{N}_i$ satisfying $\tilde{D}_i Y_i + \tilde{N}_i X_i = I$ and $\tilde{Y}_i D_i + \tilde{X}_i N_i = I$. Assume that controller K_i stabilises the real plant G in addition to both G_i and $G_{\beta,i+1}$, then G_i and $G_{\beta,i+1}$ may also be written as

$$G_i = (X - D_i \bar{R}_i)(Y + N_i \bar{R}_i)^{-1} = (\tilde{Y} + \bar{R}_i \tilde{N}_i)^{-1}(\tilde{X} - \bar{R}_i \tilde{D}_i)^{-1}$$
$$G_{\beta,i+1} = (X - D_i \bar{R}_{i+1})(Y + N_i \bar{R}_{i+1})^{-1} =$$
$$= (\tilde{Y} + \bar{R}_{i+1} \tilde{N}_{i+1})^{-1}(\tilde{X} - \bar{R}_{i+1}\tilde{D}_i)^{-1}$$

where $\bar{R}_{i+1} = \bar{R}_i + \beta R_i$. Then, controller $K_{\alpha,i+1}$ will also stabilise the real plant G if and only if $I - \bar{R}_{i+1}\alpha Q_i$ is a unit in $\mathcal{RH}_\infty$. That is, if and only if $I - \bar{R}_{i+1}\alpha Q_i$ is stable and proper with an inverse that is also stable and proper.

One simple way to ensure that $I - \bar{R}_{i+1}\alpha Q_i$ is a unit is to insist that

$$\alpha \times \|\bar{R}_{i+1}\|_\infty \times \|Q_i\|_\infty < 1$$

By choosing scalar parameters β and α that are appropriately small, this condition may be met. It should be remarked that the value of $\|\bar{R}_i\|$ is related to the achieved comparative performance of the (G, K_i) and (G_i, K_i) loops via (8.25).

These connections between caution and iterative design will be taken up again in the next chapter.

8.5 Embellishments and Conclusion

The archetypal iterative identification and control algorithm would see successive or interleaved phases. In practice, there are many other choices for the different stages, which have been discussed. Here, we shall cover just a few of these in order to provide a picture of the state of the art.

8.5.1 Two-stage (Indirect) Identification

Our system identification method considered in the development of (8.14) uses the closed-loop signals y_t and u_t directly in the computation of a least squares model, without regard for the closed-loop nature of these signals. *Closed loop* means that u_t and v_t are correlated. The consequence of this is that, if the noise model H is not known exactly, the plant model fitting is biased by the correlated noise. This can be seen in (8.14) through the plant model dependence on the second Φ_v term.

A method to yield an unbiased plant model has been suggested by Van den Hof and Schrama [147]. Here, a model is first fitted between the external reference signal r_t and the input u_t. Since the input to this fit, r_t, is independent from the noise, which is a filtered version of v_t, this yields an unbiased model. Also, because this model will not be used for anything but filtering, the model structure can be selected with very high order without penalty. We may write

$$u_t = \Psi(z)r_t + \Upsilon(z)e_t$$

Now, the artificial signal, $u_t^r = \Psi(z)r_t$, can be computed by filtering directly r_t. This new signal is the part of u_t that is caused by r_t and is automatically independent from the process disturbance v_t. A plant model is then fitted

$$y_t = \hat{G}(z)u_t^r + \hat{H}(z)e_t$$

without bias problems. There are connections between this two-stage identification, with its stable filtering of the data by a nominal closed-loop, and the dual-Youla or Hansen scheme developed in conjunction with the Windsurfer scheme of Chapter 7.

Unbiased models of the actual plant need not necessarily be the ideal outcome, however. Especially with reduced order or approximate system models, one still must ensure that the fit is connected to the ultimate application requirements. This is discussed in [108]. For the sugar mill control design problem of Chapter 13, however, because of the rather dominant nature of the disturbance process over the excitation, two-stage identification was an important step in refining the model [125].

8.5.2 Performance Robustness Versus Stability Robustness

The selection of controller tuning above sought to do two things:

- to force the designed controller to yield good achieved performance, perhaps at the expense of the designed performance;
- to keep the plant and model performances close with the same controller k_i;

Performance was the dominant agenda and no attempt was made to ensure that stability guarantees were respected.

There are different approaches to achieve this latter objective, although usually at some expense to the former.

Robustness of global objective: If one recognizes the reliance on models of these design methods, iteration notwithstanding, it is prudent to include some robustness requirement in the formulation of the global objective criterion. For LQG, this is not as simple as some other methods, but a stability robustness check ought to form part of the acceptance conditions for a controller. Connections are studied in [46].

Identification for stability robustness: Instead of selecting the performance-oriented modelling criterion of Δ^p, one might as well choose (8.1). This leads to a different choice of reference spectrum and data filter L, see [161].

Controller validation test: At each new control design phase, it is possible to perform sufficient tests to guarantee whether the controller will stabilise the real plant. The work of Vinnicombe [153] provides a measure of permissible (K_i, K_{i+1}) controller variation in terms of the currently achieved (G, K_i) performance. Both the performance and the variation are measured using frequency domain tools. This leads to the application of cautious techniques of controller update [7, 24, 25]. Other statistical tests may be performed to verify that the plant belongs to the set of systems stabilised by the controller.

8.5.3 $\mathcal{H}_\infty$ Variants

In order to achieve guaranteed robustness margins, one is tempted to transliterate these techniques from the L_2 framework to $\mathcal{H}_\infty$. However, there are still significant problems to be overcome in the formulation of $\mathcal{H}_\infty$ system identification in such a way that it links properly to the excited experiment and the control objective. This is discussed by Bayard *et al.* [20] and [161]. One feature of the $\mathcal{H}_\infty$ schemes not enjoyed by LQG–Least squares is that a truly decrescent algorithm is possible with both identification and control design having identical criteria.

8.5.4 Model-free Controller Tuning

Hjalmarsson *et al.* [78] and Kammer *et al.* [95] both consider the tuning of controllers using closed-loop experimental data but without the introduction of an explicit plant model. They use direct estimates of the gradient and Hessian of the control criterion based on the signal measurements. This has the advantage of not involving a modelling phase at all, but it is capable only of convergence to a local minimum. These methods are designed for use with restricted complexity controllers, such as PID controllers.

8.5.5 Conclusion

One view of iterative identification and control design is that it is a variety of adaptive control. Indeed, if one were to conduct adaptive control with the controller update not occurring until the end of an experiment, one would have a similar tactic to these iterative methods. Whereas adaptive control has proven surprisingly difficult to apply in many different practical problems, iterative identification and control has met with a number of early successes. One would expect that a range of profitable connections remain to be established between these related approaches.

Areas in which results still are very much needed include experiment design, where the selection of excitation spectrum Φ_r is linked to the ultimate goal, and model structure selection, where plant and noise model complexity need to be matched to the control objective. There is still a large number of open problems for this promising area. Other source material is available in [60, 69, 102, 106, 161].

Acknowledgements

The author acknowledges the valuable inputs to this work of Brian Anderson, Michel Gevers, Leonardo Kammer, Ari Partanen and Zhuquan Zang.

Research supported by US National Science Foundation under Grant ECS-0070146.

9 Identification and Validation for Robust Control

Michel Gevers

Abstract. In this chapter, we focus on the interplay between the properties of the nominal model and those of the uncertainty set of models obtained by identification on the one hand, and the stability and performance qualities of the model-based controllers that result from the identified model and its uncertainty set on the other hand. This interplay is examined first in a certainty equivalence context: this concerns the connection between the nominal model G_{mod} and a controller computed from G_{mod}, and the analysis focuses on the bias error. The analysis leads to the formulation of control-oriented identification criteria, whose minimisation yields a "control-oriented nominal model". Subsequently, we examine the interplay between the model uncertainty set $\mathcal{D}$ obtained by model validation and the corresponding controller set $\mathcal{K}$ obtained from $\mathcal{D}$ by robust control design methods. This analysis leads to the formulation of control-oriented guidelines for the design of the experimental conditions under which identification should be performed when the validated model uncertainty set, obtained by prediction error identification, serves as the basis for robust control design.

9.1 Introduction

This chapter focuses on the *interplay* between the design of the identification criterion and of the validation procedure, and the robust stability and performance of the resulting controller. The chapter elaborates on the methodologies presented in Chapters 1, 6, 7 and 8, as well as on the basic robust analysis tools presented in Chapter 4. We shall restrict our presentation to the identification and validation of parametric model sets and – except when stated otherwise – to least squares prediction error identification criteria.

For the sake of making our message as clear as possible, and even though this terminology is not universally accepted, we shall introduce the following distinction between *identification* and *validation*. We shall, in this chapter, call *identification* the task of constructing a nominal model G_{mod}. For a given data set, collected under specific experimental conditions, this nominal model is the direct result of the chosen model set and of the identification criterion. We shall call *validation* the task of constructing an uncertainty set $\mathcal{D}$ that contains the true system, perhaps at a certain probability level α. Very often, the construction of this uncertainty set $\mathcal{D}$ is also the result of a prediction error identification experiment. Thus, a single identification experiment could be used to construct both a nominal model G_{mod} and a validated uncertainty set $\mathcal{D}$. However, in identification for control, one often wants to work with a low-order model G_{mod} for control design. In such case, there are good reasons to distinguish between:

- an "identification" experiment for the construction of a control-oriented nominal model; this is typically achieved by an identification step with a low-order model structure, control-oriented experimental conditions and/or a control-oriented criterion;
- a "validation" experiment for the construction of a control-oriented uncertainty set; this can be achieved by an identification step with a full-order model structure, control-oriented experimental conditions and/or a control-oriented criterion.

First, let us recall the main motivation for studying this interplay between identification/validation and robust control, as well as the change in strategy that has been made possible by the new insights gained in this interplay. As explained in Chapter 1, the traditional strategy in model-based robust control design was to first identify the best possible model G_{mod} and construct the most reliable uncertainty set $\mathcal{D}$ around G_{mod} on the basis of prior information and available data, and to then design a controller K that achieved closed-loop stability and met the required performance with all models in $\mathcal{D}$, and hence with the unknown true system G_0. For this scenario to be successful, a very accurate (and hence complex) model G_{mod} was typically required. This not only required an important investment in identification and/or modelling; it also led to unnecessarily complex controllers.

The main lesson learned from ten years of research on the interplay between identification/validation and robust control is that one can design low-order model-based controllers that achieve high performance on the actual system. These low-order controllers are based on low-order (and hence biased) models whose bias error has been tuned for robust control. They are selected from a class of controllers $\mathcal{K}$ that achieve robust stability and robust performance with all models in an uncertainty set $\mathcal{D}$ whose shape has been tuned for robust control.

As a result, the present scenario for model-based control design resulting from these new insights can be described as follows: *on the basis of the required performance, of any knowledge of the unknown system, and of the performance achieved with the present controller (if any), design a control-oriented identification experiment to compute a low-order model G_{mod} and an uncertainty set $\mathcal{D}$; then design a new controller K that achieves closed-loop stability and meets the required performance with all models in $\mathcal{D}$, and hence with the unknown true system G_0.* If necessary, repeat this design procedure, possibly with a more demanding performance criterion. In most versions of this new scenario, one first computes a class of controllers $\mathcal{K}(G_{mod}, \mathcal{D})$ that all achieve the required performance with all models in $\mathcal{D}$; the new controller

K is then chosen within this class in such a way as to have some additional nice features (*e.g.*, low complexity).

The goal of the new scenario is to achieve the same or better performance based on models of lower complexity. The class of controllers $\mathcal{K}$ that achieve the required performance is larger because the model uncertainty set $\mathcal{D}$ is tuned towards that aim. All in all, the same or better performance is achieved with a controller that is easier to compute and of lower complexity than is possible with the traditional scenario.

In terms of the global design procedure, the main distinction between the traditional concept of robust control design and the new one is that, in the new scenario, the identification and validation steps have become part of the global control design procedure, whereas in the traditional scenario the control performance specifications played no role in the identification/validation step. In highlighting the distinction between the *identification* part of the design, which determines the nominal model G_{mod}, and the *validation* part of the design, which determines the validated uncertainty set $\mathcal{D}$, we also want to stress that the identification step focuses on the bias error distribution of the nominal model G_{mod}, whereas the validation step focuses on the variance error distribution.

The early work on the interplay between identification and robust control addressed almost exclusively the question of the nominal model. The question was: *how should the bias error of the nominal model be tuned so that the resulting controller based on this nominal model G_{mod} stabilises the unknown true system and achieves with this system a closed-loop performance that is close to the nominal closed-loop performance?*

In a nutshell, the answer to that question is that the nominal closed-loop system (G_{mod}, K) and the actual closed-loop system (G_0, K) must be "close", where the closeness is measured in a norm that is determined by the control performance criterion. A key point here is that the controller K is the controller to be designed, which itself depends on the model G_{mod}. This is what makes the problem difficult and results in the need for an iterative approach, in which a sequence of models and controllers are computed, achieving higher and higher performance. The difficulty, as already mentioned in Chapter 1, is that this procedure is not guaranteed to converge: higher performance may be achieved at the cost of lower stability margins, eventually leading to instability. Thus, such schemes must include safety checks to guarantee stability. These safety checks take the form of bounds on a measure of the distance between two successive controllers. They will be described in Section 9.4.

Even though a lot of progress has been made all through the 1990s on the development of new methods for computing uncertainty bounds on identified models, much of this work was developed independently of the control objective. Thus, the uncertainty bounds developed in this work are not "control-oriented", in the sense that their construction does not take account of the interplay between the validated set of models $\mathcal{D}$ and the set of robust controllers $\mathcal{K}$ that achieve a prescribed control performance with all models in $\mathcal{D}$.

One reason for the paucity of results on this aspect of the synergistic design problem is the difficulty of the problem. It is already hard to understand the interplay between the true system G_0, the present controller K_{id} (if any), the model G_{mod}, the designed controller K, the nominal performance $\bar{J}^{nom}$ and the achieved performance $\bar{J}^{ach}$, and to derive from this understanding the qualities that the model must possess for the nominal performance of the loop (G_{mod}, K) to be close to the achieved performance of the loop (G_0, K). It is much harder to understand the interplay between the true system G_0, the present controller K_{id} (if any), the set of validated models $\mathcal{D}$, and the set of admissible controllers $\mathcal{K}$ for which the worst-case performance with all models in the validated set $\mathcal{D}$ is acceptable.

Another reason is that, even if the interplay between validation and robust control were well understood, there would not be a unanimously accepted way of formulating a "validation for robust control" problem. In this chapter, we will review two reasonable formulations of such problem, but we are fully aware that, as the insight and the technical tools evolve, better formulations will almost certainly eventuate.

The contents of this chapter are as follows. In Section 9.2, we present the identification/control set-up, and we introduce some basic concepts from robust optimal control. Section 9.3 focuses on the interplay between true system, nominal model, achieved performance and nominal performanc. This leads to the formulation of control-oriented identification criteria, which have consequences on the shaping of the bias error distribution. In Section 9.4, we explain the need for caution in iterative model-based robust control design, and we show how such cautious steps can be implemented. Section 9.5 discusses the connection between the model uncertainty set and the corresponding controller set. Two approaches are discussed: one belongs to the realm of optimal experiment design with full-order models, the other belongs to the realm of robust experiment design. We end up with some conclusions in Section 9.6.

9.2 The Identification/Control Setup and some Basic Formulæ

For simplicity, we consider in this chapter only single-input single-output (SISO) linear time-invariant (LTI) systems, and we limit our presentation to one-degree-of-freedom controllers. The "true system" is assumed to be represented by

$$\mathcal{S}: y_t = G_0(z)u_t + v_t \tag{9.1}$$

where $G_0(z)$ is a linear time-invariant causal operator, y is the measured output, u is the control input, and v is noise, assumed to be quasi-stationary.

The control law is represented by

$$u_t = K(z)[r_t - y_t] + d_t \tag{9.2}$$

where r_t is the reference excitation and d_t is a possible disturbance acting on the system. The signal d_t can also be seen as an error between the computed control action and the actually applied control action.

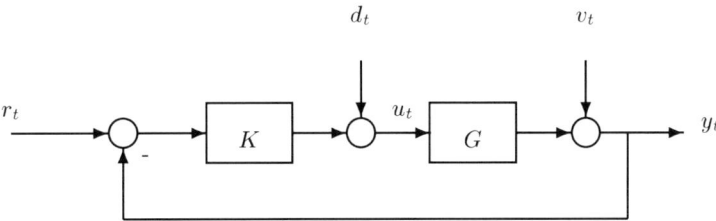

Fig. 9.1. The generic feedback configuration

Our generic feedback loop can thus be represented as in Figure 9.1. The equations of this closed-loop system can be written as:

$$y_t = \frac{GK}{1+GK}r_t + \frac{G}{1+GK}d_t + \frac{1}{1+GK}v_t \tag{9.3}$$

$$u_t = \frac{K}{1+GK}r_t + \frac{1}{1+GK}d_t - \frac{K}{1+GK}v_t \tag{9.4}$$

In analysing the identification/control interplay, it will be useful to consider – and compare – the following special cases of this feedback configuration:

- assume that, at the time of designing a new controller K, the true system G_0 is already under feedback control with a controller denoted by K_{id}. We then consider the *current feedback system* (G_0, K_{id}), on which data are typically collected in order to identify a new model G_{mod} for the design of a new controller K;
- with $G = G_{init}$, $v_t = \hat{v}_t$ and $K = K_{id}$ we have, correspondingly, the *current design loop* (G_{init}, K_{id}). Here, G_{init} is the present model of G_0, if available, while $\hat{v}_t = H_{init}e_t$, with e_t white noise, is the present noise model if available;
- with $G = G_{mod}$ (the newly-identified I/O model) and $v_t = \hat{v}_t = H_{mod}e_t$ (the newly-identified noise model, if any) we have the *nominal closed-loop system* (G_{mod}, K), where K is the new controller designed from G_{mod};
- the new controller K, designed from G_{mod}, is applied to the true system G_0, thus generating the *achieved closed-loop system* (G_0, K). If the design is robust, then the performance of this achieved closed-loop system (G_0, K) must be reasonably close to that of the nominal closed-loop system (G_{mod}, K).

For the analysis of both the stability and the performance of this closed-loop system (G, K), a key role is played by the closed-loop transfer function matrix $T(G, K)$ defined as follows:

$$T(G,K) = \begin{pmatrix} T_{11} & T_{12} \\ T_{21} & T_{22} \end{pmatrix} = \begin{pmatrix} \frac{GK}{1+GK} & \frac{G}{1+GK} \\ \frac{K}{1+GK} & \frac{1}{1+GK} \end{pmatrix} \tag{9.5}$$

It is the transfer function matrix from the external signals r and d to the loop signals y and u.

The closed-loop system (G, K) is stable (or "internally stable") if all four transfer functions in $T(G, K)$ are stable. This notion of stability guarantees that there are no unstable pole–zero cancellations in any of the paths from the external signals r, d, v to the internal signals y, u in the feedback system of Figure 9.1. This can be conveniently expressed in mathematical terms using the *generalised stability margin* $b(G, K)$ introduced by Vinnicombe [153], see Chapter 4.

The generalised stability margin of the closed-loop system (G, K) is defined as

$$b_{GK} = \begin{cases} \|T(G,K)\|_\infty^{-1} & \text{if } [G\ K] \text{ is stable} \\ 0 & \text{otherwise} \end{cases} \tag{9.6}$$

Note that this generalised stability margin takes its values between 0 and 1; the higher this value, the better the stability margin. It is also important to note that, for a given plant G, and whatever the linear controller, the generalised stability margin has a maximum value $b_{opt}(G)$ (see, e.g., [163]) given by

$$b_{opt}(G) = \sup_K b(G,K) = \sqrt{1 - \| [N\ M] \|_H^2}, \tag{9.7}$$

where $\| A \|_H$ is the Hankel norm of the operator A and $\{N, M\}$ is the normalised coprime factorisation of G; see Chapter 4 or [164] for the definitions of these concepts.

As for the performance of the closed-loop system (G, K), most of the commonly used performance criteria are defined from the following general frequency function:

$$J(G,K,W_l,W_r,\omega) = \sigma_1 \left(\overbrace{\begin{pmatrix} W_{l1} & 0 \\ 0 & W_{l2} \end{pmatrix}}^{W_l} T(G(e^{j\omega}), K(e^{j\omega})) \overbrace{\begin{pmatrix} W_{r1} & 0 \\ 0 & W_{r2} \end{pmatrix}}^{W_r} \right) \tag{9.8}$$

where $W_{l1}(e^{j\omega}), W_{l2}(e^{j\omega})$ and $W_{r1}(e^{j\omega}), W_{r2}(e^{j\omega})$ are frequency weights that allow one to define specific performance levels, and where $\sigma_1(A)$ denotes the largest singular value of A.

The frequency function $J(G, K, W_l, W_r, \omega)$ defines a template. Any function that is derived from J can, of course, also be handled. Thus, a typical optimal control problem formulation is

$$K^* = \arg\min_K \|J(G, K, W_l, W_r, \omega)\| \tag{9.9}$$

where $\| \ \|$ denotes a suitable norm. See, for example, [46] where $\| W_l T(G, K) W_r \|_\infty$ is used. The corresponding optimal cost will be denoted by

$$\bar{J}(G, K^*) = \| W_l T(G, K^*) W_r \| \tag{9.10}$$

The choice of a diagonal structure for W_l and W_r is no loss of generality, since the four transfer functions in $T(G, K)$ can all be weighted differently. For example, a common choice for the performance measure of a closed-loop system is the shape of the modulus of the frequency response of one or several of the four transfer functions defined in (9.5) (see [158]) or the ∞-norm of a frequency weighted version of one of these four transfer functions.

The optimal control problem formulation defined by (9.9) produces a controller K^* that is optimal, with respect to the control objective, for a specific system G that is assumed to be known. The controller K^* that is optimal for G may yield very poor performance with a system G_0 that is very close to G; worse still, it could destabilise G_0. This point was put very strongly in the much-studied Schrama example [136].

The whole paradigm of robust optimal control is to formulate control design objectives that deliver a controller that has a guaranteed level of performance with all systems in a given uncertainty set $\mathcal{D}$. For the formulation of such problems, one defines a *worst-case performance function*. Thus, (9.8) is replaced by

$$J_{WC}(\mathcal{D}, K, W_l, W_r, \omega) = \tag{9.11}$$

$$\max_{G(z) \in \mathcal{D}} \sigma_1 \left(\overbrace{\begin{pmatrix} W_{l1} & 0 \\ 0 & W_{l2} \end{pmatrix}}^{W_l} T(G(e^{j\omega}), K(e^{j\omega})) \overbrace{\begin{pmatrix} W_{r1} & 0 \\ 0 & W_{r2} \end{pmatrix}}^{W_r} \right)$$

A robust optimal control problem can then be formulated as

$$K^* = \arg\min_K \|J_{WC}(\mathcal{D}, K, W_l, W_r, \omega)\| \tag{9.12}$$

where $\| \ \|$ denotes a suitable norm. The corresponding optimal worst case cost is then denoted by

$$\bar{J}_{WC}(\mathcal{D}, K^*) = \|J_{WC}(\mathcal{D}, K^*, W_l, W_r, \omega)\| \tag{9.13}$$

We now show that the formulation of optimal control problems in the form of (9.9), based on a unique model G (rather than a set $\mathcal{D}$), results in optimal controllers that already possess an inherent stability robustness property. This is a feature of a control problem formulation that is based on the whole 2×2 transfer function $T(G, K)$ rather than on only one of its elements. As already stressed in Chapter 4, the main difference is that such control problem formulation is based on more than just the loop gain $L = GK$.

Suppose that a model $G_{mod} = \frac{N_{mod}}{D_{mod}}$ of a plant is given, where the stable causal transfer functions N_{mod} and D_{mod} are the normalised coprime factors of G_{mod}: see Chapter 4. Then, from a robust stability point of view, it makes sense to design a controller K that minimizes $||T(G_{mod}, K)||_\infty$. Indeed, as explained in Chapter 4, the optimising controller K^* maximises the stability robustness to errors in the coprime factors N_{mod} and D_{mod} of the model G_{mod}. More precisely, let

$$K^* = \arg\min ||T(G_{mod}, K)||_\infty \qquad (9.14)$$

and let $\gamma^* = ||T(G_{mod}, K^*)||_\infty$. Then, the closed-loop systems $(G_{mod} + \Delta G, K^*)$ are stable for all systems $G_{mod} + \Delta G = \frac{N_{mod}+\Delta N}{D_{mod}+\Delta D}$ such that

$$\left\| \begin{pmatrix} \Delta N \\ \Delta D \end{pmatrix} \right\|_\infty < \frac{1}{\gamma^*} \qquad (9.15)$$

Thus, we observe that the maximisation, with respect to a controller K, of the stability robustness to coprime factor uncertainty on a nominal model $G_{mod} = \frac{N_{mod}}{D_{mod}}$ can be formulated as the maximisation of a specific version of the performance criterion defined in (9.9), with $W_l = W_r = I$.

9.3 A Control-oriented Nominal Model

In this section, we show how the control performance objective leads to the design of a control-oriented identification criterion, whose minimisation delivers a control-oriented nominal model. As explained in the introduction of this chapter (see also Chapter 1), we assume that the model set used for identification is a restricted complexity model set $\{G(z, \theta)\}$ parameterised by some parameter vector θ, i.e., there is no θ_0 such that $G_0(z) = G(z, \theta_0)$. As a result, the model obtained by identification, $G_{mod} \triangleq G(z, \hat{\theta})$, is necessarily biased, and the focal point of the discussion in this section is therefore on the control-oriented shaping of the bias distribution of $G(z, \hat{\theta})$.

The problem can be formulated as follows. Since we shall construct our controller K on the basis of a model G_{mod} that is biased (i.e., necessarily wrong), how should we formulate the criterion in such a way that the frequency distribution of these bias errors has the smallest possible impact on the closed-loop *performance degradation*, while at the same time guaranteeing stability of the achieved closed-loop system (G_0, K). By performance degradation, we mean the difference between the performance of the nominal closed-loop system (G_{mod}, K) and the performance of the actual closed-loop system (G_0, K).

Recall first that, if we knew the true system G_0, we would compute the optimal controller as the solution of

$$K^* = \arg \min_K \| W_l T(G_0, K) W_r \| \qquad (9.16)$$

where $\| \; \|$ denotes some suitable norm, such as the $\mathcal{H}_\infty$-norm or the $\mathcal{H}_2$-norm for example. Now, one can write:

$$W_l T(G_0, K) W_r = W_l T(G_{mod}, K) W_r + W_l [T(G_0, K) - T(G_{mod}, K)] W_r \qquad (9.17)$$

By applying the triangle inequality, Schrama showed that one can then squeeze the achieved cost $\| W_l T(G_0, K) W_r \|$ between the following lower and upper bounds [136]:

$$\begin{aligned} & | \; \| W_l T(G_{mod}, K) W_r \| - \| W_l [T(G_0, K) - T(G_{mod}, K)] W_r \| \; | \\ & \leq \| W_l T(G_0, K) W_r \| \\ & \leq \| W_l T(G_{mod}, K) W_r \| + \| W_l [T(G_0, K) - T(G_{mod}, K)] W_r \| \end{aligned} \qquad (9.18)$$

As stated above, the ideal (but elusive) goal would be to compute the controller that minimises the control performance objective $\bar{J}^{ach}(G_0, K)$ on the actual system. Since G_0 is unknown, this is replaced by a model-based control design, where one computes the controller K^* that minimises the nominal performance objective:

$$K^* = \arg \min_K \| W_l T(G_{mod}, K) W_r \| . \qquad (9.19)$$

This results in the nominal cost $\bar{J}^{nom}(G_{mod}, K^*)$.

The double triangle inequality (9.18) shows that the achieved cost $\bar{J}^{ach}(G_0, K^*)$ will be close to the nominal cost $\bar{J}^{nom}(G_{mod}, K^*)$ if the performance degradation term

$$\bar{J}^{pr} = \| W_l [T(G_0, K^*) - T(G_{mod}, K^*)] W_r \| \qquad (9.20)$$

is small. This observation is at the basis of all "identification for control" schemes developed in the early part of the 1990s. It shows that what matters for a model G_{mod} to be tuned for control design is that the closed-loop transfer functions $T(G_0, K^*)$ and $T(G_{mod}, K^*)$ must be close to one another in a norm that is entirely determined by the control performance criterion. Indeed, the way in which $T(G_0, K^*)$ and $T(G_{mod}, K^*)$ must be close is exactly determined by the requirement that $\bar{J}^{pr}$ must be small; this term has often been called the *performance robustness* term: see, *e.g.*, [60]. If an $\mathcal{H}_\infty$-norm is chosen for the control performance criterion, then a model G_{mod} that is tuned for control design is one for which $\| W_l [T(G_0, K^*) - T(G_{mod}, K^*)] W_r \|_\infty$ is minimised; if an $\mathcal{H}_2$-norm is used in the control criterion, then a good model for control should minimise $\| W_l [T(G_0, K^*) - T(G_{mod}, K^*)] W_r \|_2$.

Important observations

The statements made above on the basis of the double triangle inequality require some important observations and cautionary notes.

1. We note that the estimated plant model, G_{mod}, and the controller, K^*, both influence the two terms $\bar{J}^{nom}$ and $\bar{J}^{pr}$. Thus, ideally, one should minimise the two terms jointly over the class of admissible plant models and admissible controllers. This is an impossible task in the case of restricted complexity models.[1]

2. On the other hand, minimising the nominal criterion $\bar{J}^{nom}(G_{mod}, K)$ with respect to the controller for a given model G_{mod} is a classical control design task, whereas minimizing $\bar{J}^{pr}(G, K)$ with respect to a model G for a given controller K was shown to be achievable by closed-loop identification. Therefore, an obvious suboptimal strategy is to make $\bar{J}^{nom}$ small by controller design for a given plant model, and to keep $\bar{J}^{pr}$ small by identification design for a given controller. Since $\bar{J}^{nom}$ depends on the estimated plant model, and $\bar{J}^{pr}$ depends on the designed controller, this strategy can only be applied in an iterative manner, using a succession of *local controller designs* and *local identification designs*:[2]

$$\min_K \bar{J}(G_{mod}^i, K) \longrightarrow K_{i+1}$$
$$\min_{G(\theta)\in\mathcal{M}} \bar{J}^{pr}(G(\theta), K_i) \longrightarrow G_{mod}^{i+1} \tag{9.21}$$

This idea is at the heart of the iterative identification/controller design methods developed in the early and mid 1990s [102, 136, 139, 161].

3. One important technical problem raised by these iterative identification and control schemes is that the reasoning developed on the basis of the double triangle equality assumes that the controller K is identical in all expressions. However, in the iterative schemes, G_{mod} is obtained by minimising $\| W_l[T(G_0, K_i) - T(G_{mod}, K_i)]W_r \|$ where K_i is the presently operating controller, while the new controller $K_{i+1} = K^*$ is computed from this new model G_{mod}. Hence, the controller K_i that appears in the nominal performance term $\| W_l T(G_{mod}, K_i) W_r \|$ of the triangle inequality (9.18) is not the same as the controller K_{i+1} that appears in the robust performance term $\| W_l[T(G_0, K_{i+1}) - T(G_{mod}, K_{i+1})]W_r \|$.

---

[1] In dual control, the achieved criterion is minimised jointly over the parameterised set of plant models and corresponding controllers, but the model set is assumed to contain the true system, and the minimisation leads to a tractable solution only for the very simple minimum variance control criterion: see [13].

[2] The term 'local' refers to the fact that, at each iteration, the controller design (conversely, the identification design in next phase) is performed on the basis of some present (*i.e.*, local) plant model (conversely, on the presently operating (*i.e.*, local) controller).

Thus, for the reasoning to apply, it is required that $\| W_l[T(G_0, K_{i+1}) - T(G_{mod}, K_{i+1})]W_r \|$ is very close to $\| W_l[T(G_0, K_i) - T(G_{mod}, K_i)]W_r \|$. This technical problem has generated a lot of further research, whose main result has been to introduce "caution" in the controller adjustment from K_i to K_{i+1} in such a way as to make these two quantities close to one another. A key technical tool for establishing robust stability and robust performance guarantees using caution has been the ν-gap metric introduced by Vinnicombe [153], already presented in Chapter 4, and its related robust stability and robust performance results. The most obvious use of the ν-gap is to measure a distance between two plants, say G_0 and G_{mod}, and to establish stabilisation of G_0 by some controller K based on stabilisation of G_{mod} by K and a bound on the distance $\delta_\nu(G_0, G_{mod})$ between the two systems. However, by duality, the ν-gap can also be used to measure the distance between two successive controllers, say K_i and K_{i+1}, and to establish stabilisation of G_0 by K_{i+1} on the basis of stabilisation of G_0 by K_i and a bound on the distance $\delta_\nu(K_{i+1}, K_i)$ between the two controllers. In the next section, we shall present some of these technical results and explain how they can be used to obtain prior stability guarantees when moving between successive controllers.

Before we conclude this section, we comment on the identification part of the iterative schemes. These have been formulated above as

$$G_{mod} = \arg \min_{G \in \mathcal{M}} \| W_l[T(G_0, K_i) - T(G, K_i)]W_r \| \qquad (9.22)$$

where $\mathcal{M} \triangleq \{G(z, \theta), \theta \in \mathcal{D}_\theta\}$ is a model set chosen by the user, typically a reduced order model set. Whatever the specific norm that is used, this is not a standard identification criterion. As it turns out, it has been shown that such a criterion can be minimised by closed-loop identification, with the controller K_i operating on the actual system. Several schemes have been proposed, for specific choices of norm: see, e.g., [46, 102, 113, 137, 139, 161]. An application to the control of a sugar cane crushing mill can be found in Chapter 13, while the servo design of a compact disc player based on control-relevant identification is described in Chapter 11. The need for a better control over the bias error in closed-loop identification spurred a renewed interest in closed-loop identification, leading to several new schemes. An excellent survey can be found in [148].

The fact that the identification for control criteria, in iterative identification and control, turned out to be closed-loop criteria with the most recent controller in the loop, was of course of major practical significance and explains why these schemes were quickly adopted in industry. For a process control engineer, it is indeed a lot more appealing to collect identification data under normal operating conditions than to have to perform special identification experiments, or worse still open the loop.

9.4 Caution in Iterative Design

We first present some results, based on the Vinnicombe ν-gap, that form the basis for many of the stability robustness guarantees that can be achieved in iterative identification and robust control design.

Proposition 9.1 ([155]). *Consider a plant G and two controllers K_1 and K_2, with K_1 stabilising G. Then*

1. *(G, K_2) is stable for all controllers K_2 satisfying $\delta_\nu(K_1, K_2) \leq \beta$ if and only if $b(G, K_1) > \beta$.*
2. *If $\delta_\nu(K_2, K_1) < 1$, then*

$$\arcsin b(G, K_2) \geq \arcsin b(G, K_1) - \arcsin \delta_\nu(K_1, K_2), \tag{9.23}$$

and

$$\delta_\nu(K_1, K_2) \leq \|T(G, K_1) - T(G, K_2)\|_\infty \leq \frac{\delta_\nu(K_1, K_2)}{b(G, K_1)b(G, K_2)} \tag{9.24}$$

$$= \frac{\delta_\nu(K_1, K_2)}{b(G, K_1)} \|T(G, K_2)\|_\infty$$

The importance of Part 1 of Proposition 9.1 is that it provides a sufficient condition on a new controller, K_2, that guarantees its stabilisation of the plant G:

$$\delta_\nu(K_2, K_1) < b(G, K_1) \tag{9.25}$$

The norm $\|T(G, K)\|_\infty = b^{-1}(G, K)$ is one measure of the closed-loop performance of the (G, K) loop. One upshot of these results is the intuitive property that a well-behaved controller, as measured by a small $\|T(G, K)\|_\infty$, provides greater scope for variation before striking stability or performance guarantee barriers. The following proposition is the exact dual of Proposition 9.1.

Proposition 9.2. *Consider the plants G_1 and G_2, and a controller K stabilising G_1. Then*

1. *(G_2, K) is stable for all plants G_2 satisfying $\delta_\nu(G_1, G_2) \leq \beta$ if and only if $b(G_1, K) > \beta$.*
2. *If $\delta_\nu(G_1, G_2) < 1$, then*

$$\arcsin b(G_2, K) \geq \arcsin b(G_1, K) - \arcsin \delta_\nu(G_1, G_2), \tag{9.26}$$

and

$$\delta_\nu(G_1, G_2) \leq \|T(G_1, K) - T(G_2, K)\|_\infty \leq \frac{\delta_\nu(G_1, G_2)}{b(G_1, K)b(G_2, K)} \tag{9.27}$$

$$= \frac{\delta_\nu(G_1, G_2)}{b(G_1, K)} \|T(G_2, K)\|_\infty$$

We also have the following result, derived in [6].

Proposition 9.3. *Consider two stable closed-loop plant–controller pairs* (G_1, K) *and* (G_2, K). *Then*

$$|b(G_1, K) - b(G_2, K)| \leq \delta_\nu(G_1, G_2). \tag{9.28}$$

Hence

$$|b(G_1, K) - b(G_2, K)| \leq \delta_\nu(G_1, G_2) \leq \|T(G_1, K) - T(G_2, K)\|_\infty$$
$$\leq \frac{\delta_\nu(G_1, G_2)}{b(G_1, K) b(G_2, K)}. \tag{9.29}$$

Expression (9.29) shows that the distance between the stability measures $b(G_1, K)$ and $b(G_2, K)$ is always smaller than the distance between the closed-loop transfer functions $T(G_1, K)$ and $T(G_2, K)$, measured in $\mathcal{H}_\infty$-norm. We have shown in Section 9.3 that, in identification for control, one computes the new model by minimising some norm of $\| T(G_0, K) - T(G, K) \|$, where K is the presently operating controller. If the ∞-norm is used, this implies that the stability margin of the new nominal closed-loop system (G_{mod}, K) will be close to the stability margin of the actual closed-loop system (G_0, K). We shall see that this is a key feature that will allow us to obtain prior stability guarantees.

9.4.1 Cautious Controller Adjustment

Consider now that, at some stage of an iterative model/controller design, we have arrived at a controller K_i that stabilises both the true plant G_0 and the present model G_i, and that we have computed a new model G_{i+1} by minimising some norm of $\| G_0 - G_{i+1} \|$ over a given model set. Assume also that G_{i+1} is stabilised by the present controller K_i; we shall see later how this can be guaranteed. We also consider that our control objective takes the generic form described in Section 9.2:

$$K^* = \arg\min_K \bar{J}(G_{i+1}, K), \quad \text{where} \tag{9.30}$$

$$\bar{J}(G_{i+1}, K) = \| W_l T(G_{i+1}, K) W_r \| \tag{9.31}$$

Assume now that we have an accurate estimate $\hat{b}(G_0, K_i)$ of the stability margin $b(G_0, K_i)$ of the present closed-loop system, *i.e.*, such that

$$|\hat{b}(G_0, K_i) - b(G_0, K_i)| \leq \epsilon \tag{9.32}$$

$$\Rightarrow b(G_0, K_i) > \hat{b}(G_0, K_i) - \epsilon = k\hat{b}(G_0, K_i) \text{ for some } k \in (0, 1) \tag{9.33}$$

Then, any controller K_{i+1} such that

$$\delta_\nu(K_{i+1}, K_i) < k\hat{b}(G_0, K_i) \tag{9.34}$$

is guaranteed to stabilise the true plant G_0, since K_i stabilises G_0. Thus, the new controller K_{i+1} stabilises the true plant if the adjustment from K_i to K_{i+1} is small enough, as expressed by constraint (9.34).

9.4.2 Estimation of a Bound on the Stability Margin $b(G_0, K_i)$

There are several ways of estimating the stability margin $b(G_0, K_i)$ together with a bound on its estimation error. One way is to compute a direct estimate of $\| T(G_0, K_i) \|_\infty = b^{-1}(G_0, K_i)$ from measurements on the actual closed-loop system (G_0, K_i) [24], where the selection of appropriate external excitation signals for this task is also discussed. The inverse of the estimate of $\| T(G_0, K_i) \|_\infty$ is then used as the estimate $\hat{b}(G_0, K_i)$ in (9.33), and a safety factor $k \in (0, 1)$ is introduced to account for the estimation error on $\hat{b}(G_0, K_i)$.

An alternative way is to use an H_∞ identification method for the estimation of G_{i+1} from measurements obtained on the actual closed-loop system (G_0, K_i): see, e.g., [67, 113]. These methods also deliver a bound on the ∞-norm error, i.e., they deliver a number ϵ such that

$$\| T(G_0, K_i) - T(G_{i+1}, K_i) \|_\infty \leq \epsilon \tag{9.35}$$

By Proposition 3 it then follows that

$$|b(G_0, K_i) - b(G_{i+1}, K_i)| \leq \epsilon \tag{9.36}$$
$$\Rightarrow b(G_0, K_i) > b(G_{i+1}, K_i) - \epsilon = kb(G_{i+1}, K_i) \text{ for some } k \in (0, 1) \tag{9.37}$$

Observe that $b(G_{i+1}, K_i)$ is known, since G_{i+1} and K_i are known. One can thus use $\hat{b}(G_0, K_i) = b(G_{i+1}, K_i)$ as an estimate of $b(G_0, K_i)$ in (9.33), with an error bounded by ϵ, which can again be accounted for by a safety factor $k \in (0, 1)$.

To summarise, if G_{i+1} is a new model obtained from closed-loop data on the present (G_0, K_i) loop such that G_{i+1} is stabilised by the present K_i and such that (9.35) holds, then any controller K_{i+1} such that

$$\delta_\nu(K_{i+1}, K_i) < kb(G_{i+1}, K_i) \tag{9.38}$$

with k defined as in (9.37), is guaranteed to stabilise the true plant G_0.

9.4.3 Construction of a Stabilising Controller

Condition (9.34) defines a set of controllers $\{K_{i+1}\}$ that are guaranteed to stabilise G_0, but it says nothing about how to construct such controllers. In addition, most of the controllers in that set may not achieve a good performance with G_0. We now present a procedure for the construction of controllers that satisfy the stability condition (9.34) and, at the same time, achieve better performance than was achieved with the previous controller K_i.

Consider the presently operating controller K_i, which stabilises both the unknown plant G_0 and the new model G_{i+1}. Suppose that we are given a known bound on the differences between the closed-loop transfer matrices

$$\|T(G_0, K_i) - T(G_{i+1}, K_i)\|_\infty \leq \epsilon \tag{9.39}$$

Let $K_i = U_i V_i^{-1}$ be a right coprime factorisation of K_i, with U_i and V_i stable proper transfer functions and, similarly, let $G_{i+1} = N_{i+1} D_{i+1}^{-1}$ be a right coprime factorisation of the new model G_{i+1}. Then, the set of all controllers K_{i+1} that stabilise the model G_{i+1} and that also stabilise all G_0 satisfying Condition (9.39) is given by

$$\mathcal{K} = \{K(Q) : K(Q) = (U_i - D_{i+1} Q)(V_i + N_{i+1} Q)^{-1}, Q \in \mathcal{RH}_\infty, \| Q \| \leq \epsilon^{-1}\} \tag{9.40}$$

Here, Q also belongs to the set of stable proper transfer functions. This result was established in [24]. Observe that, without the constraint on the norm of Q (i.e., $\| Q \| \leq \epsilon^{-1}$), the set $\mathcal{K}$ is the set of all controllers stabilising the new model G_{i+1}. By imposing the additional constraint on $\| Q \|$, we are limiting the distance between the old controller K_i and the new controller K_{i+1}, which is what guarantees the stabilisation of the true plant G_0.

Now, consider that the control objective $\bar{J}(G_{i+1}, K)$ of (9.30) depends in a convex manner on Q, as is the case in H_2 and H_∞ optimal control problems, and let $K_{i+1}^* = K(Q^*)$ be the solution of the unconstrained optimisation problem:

$$K_{i+1}^* = K(Q^*) = \arg\min_Q \bar{J}(G_{i+1}, K(Q)) \tag{9.41}$$

If $\| Q^* \| \leq \epsilon^{-1}$, we can take $K_{i+1} = K_{i+1}^* = K(Q^*)$. Otherwise, consider the set:

$$K(\alpha Q^*) = (U_i - \alpha D Q^*)(V_i + \alpha N Q^*)^{-1}\}, \text{ for } \alpha \in [0, 1] \tag{9.42}$$

All controllers in that set stabilise G_{i+1}. The value $\alpha = 0$ corresponds to the present controller K_i, while $\alpha = 1$ corresponds to $K_{i+1}^* = K(Q^*)$, which is not guaranteed to stabilise G_0. Let α^* be the largest value for which $\| \alpha Q^* \| < \epsilon^{-1}$. Then, the controller $K_{i+1} = K(\alpha^* Q^*)$ is guaranteed to stabilise G_0 and, in addition, we have [7]:

$$\bar{J}(G_{i+1}, K(Q^*)) \leq \bar{J}(G_{i+1}, K(\alpha^* Q^*)) < \bar{J}(G_{i+1}, K_i) \tag{9.43}$$

9.4.4 Cautious Model Adjustment

We have seen above that one of the ways of imposing a cautious controller adjustment, in order to guarantee closed-loop stability of the loop (G_0, K_{i+1}), is to limit the controller movement with respect to the nominal stability margin $b(G_{i+1}, K_i)$: see (9.38). This, of course, requires that the nominal loop formed by the present controller and the new model is stable. A convenient way to ensure that the identified model G_{i+1} is stabilised by the present controller is to parameterise the model set in the dual Youla parameterisation as follows.

Proposition 9.4. *Consider the present controller K_i that stabilises both the true system G_0 and the present model G_i. Let $K_i = U_i V_i^{-1}$ be a right coprime factorisation of K_i, with U_i and V_i stable proper transfer functions and, similarly, let $G_i = N_i D_i^{-1}$ be a right coprime factorisation of the present model G_i. Then the set of all models G_{i+1} that are stabilised by the controller K_i is given by*

$$\mathcal{M} = \{G(R) : G(R) = (N_i - V_i R)(D_i + U_i R)^{-1}, R \in \mathcal{RH}_\infty\} \qquad (9.44)$$

The subset of such models that also satisfies the condition

$$\|T(G_i, K_i) - T(G_{i+1}, K_i)\|_\infty \leq \epsilon \qquad (9.45)$$

is given by

$$\mathcal{M} = \{G(R) : G(R) = (N_i - V_i R)(D_i + U_i R)^{-1}, R \in \mathcal{RH}_\infty, \| R \| \leq \epsilon^{-1}\}$$

Thus, not only can one perform the identification in such a way that the new model is stabilised by the present controller K_i, but one can also impose that the new nominal closed-loop system, (G_{i+1}, K_i), is close to the old one, (G_i, K_i), in the sense of Condition (9.45).

9.5 Model Validation for Robust Control Design

So far, we have mostly discussed the interplay between the true system, G_0, the nominal model, G_{mod}, the controller present during identification, K_{id}, and the controller to be designed, K. In an iterative design, this corresponds to the interplay between the true system, G_0, the successive models, G_i and G_{i+1}, and the successive controllers, K_i and K_{i+1}. We have justified the use of iterative models/controllers on the basis of the control performance objective, and we have justified the need for caution on the basis of stability guarantees.

In this section, we turn to the interplay between validated model uncertainty sets (which, typically, can be obtained by identification), and corresponding controller sets. As explained in the introduction of this chapter, the analysis of this interplay is much harder. In addition, different views can be taken on what constitutes a model uncertainty set that is "tuned for robust control design". Essentially, these different views correspond to different ways of connecting a model uncertainty set $\mathcal{D}$ and a corresponding controller set $\mathcal{K}$. We shall present here two different streams of thought and, correspondingly, two different types of results.

9.5.1 Optimal Experiment Design for Certainty Equivalence Control

In this approach, an identification experiment design is called "optimal" if the controller computed from the estimated model is one that minimises the performance degradation vis-à-vis the performance that would be achieved with the ideal controller. The ideal controller is the controller that would be computed if the true system were known. In a stochastic prediction-error identification setting, the model is a random object and hence, so is the controller; thus, it is the average performance degradation that is minimised with respect to the experiment design.

To make things more precise, consider that the true unknown system is G_0 (or (G_0, H_0) when a noise model is also used in the controller design), that the performance criterion is denoted by $\bar{J}(G_0, H_0, K)$, and that the optimal controller for the true system is

$$K^* = \arg\min_K \bar{J}(G_0, H_0, K) \tag{9.46}$$

This last expression defines a certainty equivalence mapping $K^* = k(G_0, H_0)$ from the system (G_0, H_0) to the optimal controller K^*. Suppose now that an identification experiment with N data leads to a model $(\hat{G}_N, \hat{H}_N)$ of (G_0, H_0). The same mapping then defines a controller $\hat{K}_N = k(\hat{G}_N, \hat{H}_N) = \arg\min_K \bar{J}(\hat{G}_N, \hat{H}_N, K)$. As a result of the application of $\hat{K}_N$ to the real system, rather than K^*, the achieved cost will be $\bar{J}(G_0, H_0, \hat{K}_N)$, being greater or equal to $\bar{J}(G_0, H_0, K^*)$. This results in a "performance degradation"

$$\bar{J}^{deg} = \bar{J}(G_0, H_0, \hat{K}_N) - \bar{J}(G_0, H_0, K^*) \tag{9.47}$$

For a given identification experiment design, every set of N data delivers a different model $(\hat{G}_N, \hat{H}_N)$, due to the particular noise realisation, and hence a different controller $\hat{K}_N$. In a worst-case identification framework, the optimal design problem would consist of finding the experiment design that minimises the worst case performance degradation, over all possible controllers $\hat{K}_N = k(\hat{G}_N, \hat{H}_N)$, that can result from such an identification experiment. In a stochastic framework, the model $(\hat{G}_N, \hat{H}_N)$ estimated from N random data is a random variable that depends on the particular noise realisation that was present in the data. Therefore, so is the controller $\hat{K}_N$, as well as the performance degradation $\bar{J}^{deg}$ defined in (9.47). In a stochastic framework, the problem statement in optimal identification design for control, in the context described here, can therefore be phrased as follows: "Find the experimental conditions $\mathcal{X}$ that minimise the average performance degradation"[3]. In view

---

[3] We denote by $\mathcal{X}$ the set of all admissible experimental conditions that have an effect on the quality of the model estimates $(\hat{G}_N, \hat{H}_N)$, such as use of open-loop or closed-loop data, choice of input spectrum distribution, of a regulator in the case of closed-loop identification, *etc*. By "admissible" experimental conditions, we refer to experimental conditions that obey possible constraints on the signal powers or signal energies.

of (9.47), this can be formulated as follows:

$$\min_{\mathcal{X}} E\bar{J}(G_0, H_0, \hat{K}_N) = \min_{\mathcal{X}} E\bar{J}(G_0, H_0, k(\hat{G}_N, \hat{H}_N)) \tag{9.48}$$

The expected value is taken with respect to the noise, which affects the model estimate, and hence the controller estimate.

In the case of certainty equivalence control design, this is probably the most logical (and ideal) problem formulation for an *optimal identification for control design*: one searches for the experimental conditions that will deliver models $(\hat{G}_N, \hat{H}_N)$ and hence controllers $\hat{K}_N = k(\hat{G}_N, \hat{H}_N)$ which, on average, produce the least possible degradation in achieved performance vis-à-vis the optimal performance. However, there are several difficulties with this formulation:

- as with all optimal experiment design problems, the optimal experiment for the estimation of some unknown quantity depends on this unknown quantity. Thus, the optimal experiment $\mathcal{X}$ defined by (9.48) necessarily depends on the unknown system (G_0, H_0). This does not mean that such results are meaningless: they give very useful guidelines for the identification design, and they can sometimes lead to iterative schemes that converge to the optimal experiment design: see, *e.g.* [77];
- there are no known results for the problem formulated in (9.48) for the case where the control performance objective takes the general form defined in this chapter as $\bar{J}(G_0, K) = \| W_l T(G_0, K) W_r \|$. However, there are solutions for the case where the control objective reduces to minimum variance control, as well as for less ambitious choices of optimal experiment design criteria. These are briefly reviewed below;
- the solutions obtained so far are all based on the assumption that the system is in the model set, *i.e.*, they are based on variance errors only. In addition, most of the practical results obtained so far are based on the use of the asymptotic formulae for the covariance of the transfer function estimates derived in [107]. These formulae for $cov\begin{pmatrix}\hat{G}_N(e^{j\omega})\\\hat{H}_N(e^{j\omega})\end{pmatrix}$ are only approximate.

The first instance in which an optimal identification for control problem was formulated in the framework described above was in [63]. That paper formulated the problem in the more general framework of optimal design with respect to the intended model application. A special case treated in that paper was where the intended application is *minimum-variance* control design. The performance degradation measure was then, very naturally, defined as the degradation in the output variance due to the use of a model-based controller that is not the optimal controller:

$$\bar{J}_{MV} = E[y_t^{opt} - y_t^N]^2 \tag{9.49}$$

Here, y^{opt} denotes the output of the optimal system (G_0, H_0, K^*), while y_t^N denotes the output of the system $(G_0, H_0, \hat{K}_N)$ whose controller results from the identified model, *i.e.*, $\hat{K}_N = k(\hat{G}_N, \hat{H}_N)$. Both closed-loop systems are assumed to be driven by the same external reference and noise signals. The results showed that, for this application of identification to minimum-variance control design, the optimal experiment consists of performing closed-loop identification with the unknown optimal minimum-variance controller in the loop.

The results of [63] have been extended to address optimal identification experiment design in connection with a range of control design problems other than minimum- variance control [58, 77]. However, instead of formulating the optimal identification design problem as a problem of minimising the degradation with respect to the optimal cost, it has been formulated in terms of minimising the simpler Criterion (9.49):

$$\min_{\mathcal{X}} E[y_t^{opt} - y_t^N]^2 \tag{9.50}$$

The major advantage of the formulation (9.50) is that the variance of the error signal $y_t^{opt} - y_t^N$ can be expressed as a frequency-weighted integral of the controller variance $E|\hat{K}_N(e^{j\omega}) - K_0(e^{j\omega})|^2 = E|\Delta K_N(e^{j\omega})|^2$. Using a first order approximation, this controller variance can itself be expressed as a function of the covariance matrix of the transfer function estimates:

$$E|\Delta K_N(e^{j\omega})|^2 = [F_G \ F_H] \ \text{cov} \begin{pmatrix} \hat{G}_N(e^{j\omega}) \\ \hat{H}_N(e^{j\omega}) \end{pmatrix} \begin{bmatrix} F_G^* \\ F_H^* \end{bmatrix} \tag{9.51}$$

where F_G and F_H denote the sensitivity of the control design mapping $K = k(G, H)$ with respect to G and H, respectively, and $*$ denotes the adjoint operation. As a result, the optimal design problem can be formulated as follows:

$$\min_{\mathcal{X}} \int_{-\pi}^{\pi} tr[\Pi(\omega) Q(\omega)] d\omega \tag{9.52}$$

where $\Pi(\omega)$ denotes the covariance of $\begin{pmatrix} \hat{G}_N(e^{j\omega}) \\ \hat{H}_N(e^{j\omega}) \end{pmatrix}$, and where $Q(\omega)$ is a frequency weighting matrix that reflects the particular control design mapping $k(G, H)$ and that takes account of the sensitivity of this mapping with respect to model errors.

The main results achieved in optimal experiment design for certainty equivalence control can be summarised as follows:

- as stated above, it was shown in [63] that, for a minimum-variance control design criterion, the optimal identification design is to perform closed-loop identification with the *unknown* optimal minimum-variance controller acting on the system, provided the system is minimum-phase.

This result holds both for the case when the input signal variance is constrained, and for the case where the output signal variance is constrained. It was also shown in [63] that, surprisingly, addition of an external excitation signal r does not contribute to a decrease of the error variance. In fact, with the optimal identification design, the performance degradation $E[y_t^{opt} - y_t^N]^2$ becomes independent of the input signal power;

- in [77], the optimal experiment design problem was solved for the general case of a differentiable control design mapping $K = k(G, H)$. As stated already, the optimality criterion adopted in [77] was not the average performance degradation with respect to the optimal control cost, but rather the simpler suboptimal criterion $E[y_t^{opt} - y_t^N]^2$. The main outcome was that the optimal design was again to perform a closed-loop identification experiment, with a particular controller in the loop that depends on the unknown system. This optimal controller for identification is not always stabilising and proper. In the case where it is stabilising and proper, the "performance degradation" $E[y_t^{opt} - y_t^N]^2$ is again independent of the power of the excitation signal, a rather remarkable result. In the case where the optimal controller for identification is not stabilising or proper, an optimal stabilising and proper controller can be defined using the Youla parameterisation;

- in [58], the same suboptimal criterion $E[y_t^{opt} - y_t^N]^2$ was adopted as a basis for the construction of optimal experiment design problems. The main novelty of [58] was the adoption of a constraint on a mixed sum of input and output power: $\int_{-\pi}^{\pi} \alpha \Phi_u(\omega) + \beta \Phi_y(\omega) d\omega \leq 1$. The optimal design problem (9.52) was solved with respect to the design variables K_{id} and Φ_r under this signal power constraint, for various special cases of the weighting matrix $Q(\omega)$ of (9.52). Here, K_{id} is the optimal controller that operates during identification, while Φ_r is the power spectral density of the external excitation used during identification;

- it follows from the analysis of [77] and [58] that, in almost all cases, the optimal identification design for control is to perform closed-loop identification. One exception is when the controller depends only on the input-output model $\hat{G}_N$ and the power constraints apply only on the input power Φ_u. We refer the reader to [77] and [58] for more specifics.

The optimal experiment design for certainty equivalence control, developed in this section, has been formulated as a problem of minimisation of the average performance degradation. One way of viewing this approach is to consider that the set of models that are generated by a particular experiment design set-up,

$$\mathcal{D} = \{(\hat{G}_N, \hat{H}_N)\} = \{(G(\hat{\theta}_N), H(\hat{\theta}_N)), \hat{\theta}_N \in D_\theta\}$$

is matched, by the certainty equivalence control design, into a corresponding controller set:

$$\mathcal{K} = \{\hat{K}_N = k(\hat{G}_N, \hat{H}_N), \forall (\hat{G}_N, \hat{H}_N) \in \mathcal{D}\}$$

The optimal experiment design is then defined as the identification set-up for which this controller set $\mathcal{K}$ achieves the smallest possible average performance degradation with respect to the optimal cost $\bar{J}(G_0, H_0, K^*)$. Three essential features of this optimal experiment design approach are:

- the unknown optimal control cost $\bar{J}(G_0, H_0, K^*)$ is taken as the benchmark, with which the achieved costs $\bar{J}(G_0, H_0, k(\hat{G}_N, \hat{H}_N))$ are compared;
- a certainty equivalence approach is taken to design the controller from each estimated model: $\hat{K}_N = c(\hat{G}_N, \hat{H}_N)$;
- the resulting optimal experiment is a function of the unknown true system and can therefore not be computed. It can at best be approximated, perhaps by some iterative design.

9.5.2 Connecting Identification Design and Robust Control Design

In the robust control design approach to *validation for control*, one also connects a model uncertainty set $\mathcal{D}$ with a corresponding controller set $\mathcal{K}$, but the controller set is defined on the basis of robust stability or robust performance considerations rather than on the basis of a certainty equivalence principle. To be more specific, the controller set $\mathcal{K}$ is defined as the set of controllers that robustly stabilise all models in the model set $\mathcal{D}$, or that achieve a given level of performance with all models in $\mathcal{D}$. Such controller sets can be computed, and one can therefore again consider the pairs of model sets and controller sets $\{(\mathcal{D}^{(i)}, \mathcal{K}^{(i)})\}$ that are produced by each particular identification experiment design and the corresponding robust control design procedure.

The main difficulty is to define an optimal identification design criterion on the basis of such pairs of model–controller sets. Ideally, one should again call a controller set $\mathcal{K}$ optimal if it causes the smallest possible worst-case performance degradation on the actual system, but this would again involve an optimal design that is dependent on the unknown true system. Instead, one has taken a more pragmatic viewpoint, in which an experiment $\mathcal{X}^{(1)}$ is termed better tuned for robust control design than an experiment $\mathcal{X}^{(2)}$ if the controller set $\mathcal{K}^{(1)}$ obtained with the experimental conditions $\mathcal{X}^{(1)}$ is larger (in some sense) than the controller set $\mathcal{K}^{(2)}$ obtained with the experimental conditions $\mathcal{X}^{(2)}$.

In [31], a parameterised model uncertainty set $\mathcal{D}$ was described that is obtained by a validation procedure that consists in a straightforward prediction-error identification with a full order model structure. It contains the nominal model G_{mod}[4], and it also contains the true system G_0 with any prescribed probability level α. A measure of size of this uncertainty set was then adopted that is based on the ν-gap between two transfer functions. It is the *worst-case*

---

[4] For example, the centre of $\mathcal{D}$.

ν-gap between G_{mod} and all models in $\mathcal{D}$:

$$\delta_{WC}(G_{mod}, \mathcal{D}) = \max_{G_{in} \in \mathcal{D}} \delta_\nu(G_{mod}, G_{in}). \tag{9.53}$$

This allows one to define a corresponding set of controllers, every one of which stabilises each model in $\mathcal{D}$:

$$\mathcal{K}(G_{mod}, \mathcal{D}) = \{K \mid b_{G_{mod},K} > \delta_{WC}(G_{mod}, \mathcal{D})\}. \tag{9.54}$$

The stabilisation property follows directly from Proposition 9.2.

Consider now that two different validation experiments lead to two different uncertainty sets, $\mathcal{D}^{(1)}$ and $\mathcal{D}^{(2)}$. This defines two corresponding controller sets, $\mathcal{K}(G_{mod}, \mathcal{D}^{(1)})$ and $\mathcal{K}(G_{mod}, \mathcal{D}^{(2)})$ via (9.54). It is then easy to see that

$$\delta_{WC}(G_{mod}, \mathcal{D}^{(1)}) < \delta_{WC}(G_{mod}, \mathcal{D}^{(2)}) \Rightarrow \mathcal{K}(G_{mod}, \mathcal{D}^{(2)}) \subset \mathcal{K}(G_{mod}, \mathcal{D}^{(1)})$$

Thus, the worst-case ν-gap $\delta_{WC}(G_{mod}, \mathcal{D})$ can be used as a measure of how well the validated uncertainty set, in connection with the nominal model G_{mod}, is tuned for robust control design, at least insofar as robust stability is concerned. Using this particular technical tool (the worst case ν-gap) for connecting a validated uncertainty set $\mathcal{D}$ and a corresponding set of stabilising controllers $\mathcal{K}$ leads to the concept that a validation experiment that is tuned for robust control design is one that delivers a pair $(G_{mod}, \mathcal{D})$ with the smallest possible worst-case ν-gap. Preliminary optimal input design results for the minimisation of the worst-case ν-gap have been obtained recently [76].

There are two major limitations with the approach based on the worst-case ν-gap:

- the stability condition used for the definition of the controller set corresponding to a particular model uncertainty set (see (9.54)) is a sufficient condition, *i.e.*, there are other controllers outside the set $\mathcal{K}$ defined in (9.54), that stabilise all models in $\mathcal{D}$;
- more importantly, the criterion used to define the controller set is a robust stability criterion only; many of the controllers defined by (9.54) may achieve very poor performance with some of the models in $\mathcal{D}$, and possibly also with the unknown $G_0 \in \mathcal{D}$.

As a result, a new approach has recently been developed that connects a controller set $\mathcal{K}$ to a pair $(G_{mod}, \mathcal{D})$ (nominal model and validated model set) on the basis of robust performance specifications [33]. The control performance specifications are prescribed as follows:

$$(G_0, K) \text{ stable and } \quad \| J(G_0, K, W_l, W_r, \omega) \|_\infty < 1 \tag{9.55}$$

where $J(G_0, K, W_l, W_r, \omega)$ is defined as in (9.8).

Roughly speaking, the procedure can be described as follows:

1. An identification and validation experiment delivers a nominal model G_{mod} and an uncertainty set $\mathcal{D}$.
2. A controller set $\mathcal{K}(G_{mod})$ is defined such that each controller in this set achieves with the nominal model G_{mod} a performance that is slightly better than the desired performance:

$$\mathcal{K}(G_{mod}) = \{K | (G_{mod}, K) \text{ stable and } \| J(G_{mod}, K, W_l, W_r, \omega) \|_\infty < \gamma < 1\}$$

where $0 \leq \gamma \leq 1$ (*e.g.*, $\gamma = 0.9$).
3. One then checks whether all controllers in $\mathcal{K}(G_{mod})$ achieve the prescribed performance with all models in $\mathcal{D}$, *i.e.*, whether $\forall K \in \mathcal{K}(G_{mod})$ and $\forall G \in \mathcal{D}$ the conditions (9.55) hold. If so, the pair $(G_{mod}, \mathcal{D})$ is said to be tuned for robust control design.
4. If not, guidelines are given to perform a new identification/validation experiment, producing a new pair $(G_{mod}, \mathcal{D})$ that is more likely to be tuned for robust control design.

Computational tools have been developed to perform all steps of this design procedure, in the context of a parameterised model uncertainty set $\mathcal{D}$ obtained by prediction error identification: see [33] for details.

Other approaches have been developed that aim to connect model uncertainty sets and corresponding controller sets, on the basis of robust control design considerations: see, *e.g.*, [48, 69, 70, 97, 132]. As stated above, the main difficulty is that there is no obvious optimality criterion for the design of the identification/validation experiment. In addition, specific technical tools need to be developed for the combination of each identification and validation criterion with each control design criterion. Here, we have briefly sketched the results that have been achieved for the combination of standard prediction error methods with standard robust control design methods. To connect these two sets of tools required rather considerable technical work. This is just to say that the study of the interplay between model validation and robust control design is still a rather wide open field, given the complexity of the interplay and the fact that it requires a good understanding of techniques that were, until recently, studied and developed by two different communities.

One of the interesting unifying conclusions that have already emerged from the different approaches that have been studied is that, just as for the identification of the nominal model, experience indicates that the validation should be performed with closed-loop data when the goal is to design a model uncertainty set that is tuned for robust control design.

9.6 Conclusions

In this chapter, we have focused on the interplay between identification and control design. For the nominal model, we have shown that the quality of a model that is to be used for control design must be evaluated with respect to its closed-loop properties. This leads rather straightforwardly to the conclusion that the identification of the nominal model should be performed in closed-loop, and that this should be done in an iterative way, with a succession of model updates and controller updates. One way to obtain prior stability guarantees, when moving from one model/controller pair to the next one, is the use of caution, *i.e.*, small step changes in the model and/or controller updates. The tools for computing the sizes of these "small" updates have been presented.

The connection between model validation and robust control design is somewhat harder to establish because there are, at this point, no universally accepted views on what constitutes an optimal criterion for a model validation experiment design, when the objective is to perform a robust control design on the basis of such validated set. However, much progress has been accomplished in the last three years. We have sketched a recent framework for the connection of model validation in a prediction error framework and robust control design. Again, all indications so far point to the advantage of performing closed-loop validation, when the objective is robust control design. Of course, the quality of a model (or model set) identified (or validated) by closed-loop identification can always be matched by an open-loop experiment with a specific choice of input spectrum that takes the control objective into account. This input spectrum plays the role of a frequency weighting that incorporates the closed-loop quality requirements; however, this optimal input spectrum depends not only on the controller but also on the unknown system. The advantage with a closed-loop experiment is that, by using closed-loop data, the unknown system exerts a frequency weighting that approximates the optimal one.

Acknowledgements

The author wishes to thank B.D.O. Anderson, R.R. Bitmead, X. Bombois, R. Hildebrand, L. Johnston, C. Kulcsar and G. Scorletti, with whom many of the results of this chapter have been derived.

This chapter presents research results of the Belgian Programme on Interuniversity Poles of Attraction, initiated by the Belgian State, Prime Minister's Office for Science, Technology and Culture. The scientific responsibility rests with its authors.

Part II

Applications

10 Helicopter Vibration Control

Robert R. Bitmead

Abstract. The practical problem of constructing a model, from which to design a helicopter vibration control system, is presented. This is done to provide a rather extreme example of the interaction between the modelling objective and the control design. This is done to reinforce the messages of earlier chapters dealing with modelling for control design and the need to concentrate on the ultimate model application in assessing the quality of model fit.

The approach relies on the tight fitting of frequency response values across a region of interest using experimental data. This is demonstrated in the experiment design and in the data pre-filtering as was discussed in Chapter 3.

10.1 Helicopter Vibration Control Example

The identification principles for fitting models for control design developed and outlined in previous chapters will be applied and illustrated via their application to the modelling phase of a vibration control system for a helicopter. This work was originally performed with MOOG Inc [42] in an industrial setting.

Figure 10.1 shows a helicopter. It is apparent that the mounting of the

Fig. 10.1. Helicopter, showing mounting of engine, gearbox and main rotor above cabin

engine, gearbox and main rotor immediately above the cabin will lead to the introduction of vibrations into the cabin due to the passage of the rotor blades and their differential loading, which is responsible for the motion of the aircraft. The rotor speed is roughly constant, 17 Hz here, and varies ±5 per cent in flight due to loadings during manoeuvers. These vibrations are fatiguing for the pilot and limit the time available in service. A passive tuned damper system is in place, in which a set of spring-loaded plates is bolted to the airframe and is tuned to remove energy at frequencies around 17 Hz. To improve performance, it is desired to replace this passive mass-spring-damper system with an active system in which the plates are driven hydraulically. The sensors for this feedback system are accelerometers located near the pilot's head and feet, and the actuators are hydraulic rams mounted between the airframe and the damper plate.

Because the behaviour of the helicopter airframe changes with loading and configuration of the aircraft, there could be a need to adapt or adjust the controller on-line for these variations on a flight-to-flight basis.

Problem statement: *develop a model of the helicopter vibration system from hydraulic actuators to accelerometer sensors that is suited to the design of an adaptive feedback controller to reduce the amplitude of the vibration induced by the rotor blades passing at 17 Hz ±5% and which is capable of adjusting to the variations in the airframe dynamics.*

The periodic vibrations induced in the airframe also have harmonic modes present at multiples of the fundamental frequency due to non-linearities, gearbox elements, *etc.* Because rate limiters are built into the actuators, high-frequency actuation might not be feasible and so it is desired to reject as much of the fundamental component of the vibration as is possible and to ignore higher harmonics.

It is known by the designers that the airframe has a resonant mode somewhere about 40 Hz which is not intended to be part of the damper compensation. The control strategy will be to avoid actuation in this band, which further reinforces the desire to concentrate on the fundamental region.

There are two actuated damper plates installed and two accelerometers. This should allow us to reject vibration modes in two dimensions. These two problems, however, should be decoupled and so we shall concentrate on a single mode only, *i.e.*, a single-input/single-output system model, in the first instance.

10.2 A Glimpse of the Solution

The system from hydraulic actuation through the structure to accelerometer output is open-loop stable due to mechanical energy dissipation   the helicopter does not become unstable without compensation. *This is a critically important piece of plant information for the subsequent design of a controller.* It states that zero feedback control is robustly stabilising – the performance is inadequate but we do have stability for all loadings. Further, this zero controller has the feature of not exciting the mode around 40 Hz. The controller that we seek to develop will be effectively a perturbation of this robust solution. That is, we seek a controller with large action around the vibration frequency of 17 Hz and rapid roll-off before 40 Hz. One way to view this is

10 Helicopter Vibration Control

as a very low-gain (and hence robustly stabilising) controller in series with a sharp resonant peak at 17 Hz.

To understand the behaviour of such a controller, consider the transfer function relating accelerometer output to the disturbance.

$$y_t = \frac{1}{1+GK} v_t$$

To achieve rejection around 17 Hz, GK should have a large gain around 17 Hz and to avoid excitation at 40 Hz the gain should be low at frequencies near this value. The requirement can be met if a very lightly damped oscillator with centre frequency 17 Hz is present in series with a small gain in $K(s)$, because in that case $|K(j\omega)|$ will be very large at the oscillation frequency and small elsewhere. So, the controller, whatever the method used to calculate it, should have a form similar to:

$$K(s) = K^*(s) \frac{s}{(s+\epsilon)^2 + \omega_0^2}$$

with a small value of ϵ and $K^*(s)$ having no further poles near the imaginary axis and a suitably low gain. The oscillator damping acts as a design parameter for disturbance rejection "selectivity": the more the damping, the less effective the rejection but the wider the frequency band over which control is acting.

The candidate controller structure is illustrated in Figure 10.2.

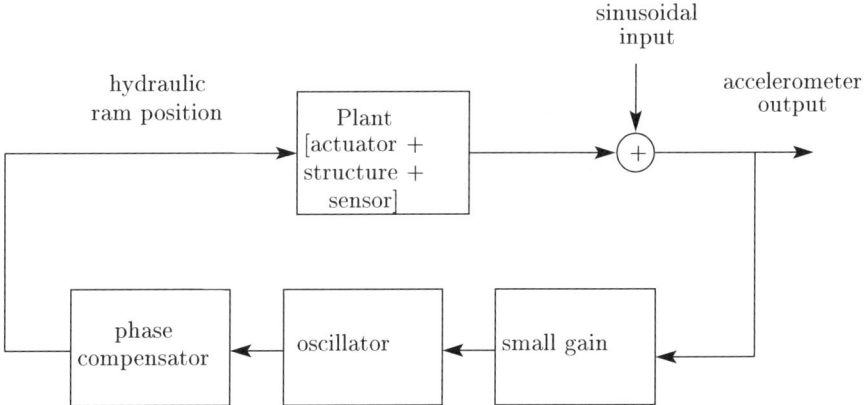

Fig. 10.2. Candidate controller structure

- the plant includes all the parts of the system from the actuator (hydraulic ram position) input to the sensor (accelerometer) output;
- the vibration disturbance is modelled as a sinusoidal additive disturbance at the output. Since the plant is presumed to be linear for modelling purposes, we may move the vibration signal to the output without loss of generality;

- the feedback controller consists of three components; the low-.gain element for robust stability and lack of excitation around 40 Hz, the oscillator for sinusoidal disturbance rejection performance, and a phase compensator for closed-loop stability.

The phase compensator is a critical component of the feedback loop. Consider the Nyquist diagram of the open loop: plant + small gain + oscillator + compensator. Because of stability, the plant, gain and compensator frequency responses will be smooth at the centre frequency of the oscillator. Thus, the magnitude of the combined frequency response will be large only around the centre frequency of the oscillator. The phase in this narrow band of frequencies will be the phase of the plant plus the phase of the compensator plus the phase of the oscillator, which moves from $-\frac{\pi}{2}$ radians to $\frac{\pi}{2}$ radians. (See Figure 10.2.) The open loop is known to be stable. For closed-loop stability, we require that the phase be such as to prevent the encirclement of the -1 point by the open-loop Nyquist diagram, and this is achieved by correct choice of the phase of the compensator at ω_0.

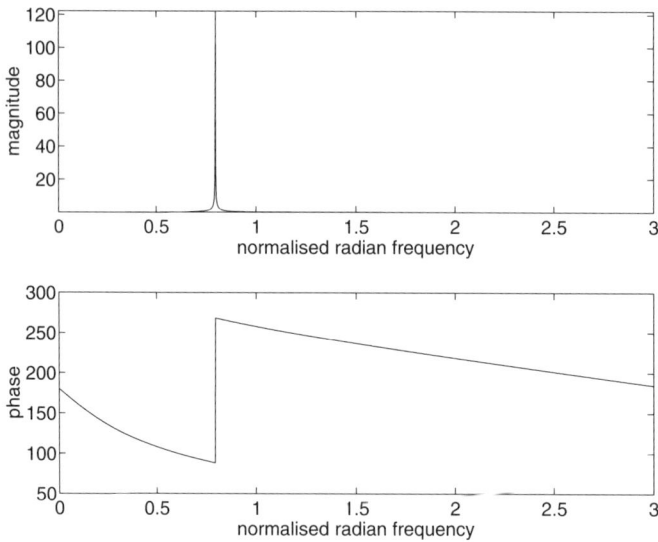

Fig. 10.3. Magnitude and phase Bode plots of a simple oscillator

Another, time-domain, way of appreciating the role of the phase compensator is to trace the signals around the loop. To cancel the sinusoidal disturbance at the output, the plant output needs to be roughly π radians out of phase with the disturbance. This requires that the phase from output measurement through the first two blocks of the controller be adjusted so that, after passage through the plant, disturbance rejection is achieved.

We note that, with this formulation of the problem, the precise gain of the controller is immaterial to stability – it affects rejection performance. Thus, the modelling for control design problem resides in effectively capturing the phase properties of the plant.

10.3 Modelling Requirements

A stable, low-order parametric model is sought that captures the phase properties of the plant system in the neighbourhood of the vibration frequency, say 15–20 Hz. This identification will be performed using experimental data from the true system.

A low-order parametric model is desired in order to introduce later the adaptation ability to accommodate system variations and frequency drifts. The modelling precision at frequencies away from the 15–20 Hz band is irrelevant because no significant control gain will be present there. This condition will be further enforced during the control design phase by using, for example, frequency-weighted LQG techniques based on the parametric model.

The stability of the model is a key property of the identification for control of a system known to be stable and for which we seek to use a high-gain feedback strategy. Allied with the discussion in Chapter 9, we see that stability plus phase matching in the controller pass band are the key factors to maintaining the achieved and designed sensitivity functions close.

10.4 Identification Experiments

The apparatus used in the identification experiment was a structural facsimile of the helicopter airframe, which demonstrated the same dynamics, with the active mass-spring-damper sets bolted on. The identification experiments were conducted to model the actuator-to-sensor transfer function. These laboratory experiments were performed to understand the modelling for control design issues prior to the development of an in-service adaptive scheme. In the on-line adaptive solution, the experimental principles outlined here for modelling based on frequency response fits still apply.

A non-parametric swept-sine experiment was performed using a standard piece of test equipment for empirical frequency response estimation. This injected a sinusoid of slowly-changing frequency into the system and compared the gain and phase of the output at the same frequency. The resulting response is shown in Figure 10.4. This reveals a number of important phenomena:

- there is a resonant mode centred at 17 Hz and extending either side for roughly 5 Hz. This is due to the tuned mechanical properties of the mass-spring-damper sets;

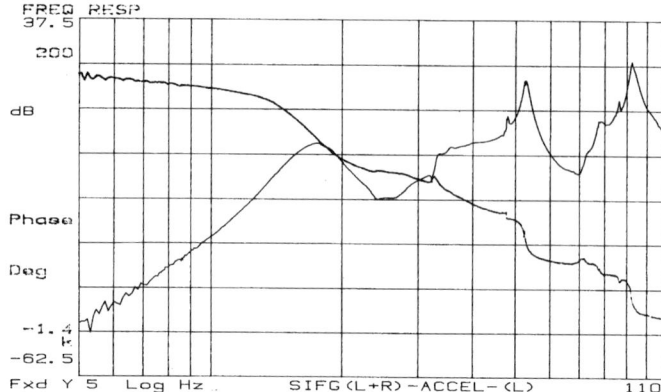

Fig. 10.4. Swept sine experiment

- there is a range of complicated resonant effects from 40 Hz and upwards with increasing amplitude. The fact that the transfer function maps a position into a velocity means that it contains two differentiators, which is evident in the high-frequency amplification;
- the double differentiation is evident at low frequencies where the response dies off at 40 dB/decade.

This initial snapshot of the system is consistent with *a priori* understanding.

We note that this swept-sine information is not really particularly useful for control design purposes. The plant transfer function is captured at a number of frequencies as gain and phase information. This is not a parametric model at this stage. A parametric model might be fitted to the data, however, using any of a number of fitting tools. The primary difficulty with these tools is that they do not ensure stability of the resulting model, nor do they give much guidance as to how to manage the complexity of the resulting model, which needs to be simple in order to admit adaptation. We shall see in the following how both these properties might be achieved by redoing the experiment in a different way.

We note that, because of the presence of rate limiters in the hydraulic actuators, it is only possible to conduct experiments with smooth signals. The textbook choice of a single experiment of pseudo-random binary noise is not acceptable because the rate limiters introduce non-linear effects, which are amplitude-dependent.

The decision was made to conduct a sequence of sinusoidally excited experiments with different frequencies – effectively a slow replica of the swept-sine data. The frequencies chosen would be concentrated around the 17 Hz range but would include outlying frequencies as well. In this way, the system identification was guided by the swept-sine experimental data. Using individual sinusoids ought not fall foul of the rate limiters.

10.5 Data Analysis and Sanitizing

Experiments were conducted with sinusoidal input signals at 20 different frequencies, with the bulk of them concentrated around the area 15–20 Hz. Figure 10.5 shows the computer-generated 15 Hz sinusoidal actuator input signal and its associated output signal. Figure 10.6 shows the magnitude of the DFT of both signals. (Note that the peak of the signal magnitude at 15Hz has been cut off by the choice of axes.)

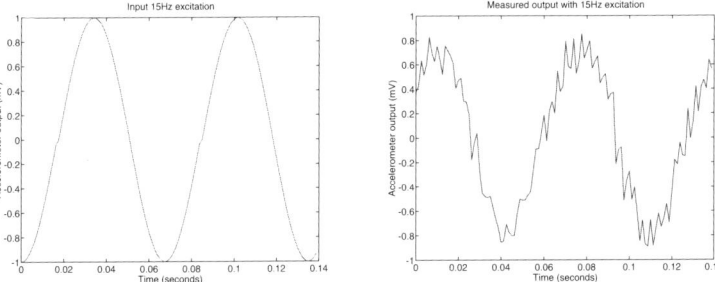

Fig. 10.5. 15 Hz actuator input signal and corresponding accelerometer output signal

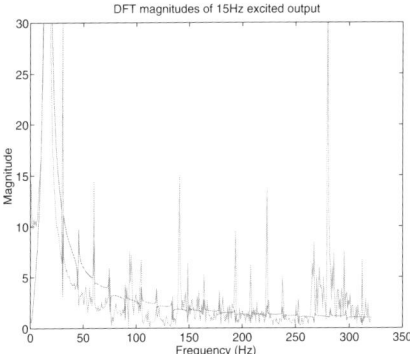

Fig. 10.6. Frequency-domain analysis of input and output signals

There are a number of immediate comments:

- there is considerable distortion introduced by the process evident in the time-domain output data;
- the frequency-domain plot reveals several things:

1. The smoother plot is the DFT magnitude of the input sinusoid. The appearance of spikes at harmonic values indicates a coding error leading to a small discontinuity in the input signal. This is just evident at the end of the cycle in the left graph in Figure 10.5.
2. The magnitude of the DFT of the output signal exhibits strong content at the same frequencies. Thus, the deviation from sinusoidal response is harmonic distortion, not just noise;

- to fit a linear model the harmonic distortion might lead to bias;
- there is no intention of controlling at the frequencies of the distortion. Indeed, control action at these frequencies will be guaranteed low in order to avoid the resonant peaks.

Thus, to fit an accurate linear model in the range of excitation frequency, we need to remove the harmonic distortion. We do this by data filtering as discussed in Chapter 3, Section 3.2.

Each experiment consisted of 1000 data samples (sampling frequency 650 Hz). Transient responses were allowed to die away before the data were recorded. The raw (distorted sinusoidal) output data obtained from each of the twenty experiments will be used as part of a transfer estimation, which needs to capture the local phase very accurately. Accordingly, we need to remove any harmonic distortion and concentrate on the information contained in the fundamental signal, centred on the driving sinusoid's centre frequency, ω_k. To do this, we calculate, for a data sequence of N points recorder after the disappearance of transients, the following values:

$$Y_i = \frac{1}{N} \sum_{k=1}^{N} y_{ik} e^{j\omega_i k} \tag{10.1}$$

$$U_i = \frac{1}{N} \sum_{k=1}^{N} u_{ik} e^{j\omega_i k} \tag{10.2}$$

The complex estimate of the plant frequency response at ω_i is then given by

$$\hat{G}(e^{j\omega_i}) = g_i = Y_i/U_i$$

The very long FIR filtering in (10.1–10.2) has very narrow pass-band centred at $e^{j\omega_i}$ and with zeros at harmonically-related frequencies. This helps to remove the harmonic distortion from both input and output. The 20 derived estimates $\{g_i : i = 1, \ldots, 20\}$ are shown in Figure 10.5. Note the close similarity to the empirical (non-parametric) frequency response values of the swept-sine data of Figure 10.4. Note also that the experiment design has seen us concentrate on the fitting of frequency response values in the 15-20 Hz region of interest. Recall that this curve is simply a collection of points of frequency response values. It is not yet a dynamical model.

The use of these filters has the effect of focusing the fit precisely onto the frequencies of interest. Thus, the estimated complex gains $\{g_i\}$ are very

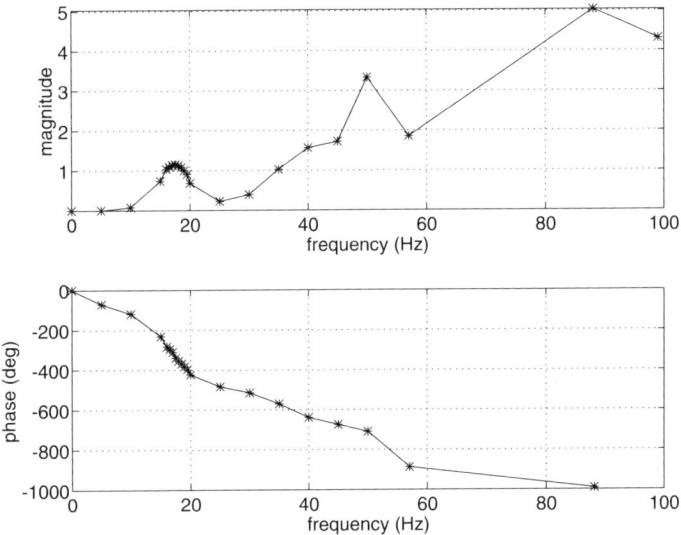

Fig. 10.7. Frequency response points fitted using individual experiments

accurate estimates of the plant frequency response – a refined version of the swept-sine data. We may now use these estimates to reconstruct data with no harmonic distortion, so-called *sanitised data*.

$$\bar{u}_{ik} = \cos(\omega_i T k) \tag{10.3}$$
$$\bar{y}_{ik} = |g_i| \cos(\omega_i T k + \arg g_i) \tag{10.4}$$

In addition to the sanitised data above, one more set was generated from a fictitious experiment at DC; $\{u_{0k}, y_{0k}\} = \{1, 0\}$. That is, we replicated an experiment in which a step in the hydraulic ram was matched in steady-state with zero acceleration. This now gives us a set of 21 sets of sinusoidal input-output sanitised data. We now focus on the derivation of a parametric time-domain dynamical model from these sanitised frequency-domain data.

10.6 Parametric System Identification

Recall that our requirements of the identified model are that it should be stable, parametric, low-order, and capture the phase of the plant in the frequency range of interest. We use time-domain model-fitting techniques to perform the fit satisfying these conditions. To do this, we need to create a time-domain input/output signal pair. We do this by forming a linear combination of the sanitised signals $\{\bar{u}_{i,k}, \bar{y}_{ik}\}$ with the coefficients of the combination reflecting our experiment design.

We select

$$\mathcal{U}_k = \sum_{i=0}^{20} w_i \bar{u}_{ik} \tag{10.5}$$

$$\mathcal{Y}_k = \sum_{i=0}^{20} w_i \bar{y}_{ik} \tag{10.6}$$

The weights are chosen as follows:

$$w_i = \begin{cases} 0.01 & w_i < 10 \text{ Hz} \\ 0.1 & 10 \text{ Hz} \leq w_i < 15 \text{ Hz} \\ 1.0 & 15 \text{ Hz} \leq w_i < 25 \text{ Hz} \\ 0.1 & 25 \text{ Hz} \leq w_i < 40 \text{ Hz} \\ 0.01 & 40 \leq w_i \end{cases}$$

Because of linearity, hypothetically at least, the output $\mathcal{Y}$ would result from the input $\mathcal{U}$, even though this is a fictitious experiment. An advantage of this fictitious experiment is that no problems arise with rate limiters and the harmonic distortion of the low-frequency ranges do not contaminate the frequency response estimates at higher frequencies.

With this single time-domain set of manufactured data, a good signal-to-distortion ratio has been achieved together with appropriate frequency weighting for our control purpose. It remains to fit the stable, low-order parametric model. For this, we elect output error (MATLAB® oe) fitting of a second-order system. Output error methods, because they rely on simulation of model outputs, always yield stable models from bounded data. Given the absence of noise in the data, the model fit will be a direct frequency-weighted minimisation as described in [108]. The resulting fit of the second order model is shown in Figure 10.8.

We can offer the following comments:

Complexity: the model is second order and captures approximately the resonant mode about the 17 Hz region.

Linearity: the model is linear and was fitted using sanitised data uncontaminated by harmonic distortion.

Stability: the second order model is stable by construction by dint of the output error identification algorithm.

Magnitude fit: the fit about 17 Hz is not close but is manageable. The model fails to capture any of the high-frequency resonances – it is not complex enough for this. It also fails to capture the double zero at zero frequency.

Phase fit: the phase fit across the 15–20Hz region is excellent. It is poor outside this region.

Suitability: this model is almost perfect for control design for sinusoidal disturbance rejection. It captures the important features of open-loop stability and phase in the disturbance band. It is simple, which permits the

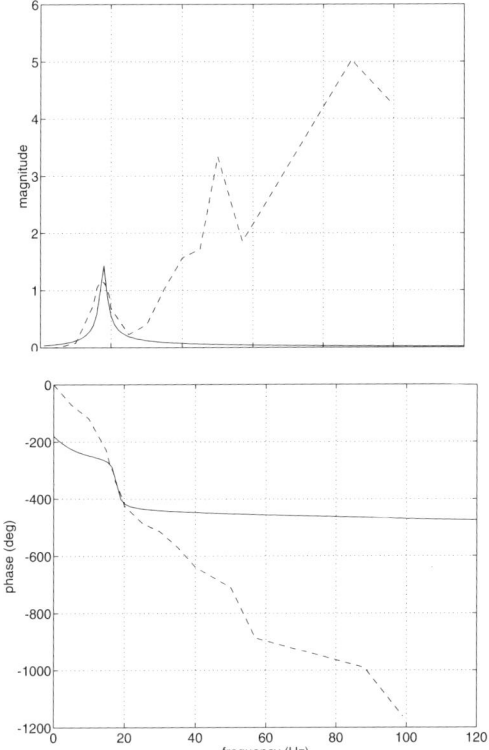

Fig. 10.8. Final model fit magnitude and phase plots versus the frequency response samples

accommodation of model mismatch through, say, frequency-weighted control design. Because of the low-gain control strategy, the missed zeros at DC are immaterial.

Control Design Methodology

Once the model is available, an appropriate low-gain-plus-oscillator controller is designed using frequency-weighted LQG design based on the parametric identified model above. Unsurprisingly, this design follows closely the prescription glimpsed earlier. The use of frequency weighting is mandated by the need to ensure that excitation of the high-frequency resonances is avoided; those resonances are evident in the experimental data in Figures 10.4 and 10.5.

Once a controller is developed, based on the reduced order model, its tentative closed-loop stability performance with the "full order" detailed system can be tested by computing the frequency-by-frequency values of KG, using the experimental frequency response data for G.

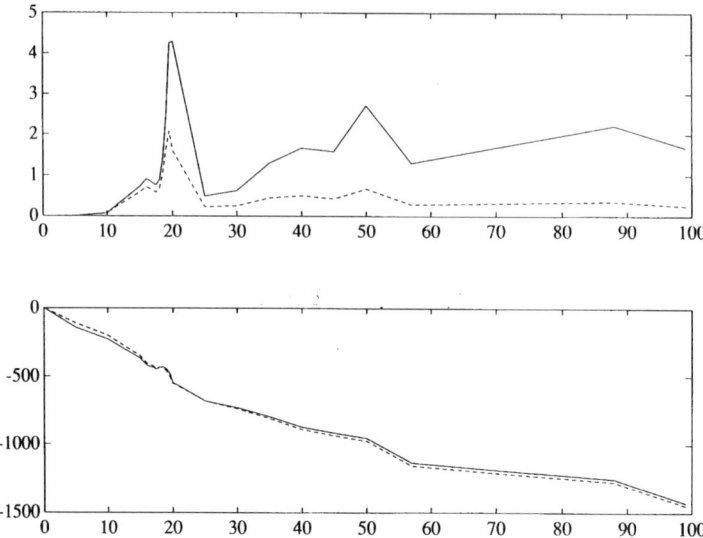

Fig. 10.9. Magnitude and phase plots of KG versus frequency for two different frequency-weighted LQG controllers based on a second order model $\hat{G}$

Figure 10.9 shows a preview of the magnitude and phase of KG for two distinct controllers designed from the model $\hat{G}$. The solid curve indicates a controller that might lead to stability problems with G because the combined magnitude exceeds 1 as the phase crosses 540°. It also preserves too great a gain at the high-frequency resonances. This controller is unlikely to stabilise G. By comparison the second controller, for which KG is plotted with a dashed line, should yield robust stability. This controller was designed using more stringent frequency-weighting penalties.

This latter controller works!

10.7 Conclusions

The example in this chapter has presented a rather extreme demonstration of modelling for control. Because the problem is one of narrowband disturbance rejection, and because we know the open-loop system to be stable, the modelling task becomes one of very precise phase matching in the bandwidth of the disturbance. The magnitude matching is relatively unimportant. What this shows about the problem of modelling for control is that an understanding of the nature of the controller to be used can very strongly influence the interpretation of the quality of the model. This is designed to reinforce the statements of Chapters 1, 4 and 9.

This example provides very strong evidence of the interplay between modelling and control. The main conclusion is that these are not separate issues so that information regarding the ultimate control objective and *a priori* knowledge of the process are important elements to be included in developing a good model for control. The example of the helicopter vibration control problem is a rather extreme demonstration of these principles in an area of practical relevance.

Depending on its purpose, models that produce a close open-loop response can be useless for control, and *vice versa*: robust controllers can be built from models that would be rejected if open-loop criteria were to be applied. The reason is that, in the closed-loop with a controller, the signals impinging upon the system provide a different frequency excitation from that explored in open loop. The helicopter vibration example provides almost a pathological situation, where the fit displayed in Figure 10.8 would be unacceptable from any open-loop perspective. Its acceptability is moderated by our understanding of the vibration control objective.

Once the model is available, an appropriate low-gain phase-lead plus band pass filter can act as a controller, or either a frequency-weighted LQG design. The effort in modelling with the control requirements in mind has obtained a good model valid for control design, while its (actually bad) accuracy for open-loop prediction is irrelevant.

As stated at the outset, the sinusoidal disturbance rejection problem leads to very specific requirements of the model: stability, phase matching and low-order. This latter property is used to develop an adaptive version of the control design – adaptive in both estimation of the disturbance frequency (since ω_0 can drift ± 5 per cent) and estimation of the plant phase response (due to loading and configuration changes). The adaptive problem is discussed in more detail elsewhere [135, 162].

Acknowledgements

The author acknowledges the contribution to this work of Allan Connolly, Sam Crisafulli, Rick Johnson and Gonzalo Rey.
Research supported by US National Science Foundation under Grant ECS-0070146.

11 Control-relevant Identification and Servo Design for a Compact Disc Player

Paul M.J. Van den Hof and Raymond A. de Callafon

Abstract. The application of control-relevant modelling of a radial servo loop in a CD (compact disc) player is presented in this chapter. The modelling is based on an identification algorithm that uses experimental data obtained from closed-loop experiments to find a low-order feedback relevant linear discrete time model for model-based servo control design. In this procedure, a plant model is identified in terms of a normalised coprime factorisation. Enhanced performance controllers are designed and implemented, and it is shown how estimated model uncertainty bounds can be used to verify robust stability of the designed controller prior to implementation.

11.1 Introduction

The control-relevant identification approach discussed in this chapter illustrates the development of a dynamical model of a CD (compact disc) radial positioning mechanism. The intention of this application is to focus on the closed-loop approximate identification of the radial actuator in a CD player and on improved servo track-following properties by designing a high-performance servo controller.

An increasing amount of rotational data storage applications such as hard disk drives or CD players are used in portable applications subjected to shock disturbances. The shock resistance of such a storage device operating can be improved by designing a high-performance track-following servo controller. As a result, the application of a control-relevant identification algorithm to the radial servo loop in the CD player discussed in the application of this chapter is faced with several interesting challenges:

- the radial actuator exhibits a marginally stable open loop behaviour and has multiple lightly-damped mechanical resonance modes;
- due to the operating conditions of the CD player, the identification procedure is required to estimate models on the basis of closed-loop experiments only;
- finally, the models of the CD radial actuator are required to be suitable for servo control design.

The first two items illustrate the nature of the system to be modelled and controlled. More details on the radial mechanism in the CD player are given in Section 11.2. The latter item illustrates that models used for control design are necessarily approximative. On the one hand, exact modelling can be impossible or too costly; on the other hand, control design methods can become unmanageable if they are applied to models of high complexity. For example, a highly-complex (and high-order) finite element model is useful for mechanical design considerations. However, for servo design purposes, a low-order (approximate) model is needed for a manageable feedback control design.

In the application of the CD radial servo mechanism discussed in this chapter, the above-mentioned challenges are tackled by adopting a control-relevant identification framework based on fractional model representations. In line with the idea of iterative identification and control [60, 148], as reported in Part I of this book, the aim is to identify a limited-order approximate model to be used as a basis for enhanced control design, on the basis of experiments performed under closed-loop conditions. More details on this framework are given in Section 11.6. First, a description of the CD radial servo loop is given in the following section.

11.2 The Compact Disc Mechanism

The CD mechanism considered in this application is a Philips CDM9 mechanism. The CDM9 consists of a turntable DC motor for the rotation of the compact disc and a radial actuator with an optical pick-up unit (OPU) that emits a laser spot for data reading. A schematic impression of the CD mech-

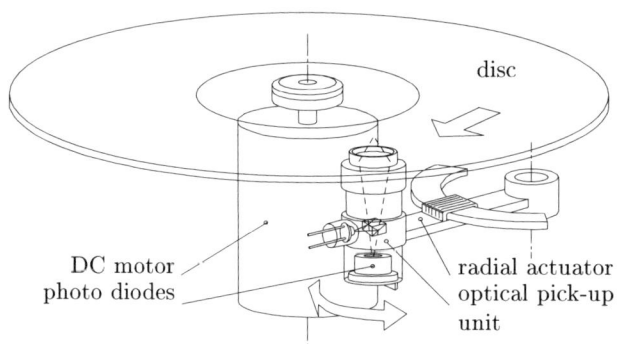

Fig. 11.1. Schematic view of CDM9 CD servo mechanism

anism is given in Figure 11.1.

A diode generates a laser beam that passes through a series of optical lenses in the OPU to give a spot on the disc surface. The light reflected from the disc is measured on an array of photo diodes, mounted in the bottom of the OPU, yielding the signals required for position error information of the laser spot on the compact disc. The light intensity of the reflected beam is directly related to the percentage of the spot area that covers a pit. A schematic depiction of a track on an audio disc is shown in Figure 11.2.

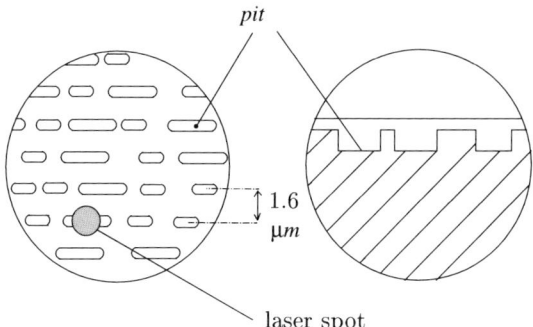

Fig. 11.2. Audio disc surface

Intensity of the reflected light is measured by a photo diode, which converts light intensity into an electrical current. The audio signal is obtained through sampling at 44.1 kHz and adjusting the sampled signals for missing bits.

Following the track on the compact disc involves basically two control loops. First, a radial control loop using a permanent magnet/coil system mounted on the radial arm, in order to position the laser spot in the radial direction orthogonal to the track. Secondly, a focus control loop using an objective lens suspended by two parallel leaf springs and a permanent magnet/coil system, with the coil mounted in the top of the OPU to focus the laser spot on the disc. In Figure 11.3, a block diagram of the two control loops is shown. Here, $G_a(q)$ denotes the transfer function of radial and focus actuator, K_{opu} the OPU, $K(q)$ the controller and $G_o(q) = -K_{opu}G_a(q)$. The variable q is the forward shift operator, yielding $x(t+1) = qx(t)$.

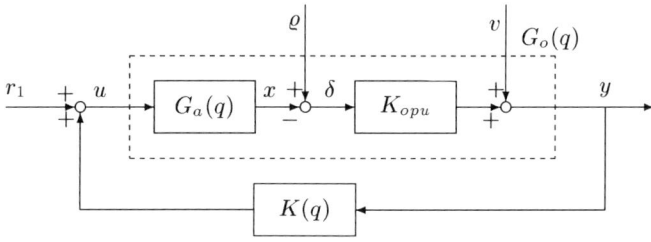

Fig. 11.3. Block diagram of the compact disc mechanism

The spot position error $\delta(t)$, which is the difference between the track position $\varrho(t)$ and the actuator position $x(t)$ in radial and focus direction, generates a (disturbed) error signal $y(t)$ via K_{opu}. This error signal $y(t)$ is led into the controller $K(q)$ and feeds the system $G_a(q)$ with the input $u(t)$. The signal $v(t)$ reflects the disturbance on the error signal $y(t)$. The absolute track position $\varrho(t)$ and actuator position $x(t)$ cannot be measured directly. Only the error signal $y(t)$ and the input $u(t)$ are available. For identification purposes, an additional and known excitation signal $r_1(t)$, uncorrelated with the additive noise $v(t)$ will be injected into the control loops, as illustrated in Figure 11.3. In Figure 11.4, the amplitude frequency response of the system G_o is sketched.

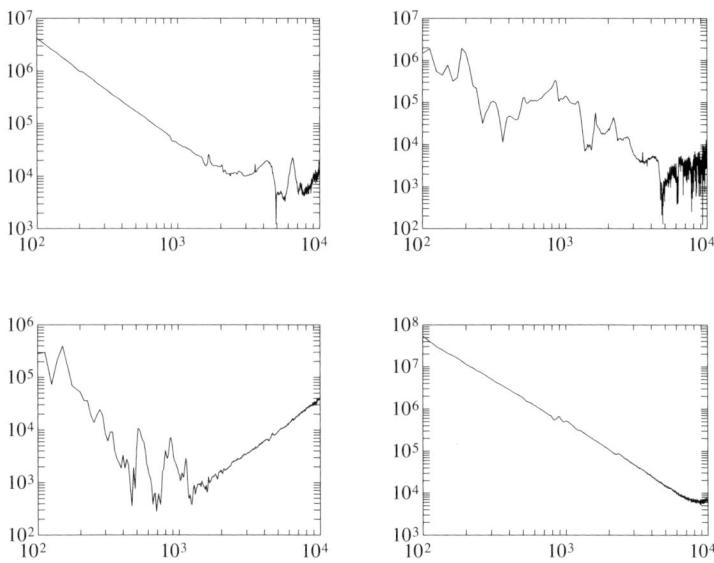

Fig. 11.4. Amplitude of spectral estimate of G_o; left upper figure is radial transfer, right lower figure is focus transfer; frequency axes (horizontal) are in Hz

The radial actuator uses a conventional voice-coil motor to follow the spiral track recorded on the track of the reflective disc. As a result, the dominant dynamical behaviour between voice-coil current and radial laser spot position can be described by a relatively simple fourth order model that consists of a double integrator and the lateral bending mode of the radial arm [55]. This model adequately describes the dynamics of the radial actuator to design a conventional PID controller for track following. Traditionally, this PID controller is designed to achieve a closed-loop bandwidth of the compact disc radial servo loop of approximately 450 Hz, being a compromise between several conflicting factors [146].

Both control loops make the CD mechanism a 2-input 2-output system [47, 55]. However, for the illustration purposes of this chapter, only the identification and control of the radial servo mechanism is considered here.

11.3 Control-relevant Modelling of Radial Actuator

The performance of the CD mechanism is determined mostly by the radial track-following properties of the CD player. An increase of the radial closed-loop bandwidth can achieve increased shock disturbance rejection for better track-following performance. For example, by trying to double the closed-loop bandwidth, track-following errors can be reduced by an order of magnitude. In the case of the CDM9 mechanism discussed in this chapter, a redesign of the traditional PID controller is needed to nearly double the radial closed-loop bandwidth to 800 Hz.

Unfortunately, the relatively simple fourth order model that consists of a double integrator and the lateral bending mode of the radial arm does not provide accurate enough information to redesign the PID controller at a higher radial closed-loop bandwidth. A straightforward redesign of the PID controller for a higher closed-loop bandwidth leads to ringing or even instability of the radial servo loop. This is illustrated in Figure 11.5, where excessive peaking in the measurement of the frequency response of the closed-loop sensitivity function can be observed.

In order to design an improved controller for the radial servo loop, a more accurate model of the radial actuator is needed. Additionally, for manageable control design, the model should be kept low in complexity and should capture only those features in the radial actuator that are essential for the control design. Specifically, the radial actuator dynamics that cause the excessive peaking of the sensitivity function in Figure 11.5 needs to be accurately modelled. This can be achieved by a control-relevant approximate identification.

However since the to-be-designed controller is still unknown in the modelling stage, it has been advocated to use an iterative scheme of identification and control design, using the controller of step $i-1$, denoted by K to estimate a model $\hat{G}$ for step i and to design an improved controller $K_{\hat{G}}$ based on this model $\hat{G}$. In this chapter, one step in such an iterative scheme will be illustrated involving an identification of a model $\hat{G}$ of the radial actuator G_o using an experimental situation where a known controller K is used. On the basis of the control-relevant estimated model $\hat{G}$, we consider a redesign of the controller K and show the significance of the modelling procedure for the control design. Estimated model uncertainty bounds will be used to verify robust stability properties of the designed controller prior to its implementation.

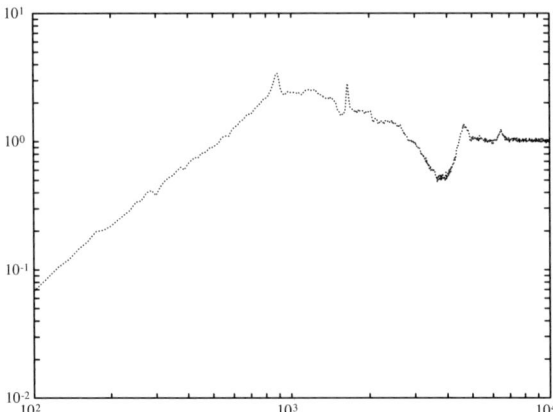

Fig. 11.5. Bode magnitude plot of measured sensitivity function $(1+KG_o)^{-1}$ of the radial transfer for a PID controller design aiming at a radial closed-loop bandwidth of 800 Hz

11.4 Preliminaries and Notations

The closed-loop system of the radial servo loop in a compact disc player is written into the general feedback system $T(G_o, K)$[1] given in Figure 11.6, which will be used throughout this chapter.

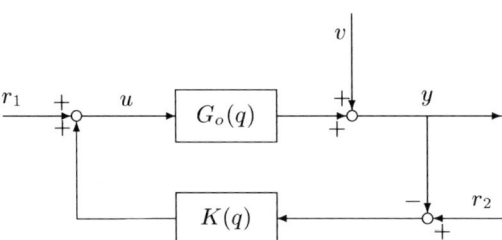

Fig. 11.6. General feedback system $T(G_o, K)$

In Figure 11.6, the signal $v(t)$ reflects the additive noise on the output $y(t)$ of the plant $G_o(q)$, which is assumed to be uncorrelated with the externally applied reference signals $r_1(t)$ and $r_2(t)$. From an identification point of view, the signals $u(t)$ and $y(t)$ are measured, $v(t)$ is unknown and $r_1(t)$ and $r_2(t)$ are possibly at our disposal for experiment design purposes.

---

[1] For notational convenience, the time shift operator q will generally be omitted.

Using Figure 11.6, the data from the closed-loop system $T(G_o, K)$ is described with the following equations

$$\begin{bmatrix} y \\ u \end{bmatrix} = T(G_o, K) \begin{bmatrix} r_2 \\ r_1 \end{bmatrix} + \begin{bmatrix} I \\ -K \end{bmatrix} [I + G_o K]^{-1} v \tag{11.1}$$

where $T(G_o, K)$ reflects the general feedback matrix

$$T(G_o, K) = \begin{bmatrix} G_o \\ I \end{bmatrix} [I + K G_o]^{-1} \begin{bmatrix} K & I \end{bmatrix} \tag{11.2}$$

This 2×2 transfer function reflects all feedback-relevant properties of the closed-loop system, such as sensitivity function (right lower part) and complementary sensitivity (left upper part).

A (possibly unstable) plant G can be described as a ratio of two stable transfer functions. In the theory of fractional representations [152], a pair of stable transfer functions (N, D) is called a *coprime factorisation (cf)* of G if $G = ND^{-1}$ and N and D do not have unstable zeros that cancel in ND^{-1}. Additionally a coprime factorisation (N_n, D_n) is called *normalised (ncf)* if it satisfies

$$N_n^* N_n + D_n^* D_n = I$$

where $*$ denotes the complex conjugate transpose. A straightforward example of a (coprime) factorisation of a plant G is

$$G = \frac{G(1 + KG)^{-1}}{(1 + KG)^{-1}}$$

showing that a possibly unstable plant G can be represented by a fraction of two transfer functions that each are stable, provided that K is a stabilising controller.

Normalised coprime factorisations have the particular property that

$$|N_n(e^{i\omega})| \leq 1, \quad |D_n(e^{i\omega})| \leq 1.$$

Additionally, it can be shown that for a normalised coprime factorisation $G = N_n D_n^{-1}$, the McMillan degree of G is equal to the McMillan degree of N_n and D_n, in other words, the quotient operation of the two transfers does not lead to an increasing system order.

Coprime factorisations are unique up to multiplication with a stable and stably invertible transfer function. Normalised coprime factorisations are unique up to multiplication with a unitary matrix.

In the following lemma, the set of all coprime factorisations of a plant is described:

Lemma 11.1. *Let the plant G_o be controlled by a stabilising controller K. Then, all coprime factorisations (N, D) of G_o can be written as*

$$N = G_o[I + KG_o]^{-1}[I + KG_x]D_x$$
$$D = [I + KG_o]^{-1}[I + KG_x]D_x$$

with G_x being any auxiliary system that is stabilised by K, having coprime factorisation (N_x, D_x).

Proof. See [150].

11.5 Common Objectives in Identification and Control

The general feedback transfer function matrix $T(G, K)$ has been recognised as an important feedback property of the closed-loop system. It induces a feedback relevant topology, meaning that if two such operators are alike, the corresponding feedback controlled systems will have similar performances. Moreover, control design methods can be formulated on the basis of the minimisation of the ∞-norm or 2-norm of the (frequency-weighted) $T(G, K)$ matrix, [114, 164].

The feedback transfer function matrix $T(G, K)$ encompasses both the information of the dynamics of the plant G and the controller K used to create the feedback connection. Therefore, for a given controller K, the (weighted) difference between $T(G_o, K)$ and $T(\hat{G}, K)$ forms a so-called feedback-relevant mismatch, caused by the difference between the nominal model $\hat{G}$ and plant G_o.

This can be understood by considering a triangle inequality applied to $\|T(G_o, K)\|$ with $\|\cdot\|$ denoting any norm or distance function, yielding [148]:

$$\left| \|T(\hat{G}, K)\| - \|T(G_o, K) - T(\hat{G}, K)\| \right| \leq \|T(G_o, K)\| \tag{11.3}$$

$$\|T(G_o, K)\| \leq \|T(\hat{G}, K)\| + \|T(G_o, K) - T(\hat{G}, K)\| \tag{11.4}$$

From (11.3) and (11.4), we see that by imposing the requirement

$$\|T(G_o, K) - T(\hat{G}, K)\| \ll \|T(\hat{G}, K)\| \tag{11.5}$$

similar performances for the controlled plant G_o and the controlled model $\hat{G}$ can be guaranteed. Therefore, constructing a model by minimising the difference $\|T(G_o, K) - T(\hat{G}, K)\|$ can be seen as a feedback-relevant identification of the plant G_o.

The difference $\|T(G_o, C) - T(\hat{G}, K)\|$ can be expressed in terms of the coprime factors of the plant G_o and the model $\hat{G}$. A direct identification of stable coprime factors then enables a unified approach to estimate stable and unstable models of plants operating under feedback controlled conditions. The results are summarised in the following section.

11.6 Estimation of Coprime Factorisations

11.6.1 Motivation

The framework for identification used in this chapter is based on the algebraic theory of fractional representations by directly estimating a coprime factorisation of the plant G_o using standard prediction error techniques. The motivation for using fractional representations from an *identification point of view* can be summarised as follows:

- the framework unifies the problem of dealing with unstable plants G_o and controllers K;
- the closed-loop identification problem can be recasted into an equivalent open-loop identification of coprime plant factors. Consequently, the results in approximate open-loop identification, based on prediction error methods like in [108], can be exploited;
- the fractional representation allows the formulation of an identification criterion that asymptotically matches a control-relevant plant–model mismatch $\|T(G_o, K) - T(\hat{G}, K)\|$, as described in the previous section.

Clearly, the first item is evident since coprime factorisations are defined to be stable. The last two items will be illustrated in subsequent sections.

11.6.2 Equivalent Open-loop Identification

By using the equation of the data generating system given in (11.1) and assuming without loss of generality that $r_2 = 0$, so the notation r stands as shorthand for r_1, we have

$$r = u + Ky \tag{11.6}$$

Using this signal r, (11.1) can be simplified to

$$\begin{bmatrix} y \\ u \end{bmatrix} = \begin{bmatrix} G_o(I + KG_o)^{-1} \\ (I + KG_o)^{-1} \end{bmatrix} r + \begin{bmatrix} (I + KG_o)^{-1} \\ -K(I + KG_o)^{-1} \end{bmatrix} v$$

If the controller K stabilises the plant G_o, all elements of the matrix $T(G_o, K)$ will be stable. Hence, both $G_o(I + KG_o)^{-1}$ and $(I + KG_o)^{-1}$ will be stable and can be considered to be a coprime factorisation (N_o, D_o) of the plant G_o. Moreover, the signal r defined in (11.6) is uncorrelated with the noise v of the closed-loop system given in Figure 11.6. This gives rise to an equivalent open-loop identification problem by estimating a stable coprime factorisation of the plant using r as input and $[y \ u]^T$ as output.

However, a coprime factorisation is not unique and one can incorporate this freedom similarly as in [150], by introducing an additional filtering of the signal r, with $x := Fr$. Again, using (11.1) and considering x as input signal in the identification, this yields a cf $(N_{o,F}, D_{o,F})$ of the plant G_o, with

$$\begin{cases} N_{o,F} = G_o(I + KG_o)^{-1}F^{-1} \\ D_{o,F} = (I + KG_o)^{-1}F^{-1} \end{cases} \tag{11.7}$$

where the filter F denotes the additional freedom in the cf of the plant G_o. According to the result of Lemma 11.1, the freedom in F can be characterised by

$$F = D_x^{-1}[I + KG_x]^{-1} = [D_x + KN_x]^{-1} \tag{11.8}$$

with $G_x = N_x D_x^{-1}$ denoting any auxiliary model that is stabilised by K. The identification set-up is depicted in Figure 11.7, where $S_o = (1 + KG_o)^{-1}$.

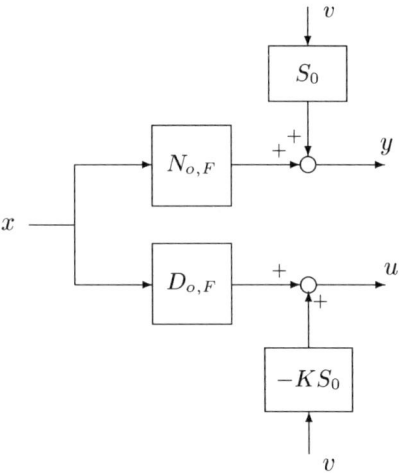

Fig. 11.7. Identification of coprime factors from closed-loop data

Since $x = Fr$ is uncorrelated with the noise v, the identification of the plant G_o from closed-loop measurements u and y is equivalent to an open-loop identification of $N_{o,F}$ and $D_{o,F}$, considering x as input signal and $(y, u)^T$ as output. Constructing a (filtered) one-step ahead prediction error for an output error (OE) type of model structure [108], we obtain:

$$\varepsilon(t, \theta) = \begin{bmatrix} y(t) \\ u(t) \end{bmatrix} - \begin{bmatrix} N(q, \theta) \\ D(q, \theta) \end{bmatrix} x(t) \tag{11.9}$$

$$\varepsilon_f(t, \theta) = L(q) I_{2 \times 2} \varepsilon(t, \theta) \tag{11.10}$$

where $L(q)$ is an additional user-defined filtering of the prediction error. By applying a least squares identification criterion

$$\hat{\theta}_N = \arg\min_{\theta} \frac{1}{N} \sum_{t=0}^{N-1} \{\varepsilon_f^T(t,\theta)\varepsilon_f(t,\theta)\} \tag{11.11}$$

it can be shown that for $N \to \infty$ and under the usual regularity conditions applied in prediction error identification, $\hat{\theta}_N \to \theta^\star$ with probability 1, with $\theta^\star = \arg\min_{\theta} \bar{V}(\theta)$ and

$$\bar{V}(\theta) = \frac{1}{2\pi} \int_{-\pi}^{\pi} \begin{bmatrix} N_{o,F}(e^{i\omega}) - N(e^{i\omega},\theta) \\ D_{o,F}(e^{i\omega}) - D(e^{i\omega},\theta) \end{bmatrix}^* \cdot \tag{11.12}$$

$$\begin{bmatrix} N_{o,F}(e^{i\omega}) - N(e^{i\omega},\theta) \\ D_{o,F}(e^{i\omega}) - D(e^{i\omega},\theta) \end{bmatrix} |L(e^{i\omega})|^2 \Phi_x(\omega) \, d\omega \tag{11.13}$$

In this expression, $\Phi_x(\omega)$ is the auto spectral density of the signal $x(t)$. The interpretation of the frequency domain representation will be scrutinised in the following subsection.

11.6.3 Feedback-relevant Identification

The difference $\Delta T(G_o, \hat{G}, K) := T(G_o, K) - T(\hat{G}, K)$, introduced in Section 11.5, can be seen as a feedback-relevant mismatch between the plant G_o and the nominal model $\hat{G}$. This mismatch can be expressed in terms of coprime factors. In [150], it has been analysed that under the condition that $N_{o,F}$ and $D_{o,F}$ are normalised coprime factors, $\Delta T(G_o, \hat{G}, K)$ can be written as

$$\Delta T(G_o, \hat{G}, K) = \begin{bmatrix} N_{o,F} - \hat{N} \\ D_{o,F} - \hat{D} \end{bmatrix} F[K \ I]. \tag{11.14}$$

Apparently, the feedback-relevant mismatch between G_o and $\hat{G}$ is a linear function of the difference between the cf of the model $\hat{G}$ and the corresponding cf of the plant G_o. By replacing the norm operator $\|\cdot\|$ by the $\mathcal{H}_2$-norm [112], the following quadratic feedback-relevant performance criterion $J_f(\theta)$, based on (11.14), can be defined

$$J_f(\theta) := \frac{1}{2\pi} \int_{-\pi}^{\pi} \begin{bmatrix} N_{o,F}(e^{i\omega}) - N(e^{i\omega},\theta) \\ D_{o,F}(e^{i\omega}) - D(e^{i\omega},\theta) \end{bmatrix}^* \cdot$$

$$\begin{bmatrix} N_{o,F}(e^{i\omega}) - N(e^{i\omega},\theta) \\ D_{o,F}(e^{i\omega}) - D(e^{i\omega},\theta) \end{bmatrix} |F|^2 (1 + |K|^2) \, d\omega \tag{11.15}$$

Comparing the feedback-relevant performance criterion $J_f(\theta)$ in (11.15) with the frequency domain representation of the least squares OE minimisation given in (11.13), it follows that the two criteria are equivalent in the case

$$|L(e^{i\omega})|^2 \Phi_x(\omega) = |F(e^{i\omega})|^2 \cdot (1 + |K(e^{i\omega})|^2)$$

or equivalently,

$$|L(e^{i\omega})|^2 = c_1 \frac{1 + |K(e^{i\omega})|^2}{\Phi_r(\omega)}$$

where $c_1 \neq 0$ is an arbitrary constant and $\Phi_r(\omega)$ is the auto spectral density of the excitation signal r.

As mentioned above, for the validity of (11.14), the accessed plant factors $N_{o,F}$, $D_{o,F}$ have to be normalised. This can be (approximately) achieved by using a normalised cf (N_x, D_x) of an auxiliary model G_x that approximately equals G_o. From (11.7) and (11.8), it then follows that $N_{o,F}$, $D_{o,F}$ become normalised too; a more detailed explanation can be found in [150].

11.7 Application to CD Radial Actuator

11.7.1 Data Acquisition

Measurements of the compact disc radial servo loop have been obtained from an experimental set up of a compact disc player provided by Philips Research Laboratories in Eindhoven, the Netherlands. This experimental set-up is used to gather a closed-loop measured time sequence of 8192 data points of the voltage input signal $u(t)$ to the radial actuator and the track position error signal $y(t)$.

For the closed-loop experiments, the same (and known) PID controller that caused the excessive peaking in the sensitivity function indicated in Figure 11.5 was used. During the closed-loop experiment, a periodic reference signal $r(t) = r_1(t)$ was added to the input signal u constructed by taking one realisation of 1024 data points from a white noise stochastic process, and concatenating this signal eight times. The gathered time sequence of $u(t)$ and $y(t)$ and the knowledge of the controller K were used to perform a control-relevant estimation of a coprime factorisation of the radial actuator G_o.

11.7.2 Estimation of Coprime Factors

To access a coprime factorisation of G_o from a closed-loop experiment, first a filter F and the signal x must be created. Using a relatively simple model that consists of a double integrator and the lateral bending mode of the radial arm, a fourth order auxiliary model G_x with a normalised cf (N_x, D_x) is created. The coprime factor model (N_x, D_x) is used to create the filter F and the signal x, according to $x = Fr$ with F given by (11.8).

Subsequently, a relatively high-order (order 32) model of the coprime factors of the radial actuator is estimated. This relatively high-order coprime factor model is used to update the knowledge of the coprime factorisation

(N_x, D_x) of the auxiliary model G_x. An amplitude Bode plot of the updated high-order coprime factor model (N_x, D_x) is given in Figure 11.8 (left). The amplitude Bode plot of the model (N_x, D_x) is compared with a measured frequency response using the signals x and (y, u).

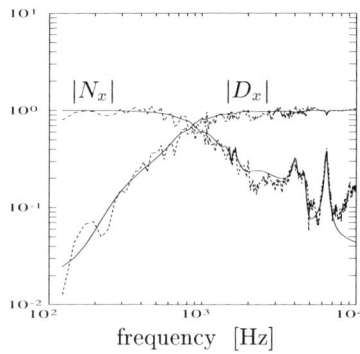 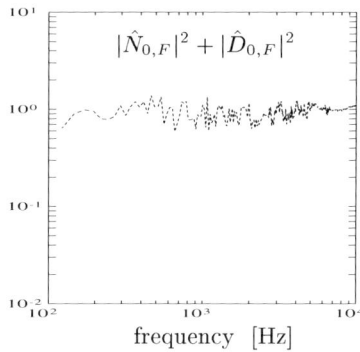

Fig. 11.8. *Left*: Bode magnitude plots of coprime factors (N_x, D_x) of auxiliary model G_x; spectral estimate (dashed) and thirty-second order parametric estimate (solid); *right*: $|N_{0,F}|^2 + |D_{0,F}|^2$ induced by the prefilter $F = (D_x + KN_x)^{-1}$

It should be noted that the relatively high-order (32) coprime factor model (N_x, D_x) depicted in Figure 11.8 is not the actual low-order control-relevant model of the radial actuator that needs to be estimated. The high-order factorisation (N_x, D_x) is used *only* to create the filter $F = (D_x + KN_x)^{-1}$ and therefore the order is not of crucial importance. With the updated information of the auxiliary model G_x in the form of the high-order coprime factor model (N_x, D_x), a new filter F and signal x can be constructed for identification purposes. With the updated and more accurate auxiliary model, the filter F will simplify the dynamics of the coprime factors of the plant in (11.7) and facilitate the minimisation of the control-relevant criterion given in (11.14). In the right picture of Figure 11.8, it is indicated that with the chosen filter F, the accessible coprime plant factors $(N_{o,F}, D_{o,F})$ are (almost) normalised. The verification of $|N_{o,F}|^2 + |D_{o,F}|^2 = 1$ is being done on the basis of spectral estimates.

For approximation and control applications, a low-order and control-relevant coprime factor model $(\hat{N}, \hat{D})$ needs to be estimated. A Bode plot of a relative low (tenth) order coprime factor model $(\hat{N}, \hat{D})$ is given in Figure 11.9. In these plots, the model $(\hat{N}, \hat{D})$ is compared with a measured frequency response using the signals x and (y, u).

Compared with the spectral estimate of the coprime factorisation of the radial actuator, it can be observed from Figure 11.9 that the essential dy-

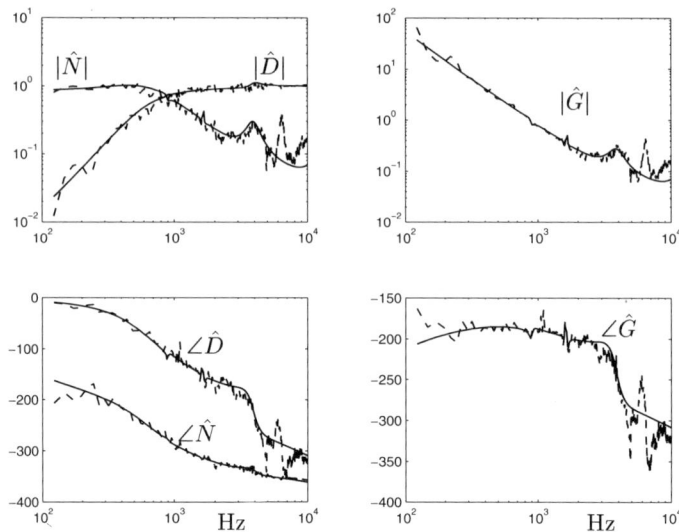

Fig. 11.9. Amplitude (upper) and phase (lower) Bode plots of spectral estimate (dashed) and parametric estimate (solid) of low (tenth) order coprime factors $(\hat{N}, \hat{D})$ (left) and model $\hat{G} = \hat{N}/\hat{D}$ (right)

namics that would explain the excessive peaking of the sensitivity function in Figure 11.5 have been captured well in the tenth order coprime factor model.

11.8 Control Design and Stability Robustness

The improved knowledge of the radial actuator in the form of the model $\hat{G} = \hat{N}/\hat{D}$ can be used to redesign the controller K. This controller synthesis is carried out by means of the weighed optimisation:

$$K_{\hat{G},\nu_b} = \arg_{\tilde{K}} \min \left\| \begin{bmatrix} W_{\nu_b}\hat{G} \\ 1 \end{bmatrix} [1 + \tilde{K}\hat{G}]^{-1}[\tilde{K}/W_{\nu_b} \ \ 1] \right\|_\infty \quad (11.16)$$

where W_{ν_b} is a filter, designed for a nominal design bandwidth of ν_b Hz. The controller synthesis is an $\mathcal{H}_\infty$ loop shape design according to McFarlane and Glover [35, 114], which is known to optimise robustness against additive perturbations of the coprime factors of the model. The weighting filter W_{ν_b} is designed to shape the loop transfer $W_{\nu_b}\hat{G}$ to achieve a pre-specified bandwidth, and additional robustness properties are achieved by solving the optimisation problem (11.16). The weighting function applied here is constructed according to:

$$W_{\nu_b} := \frac{K}{s} \cdot \frac{\tau_1 s + 1}{\tau_2 s + 1}$$

with K, τ_1, τ_2 appropriately chosen to achieve a nominal design bandwidth of ν_b. W_{ν_b} incorporates a phase lead in the crossover frequency region, which allows the stabilisation of a double integrator; it incorporates an integrator to guarantee good disturbance attenuation for low frequencies; additionally there is a controller roll-off for high frequencies. This control design procedure was used to design a new (improved) feedback controller $K_{\hat{G}}$ of order 8 with a nominal design bandwidth of $\nu_b = 800$ Hz. An amplitude Bode plot is depicted in Figure 11.10.

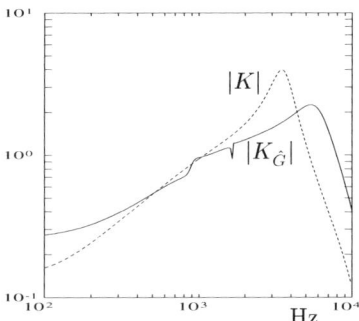

Fig. 11.10. Amplitude plots of traditional PID controller K (dashed) and newly-redesigned eigth order radial servo controller $K_{\hat{G}}$ (solid)

Before the newly-designed controller can be safely implemented, it has to be verified whether it will stabilise the radial control loop. One of the implemented procedures is directed towards estimating an upper bound on the additive error of the normalised coprime factors $(\hat{N}, \hat{D})$ being estimated. Denoting:

$$N_{o,F}(e^{i\omega}) = \hat{N}(e^{i\omega}) + \Delta_N(e^{i\omega}) \tag{11.17}$$
$$D_{o,F}(e^{i\omega}) = \hat{D}(e^{i\omega}) + \Delta_D(e^{i\omega}) \tag{11.18}$$

the uncertainty bounding procedure of [48] is a mixed deterministic/probabilistic approach that is fully compatible with the prediction error identification framework, delivering upper bounds on the additive model errors $|\Delta_N(e^{i\omega})|$ and $|\Delta_D(e^{i\omega})|$. It provides frequency-dependent upper bounds with an *a priori* chosen probability level per frequency:

$$\left. \begin{array}{l} |\Delta_D(e^{i\omega})| \leq \gamma_1(\omega) \\ |\Delta_N(e^{i\omega})| \leq \gamma_2(\omega) \end{array} \right\} \text{ w.p.} \geq \alpha \tag{11.19}$$

where α is a prespecified probability. The uncertainty bounding procedure combines a deterministic (worst-case) bound on the undermodelling (bias) error, with a stochastic confidence interval for noise-induced uncertainty. The bounding procedure is performed on the basis of periodic excitation signals.

Results of this procedure applied to the estimated coprime factors are depicted in Figure 11.11, showing three upper bounds on the additive errors, relating to three different choices of α.

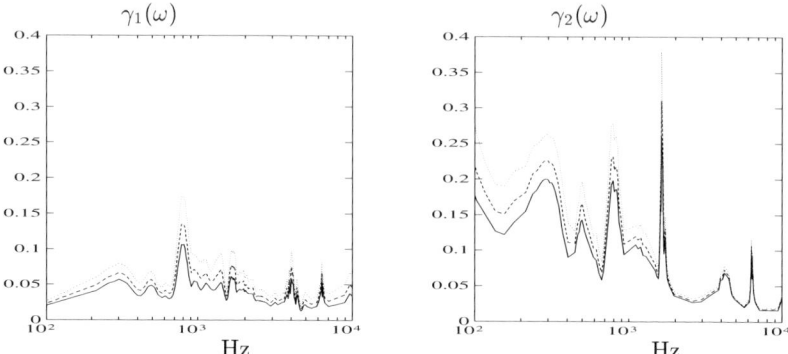

Fig. 11.11. Upper bounds $\gamma_1(\omega)$ and $\gamma_2(\omega)$ on additive model error of normalised coprime factors $|\hat{D}(e^{i\omega})|$ (left) and $|\hat{N}(e^{i\omega})|$ (right) for $\alpha = 90\%$ (—), 99% (- -) and 99.9% ($\cdots$)

Using this knowledge of an upper bound on (Δ_N, Δ_D), we can check stability robustness properties of the closed-loop system by employing the following result. Provided that $\hat{G} = \hat{N}\hat{D}^{-1}$ is stabilised by $K = D_k^{-1}N_k$, the set of plants $G = ND^{-1}$ determined by

$$N(e^{i\omega}) = \hat{N}(e^{i\omega}) + \Delta_N(e^{i\omega}), \quad |\Delta_N(e^{i\omega})| \leq \gamma_1(\omega) \tag{11.20}$$
$$D(e^{i\omega}) = \hat{D}(e^{i\omega}) + \Delta_D(e^{i\omega}), \quad |\Delta_D(e^{i\omega})| \leq \gamma_2(\omega) \tag{11.21}$$

is stabilised by K if [35]: $\|[N_k \ D_k] \begin{bmatrix} \Delta_N \\ \Delta_D \end{bmatrix} \hat{\Lambda}^{-1}\|_\infty < 1$,

with $\hat{\Lambda} = \tilde{D}_k \hat{D} + \tilde{N}_k \hat{N}$.

This condition can be checked by verifying whether

$$\frac{1}{|\hat{\Lambda}(e^{i\omega})|}[|N_k(e^{i\omega})|\gamma_1(\omega) + |D_k(e^{i\omega})|\gamma_2(\omega)] < 1 \quad \text{for all } \omega \tag{11.22}$$

This sufficient condition for robust stability has been implemented and its result is depicted in Figure 11.12. It shows that even for an α-level of larger than 99 per cent robust stability can be guaranteed. If α is chosen 99.9 per cent, the sufficient condition is not satisfied. However, this result is assessed to be sufficiently promising to allow implementation of the controller. Successful implementation of the controller $K_{\hat{G}}$ using a DSP environment yields a high-performance closed-loop system with a bandwidth of approximately 800 Hz without excessive peaking of the sensitivity function, leading to improved disturbance rejection. This has been illustrated in Figure 11.13.

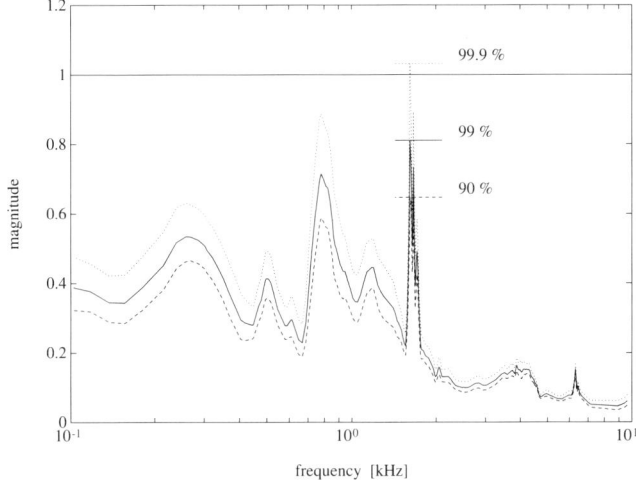

Fig. 11.12. Stability robustness tests of the new controller for point wise confidence intervals on the estimated coprime factors at three probability levels

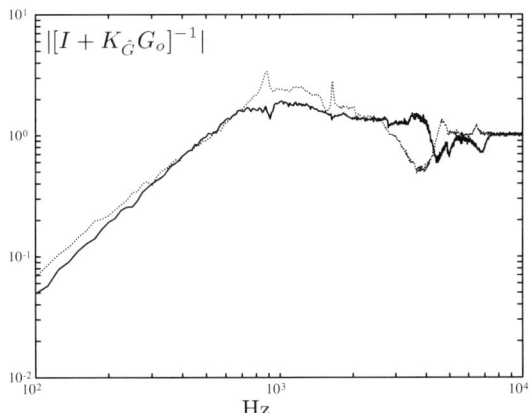

Fig. 11.13. Amplitude Bode plots of sensitivity function with traditional PID controller (dashed) and newly-redesigned eigth order radial servo controller $K_{\hat{G}}$ (solid)

11.9 Summary

In this chapter, a control-relevant parametric identification scheme is applied to a compact disc radial servo system, using the well-known prediction error methods. In the identification methodology, the problem of approximate (low-order) identification and closed-loop experiments are addressed. The problems arising from the closed-loop and approximate identification have been handled using an identification based on fractional representations. By appropriate filtering, this yields a manageable control-relevant identification of the plant, while pertaining to low-order models. Additionally, uncertainty bounding techniques have been shown to allow for a practically applicable stability robustness test that is performed prior to controller implementation. The designed and implemented controller has been shown to lead to the desired improved performance. A similar identification approach extended to the multivariable case, and in combination with a *robust* control design procedure, is incorporated in Chapter 12 of this book.

Acknowledgements

The authors acknowledge the contributions of Peter Bongers, Douwe de Vries, Hans Dötsch, Okko Bosgra and Maarten Steinbuch to the results presented here, and the support of Philips Research in providing the compact disc hardware.

12 Control-relevant Identification and Robust Motion Control of a Wafer Stage

Raymond A. de Callafon and Paul M.J. Van den Hof

Abstract. This chapter illustrates the identification of control-relevant models and the subsequent robust control design applied to a wafer stage. A wafer stage is part of a wafer stepper and used in chip manufacturing processes for accurate positioning of the silicon wafer on which the chips are to be produced. Accurate and fast positioning requires a robust and high-performance multivariable servo controller that enables a fast throughput of silicon wafers. An advanced servo controller is developed by an iterative procedure of control-relevant model identification including model uncertainty bounding, and robust control based on worst-case performance optimisation. Both stability and performance robustness can be monitored, enabling the possibility of guaranteeing performance improvement in a single step of the iteration. This is shown to lead to a successful design and implementation on a wafer stage set-up.

12.1 Introduction

The overlapping theme in the application discussed in this chapter is the robust and enhanced control of a mechanical positioning mechanism. The methodology presented and illustrated in this chapter is applicable to many mechanical systems that are subjected to high-performance servo positioning demands. The dedicated servo positioning mechanism considered in this chapter can be found in a wafer stepper. Wafer steppers combine a high-accuracy positioning and a sophisticated lithographic process to manufacture integrated circuits (chips) via a fully automated process.

By means of a photolithographic process, the chip architecture is exposed on the surface of a wafer, a silicon disc covered with a photoresist. Typically, the wafer can carry more than 80 chips and in order to expose the surface of the wafer, each chip has to be processed sequentially. Such a sequential process is needed as only one mask of the chip layout is available during the exposure phase of the photolithographic process. In order to enable sequential processing, the wafer is mounted on a wafer stage that needs to be accurately moved (stepped) in three degrees of freedom (3-DOF).

Both the accuracy and the speed of the wafer stage will influence the success and throughput of the production process of the chips on the wafer. Sophisticated design and control of this multivariable servo mechanism can help in achieving a required throughput. An important design step is the construction of a servo controller that is able to satisfy high-performance requirements: fast positioning to within an accuracy of ± 52 nm. The inherent multivariable (3-DOF) nature of the wafer stage and the high positioning performance requirements require a servo design approach in which the dynamic behaviour of the positioning mechanism is characterised accurately.

The required accuracy of the dynamical model will be dictated by the performance requirements imposed on the control system; additionally low-complexity (approximate) models are desired in order to limit the complexity of the subsequent control design and implementation. As a result, the modelling procedure will be subjected to the following basic requirements.

- a (nominal) model should be of low complexity and approximate the dominant dynamical behaviour that is relevant for high-performance control design;
- to address stability and performance robustness and facilitate the design of a robust performing controller, a nominal model should be accompanied with a bound on its uncertainty;
- closely related to the development of a nominal model, the size and shape of this model uncertainty is implicitly bounded by the requirements for high-performance control design.

Inevitably, low-complexity (nominal) models and a small model uncertainty are conflicting requirements. For high-performance servo control requirements, the trade-off between approximation and uncertainty provides a challenging modelling procedure. In the application of the wafer stage mechanism discussed in this chapter, the above-mentioned requirements are tackled by adopting an identification framework that is close to the framework discussed in the previous chapter. However besides an extension to the multivariable case, now an identified uncertainty set of models is used to guide a *robust* control design, rather than only veryfing stability robustness prior to controller implementation. The model uncertainty set is used to design a controller that has the best worst-case performance.

12.2 The Wafer Stepper Positioning Mechanism

12.2.1 Description of Servo Vechanism

The wafer stage servo mechanism is an integral part of a silicon repeater third generation (SIRE3) wafer stepper. The wafer stage is used to accurately position the silicon wafer on which the chips are to be produced. A schematic picture of the stage is shown in Figure 12.1.

The wafer chuck is part of the wafer stage and is equipped with an air bearing and placed on a large suspended granite block to reduce the effect of external vibrations. The wafer chuck is used to position the wafer in three degrees Of freedom along the surface of the granite block. As part of the mechanical positioning mechanism, three linear voice coil motors are mounted

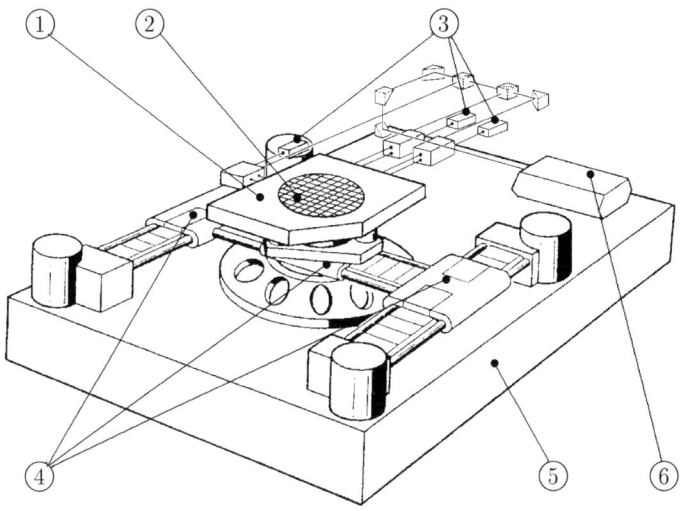

Fig. 12.1. Schematic view of a wafer stage; 1: mirror block, 2: wafer chuck, 3: laser interferometers, 4: linear motors, 5: granite block, 6: laser

in an H-shape on the granite block. The three linear motors are used to position the wafer chuck in 3-DOF along the surface of the granite block. Independent steering of the motors enables free surface translation with a relatively small rotational freedom.

The position in the horizontal plane is measured by means of laser interferometry. Relative movements are measured by determining the phase shift of the laser beams reflected on the mirror block depicted in Figure 12.2. As the horizontal plane allows three degrees of freedom, three laser measurements uniquely determine the horizontal position of the wafer.

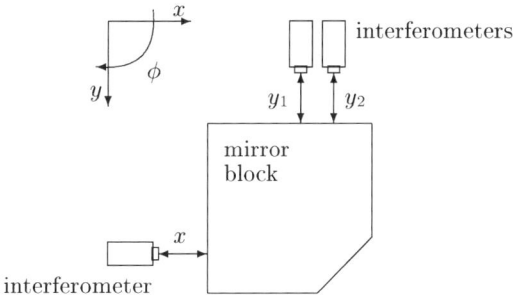

Fig. 12.2. The three position measurements on the mirror block

The three position measurements can be used to reconstruct the position of the stage in x-, y- and ϕ-direction. From the three (relative) position measurements given in Figure 12.2, the three degrees of freedom of the stage in the x-, y- and ϕ- direction can be obtained via

$$\begin{bmatrix} x \\ y \\ \phi \end{bmatrix} = T \begin{bmatrix} x \\ y_1 \\ y_2 \end{bmatrix} \quad \text{with } T := \begin{bmatrix} 1 & 0 & 0 \\ 0 & 1/2 & 1/2 \\ 0 & 1/2 & -1/2 \end{bmatrix}.$$

The transformation matrix T statically decouples the two measurements y_1 and y_2 in the y-direction. However, note that $\phi = (y_1 - y_2)/2$ is not the actual rotation of the stage, as this would require the computation involving a sinusoidal term. However, the rotation of the stage is limited and for a small difference between y_1 and y_2, e.g., small rotations, the value of ϕ is proportional to the angular rotation of the stage.

The three independent linear motors and the laser position measurements make the servo mechanism of the wafer stepper a multivariable system. The inputs reflect the currents to the three linear motors, whereas the outputs are the positions of the wafer chuck both in the x-, y-direction (translation) and in ϕ-direction (rotation).

12.2.2 Experimental Set-up

In order to gather experimental data for system identification purposes and to test the servo control of the wafer stage, an experimental set-up has been provided by Philips Research Laboratories. A close-up photograph of the set-up is provided in Figure 12.3; in it the granite block and the wafer stage are protected by a glass surface on top.

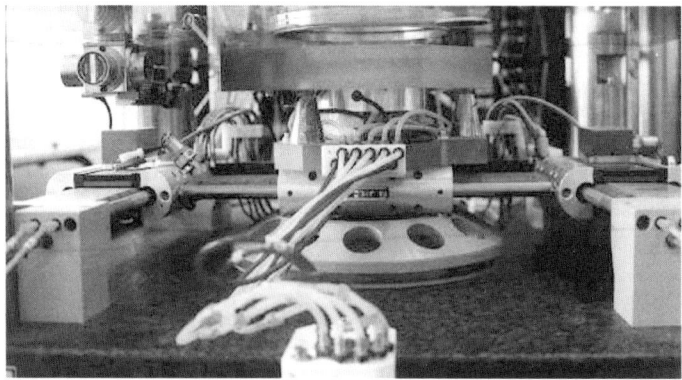

Fig. 12.3. Close-up of wafer stage displaying the three linear motors and the wafer chuck with air bearing

The experimental set-up is equipped with a computer interface to measure the position in x-, y- and ϕ-direction of the wafer chuck on discrete time samples via a digital signal processor. Due to safety requirements and operating conditions of the laser interferometers, the signals can be measured only if a digital controller is used to control the positioning of the wafer chuck. Such a digital controller can be implemented using the same digital signal processor. A schematic overview of the signals that can be accessed is depicted in the block diagram of Figure 12.4.

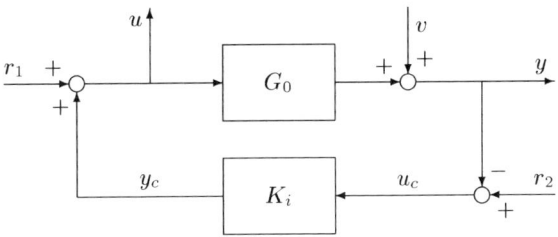

Fig. 12.4. Block diagram of experimental set-up of feedback controlled positioning mechanism

As indicated in Figure 12.4, the positioning mechanism of the wafer chuck is denoted by G_0, while the (initial) feedback controller currently used to control G_0 is denoted by K_i. For notation purposes, the feedback connection of G_0 and K_i is denoted by $\mathcal{T}(G_0, K_i)$.

The input u is used to indicate the three input signals to the linear motors for position actuation in x-, y- and ϕ-direction. In a similar way, the output signal y consists of the three position signals in x-, y- and ϕ-direction of the wafer chuck. Besides providing sufficient excitation of the feedback system, the reference signals in Figure 12.4 can be used to move or step the wafer chuck in a desired direction. As such, the signals r_1 and r_2 can be used to evaluate the performance of the feedback controlled plant by applying a reference signal r_2 and a feedforward signal r_1 in order to track a certain desired 3-DOF position signal y of the wafer chuck. The input signal u_c to the controller K_i reflects the servo error between a desired reference r_2 and the actual desired position y.

12.2.3 Modelling and Servo Performance of the Positioning Mechanism

With the previous conventional closed-loop servo control strategy, the servo controller K_i consists of three parallel PID controllers. In this situation, the interaction in the 3-DOF positioning mechanism is clearly neglected, and the PID controllers are able to achieve only moderate control performance.

Servo performance requires a minimisation of the servo error, while moving the chuck as fast as possible. The design specification for the SIRE3 wafer stepper is to bring the servo error within a bound of 52 nm (four times the measurement resolution) as soon as possible after a step trajectory has been performed. This is due to the fact that the chuck must be kept in a constant position before a chip can be exposed on the surface of the wafer.

In order to compare the servo performance of (newly-designed) feedback controllers, the reference signals r_2 and r_1 are fixed to some pre-specified desired trajectory for servo performance evaluation. This pre-specified trajectory is based on the dominating open-loop dynamical behaviour of the wafer stage G_0 that is given by a double integrator, relating the force generated by the linear motors to the position of the wafer chuck. Based on this relatively simple model, r_2 will denote a desired position profile, allowing a maximum speed and a maximum jerk (derivative of acceleration), whereas r_1 denotes (a scaled) acceleration profile obtained by computing the second derivative of r_2. A typical shape of the reference signal r_2 and the feedforward signal r_1 to position the wafer chuck in either the x- or y- direction over 1cm is depicted in Figure 12.5.

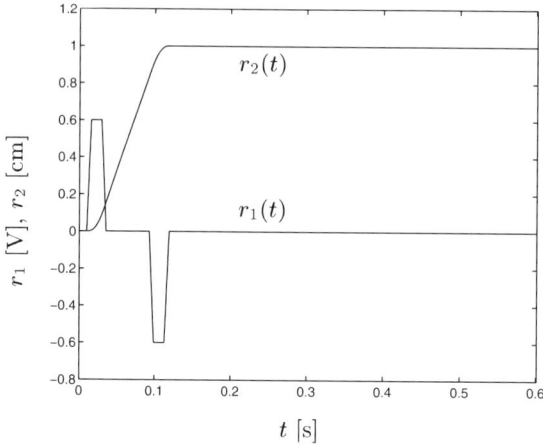

Fig. 12.5. Shape of reference signal r_2 and feedforward signal r_1 for servo performance evaluations

Although optimal reference signals can be designed for finite time optimal control problems, it should be noted that stepwise reference signals are being used here *only* to compare the performance of the servo controllers being designed and actually implemented on the wafer stage.

The specified reference signals r_1 and r_2 in Figure 12.5 can be used to create a step in the x-direction for the evaluation of the performance of the three parallel PID servo controllers. The resulting servo error $u_{c,x}$ for a step in the x-direction is depicted in Figure 12.6.

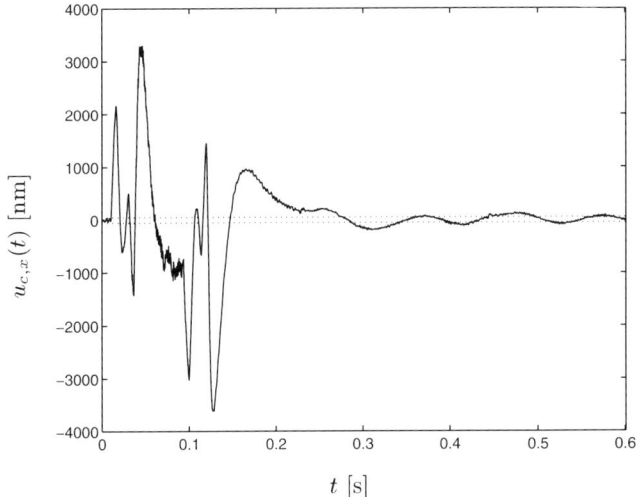

Fig. 12.6. Servo error response to a step in the x-direction using conventional parallel PID controllers

It can be observed from Figure 12.6 that the servo error $u_{c,x}$ is hardly within the bounds of 52 nm indicated by the dotted lines, even after 0.6 s. Furthermore, $u_{c,x}$ exhibits a low-frequency vibration after the step has ended (after 0.12 s). As a result, the settling time of the positioning step is strongly influenced and both an improvement of the speed of decay and a reduction of the low-frequency vibration of the servo error is desired to improve the behaviour of the servo mechanism. Clearly, the process under consideration exhibits dynamic phenomena additional to the simple model of a double integrator. Characterisation of these dynamics in a way relevant for control design is essential in achieving maximum control performance.

For both the design of the feedback controller K_i and appropriate reference signals, a dynamic model of the wafer stage is required. In this chapter, the attention is focused on the construction and use of dynamical models for the design of a feedback controller K_i.

12.2.4 Experiment Design

Experimental data from the wafer stage are gathered from the feedback connection $\mathcal{T}(G_0, K_i)$ depicted in Figure 12.4. As mentioned in Section 12.2.3, the reference signals r_1 and r_2 are used to specify respectively an acceleration reference and a position reference profile to evaluate the servo performance of a feedback controller. However, from an identification point of view, the reference signals are used to excite the closed-loop system $\mathcal{T}(G_0, K_i)$ to avoid problems associated with closed-loop identifiability.

Although the signals r_1 and r_2 can be used to conduct identification experiments, mostly low frequent information is contained in the signals depicted in Figure 12.5. To get enough information on the system in the frequency range of interest, different reference signals have to be used. Furthermore, the flexibility of the experimental set-up and the speed of the mechanical system allow the use of excitation signals to obtain a frequency response estimate of the positioning mechanism in the frequency range from approximately 10 Hz till 1 kHz.

For identification purposes, the reference signals are specified as periodic signals constructed via a sum of sinusoids

$$r(t) := \sum_{i=1}^{l} sin(\omega_i t + \phi_i) \tag{12.1}$$

specified at a pre-defined frequency grid $\Omega = \{\omega \mid \omega = \omega_i,\ i = 1, 2, \ldots, n\}$. The phase shifts ϕ_i in the sequence $\{\phi_i\}$, $i = 1, 2, \ldots l$ of (12.1) are chosen independently from a uniform distribution over the interval $(-\pi, \pi)$. In this way, the six[1] reference signals will be uncorrelated and a random phased sequence of sinusoids is generated with favourable properties [128].

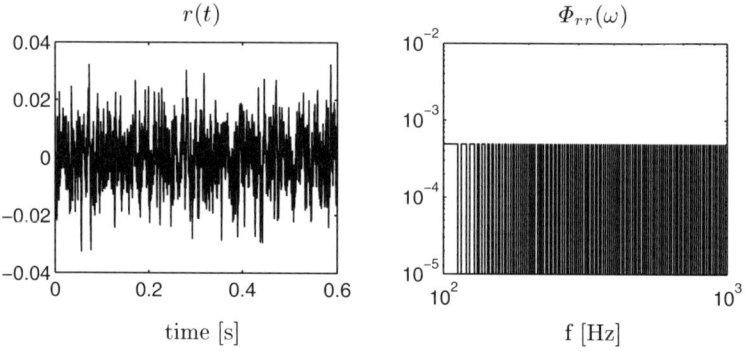

Fig. 12.7. Time domain plot (left) and spectrum (right) of the reference signal r_1 in x-direction configured as a sum of 200 sinusoids with random phase

The designed excitation signals are periodic with a period length of 2048 data points sampled at $T_s = 0.3$ ms. As a result, the frequency resolution $\Delta f = 1/(2048 \cdot 0.003) \approx 1.628$ Hz. They are each composed of 200 sinusoids distributed (approximately logarithmically) between $9 \times \Delta f \approx 14.65$ Hz and $714 \times \Delta f \approx 1162.11$ Hz.

---

[1] Both r_1 and r_2 consist of three reference signals, respectively in x-, y- and ϕ-direction.

For illustration purposes, a time-domain plot and the spectrum density in the frequency range between 100 Hz and 1 kHz of one of the six references signals is shown in Figure 12.7. It can be observed that the signal has a well-bounded amplitude. Although the signal looks like a noise in the time-domain, the spectrum is very well defined for the 200 frequency points in the frequency grid Ω. With these reference signals, the identification experiments have been performed. They will be further explained and analysed in the sequel of this chapter.

12.3 Iterative Model Set Estimation and Control Design

12.3.1 Motivation of Model Set Estimation

The motivation to apply an iterative procedure of identification and control is induced by the fact that a simultaneous (off-line) optimisation of both an identification and a model-based control design criterion can be highly non-linear [20]. Although convergence and optimality of iterative schemes cannot be guaranteed in general, countless numerical simulation examples presented in the literature show promising results [19, 69, 102, 138, 161].

From a practical point of view, it is valuable to monitor and guarantee performance improvement of the servo control system, while performing a step of subsequent identification and model-based control design. In this way, effort put into the steps of an iterative scheme can be justified by assuring an improvement of the feedback controlled plant.

Guaranteeing an improvement of the performance of a feedback controlled plant cannot be achieved by iteratively trying to improve nominal performance specifications alone. As the resulting model is just an approximation of the system to be identified, the controller based on the model has to be robust against any dissimilarities between the model and the system. This has been a motivation for the development of identification techniques that estimate an upper bound on the model uncertainty (model error), see, for example, the survey [117] and the more recent contributions [31, 70]. The resulting model uncertainty bounds constitute an allowable perturbation around a nominal model being estimated and define a set of models $\mathcal{G}$ of which the actual system G_0 is assumed to be an element. Subsequently, a robust controller can be designed on the basis of this set of models [164]. In this approach, stability and performance requirements are guaranteed for the complete set of models (which includes the actual system to be controlled).

12.3.2 Enhancement of Control Performance

To formalise the notion of a set of models and the characterisation of robust control performance for them, the following notation is considered. The notation G will be used to denote a linear time-invariant system that may represent the actual wafer stage denoted by G_0 or a (nominal) model of the wafer stage denoted by $\hat{G}$. Furthermore, let $\mathcal{G}$ be used to denote a set of models and K to represent a feedback controller. To indicate the progress in the iterative scheme of subsequent approximate modelling and model-based control design, the subscript i will be applied to the variables $\hat{G}$, $\mathcal{G}$ or K to indicate that the variable depends on the i-th step of the iterative scheme. Finally, a control objective function is denoted by $J(G, K)$ and the notion of control performance will be characterised by the value of a norm $\|J(G,K)\|$; a smaller value of $\|J(G,K)\|$ indicates better closed-loop control performance.

Examples of commonly used control objective functions may include mixed sensitivity, as well as LQG and IMC type of control objectives, see, e.g., [148]. Throughout this chapter, the control objective function $J(G,K) \in \mathcal{RH}_\infty$ and is restricted to $\mathcal{H}_\infty$-norm based performance specifications, allowing for worst-case and robust performance control design methodologies to be used.

Guaranteeing an improvement of the servo performance of a feedback controlled plant can be formalised by considering a feedback connection of the wafer stage plant G_0 and a controller K_i that satisfies a performance specification $\|J(G_0, K_i)\|_\infty \leq \gamma_i$. To improve control performance, a new and improved controller K_{i+1} has to be designed such that the performance $\|J(G_0, K_{i+1})\|_\infty$ satisfies

$$\|J(G_0, K_{i+1})\|_\infty \leq \gamma_{i+1} < \gamma_i \qquad (12.2)$$

To make the design problem in (12.2) tractable for an unknown plant G_0, basically two main items should be considered. Firstly, a procedure must be found to access the performance γ_i for $\|J(G_0, K_i)\|_\infty$ *a posteriori*, *i.e.*, when the controller K_i is implemented on the plant G_0. Secondly, the synthesis of a controller K_{i+1} must be formulated that satisfies the performance condition (12.2) *a priori* before implementing the controller K_{i+1} on the plant G_0. To accomplish both aspects, a set of models $\mathcal{G}$ will be identified and used for stability and performance robustness assessment. The following general procedure is followed.

Procedure 12.1. Let a plant G_0 and an initial controller K_i form a stable feedback connection. To improve control performance undertake the following steps:

(a) **Performance assessment by model set identification**
Use experimental data and prior information on both the data and the plant G_0 to estimate a set of models $\mathcal{G}_i$ such that $G_0 \in \mathcal{G}_i$ and determine

$$\gamma_i = \sup_{G \in \mathcal{G}_i} \|J(G, K_i)\|_\infty. \qquad (12.3)$$

Subsequently, consider the following subsequent steps for performance robustness improvement of the feedback controlled plant G_0;

(b) **Robust control design on the basis of identified model set**
Design a new controller K_{i+1} such that

$$\|J(G, K_{i+1})\|_\infty \leq \gamma_{i+1} < \gamma_i \ \forall G \in \mathcal{G}_i \quad (12.4)$$

and, when achieved, implement the controller.

(c) **Validate robust controller and identify a new model set**
Use (new) experimental data and prior information on both the data and the plant G_0 to estimate a set of models $\mathcal{G}_{i+1}$ such that $G_0 \in \mathcal{G}_{i+1}$ and

$$\|J(G, K_{i+1})\|_\infty \leq \gamma_{i+1} \ \forall G \in \mathcal{G}_{i+1}. \quad (12.5)$$

The formulation of Procedure 12.1 is a rather general set-up to generate a sequence of model-based controllers that will satisfy (12.2). Within this set-up, step (b) reflects the design of a robust controller in order to ensure (12.2). Both steps (a) and (c) contain the estimation of a set of models $\mathcal{G}$. These steps will constitute an identification problem to estimate the set $\mathcal{G}$ and/or a model (in)validation problem [143] in order to guarantee $G_0 \in \mathcal{G}$. However, both the identification problem and the model (in)validation problem should be control-relevant. This is reflected by the fact that the quality of the models G within a set $\mathcal{G}$ is evaluated by the performance specification $\|J(G, K)\|_\infty$, where steps (a) and (c) differ only in the feedback controller K being used.

By iterating on the subsequent steps (b) and (c), an iterative scheme of identification and control is formulated, where (12.4) and (12.5) reflect respectively a controller and a model closed-loop *validation test* in order to enforce (12.2). Starting from step (a), where K_i is the controller (initially) implemented on the plant G_0, (12.3) can be viewed as an initial closed-loop performance assessment test to evaluate $\|J(G_0, K_i)\|_\infty$ *a posteriori*. In the robust control design of step (b), Equation 12.4 is needed to ensure (12.2) *a priori*. In this way, both performance robustness and improvement of the upper bound on the closed-loop performance can be guaranteed for K_{i+1}. The performance $\|J(G_0, K_i)\|_\infty$ can be evaluated *a posteriori*, by implementation of K_{i+1} on the plant G_0 and estimating a new set $\mathcal{G}_{i+1}$. If indeed (12.5) is satisfied, in step (b) again a new controller can be designed on the basis of $\mathcal{G}_{i+1}$.

Although the problem formulation in Procedure 12.1 is fairly general and somehow trivial, it does provide a monotonic non-decreasing sequence of γ_i. A similar idea was proposed in [20]. Obviously, to provide a feasible procedure for handling Procedure 12.1, the choice of structure of the set of models $\mathcal{G}$ should be addressed [149].

Summarising, the following items will play an important role in the realisation of an iterative scheme as proposed here:

- the control objective function $J(G, K)$. The objective function plays a crucial role in the characterisation of performance robustness and the way in which a controller is designed;

- evaluation of the closed-loop performance (12.3) and the closed-loop validation test of (12.4) and (12.5) in a non-conservative way. A proper choice for the structure of the set of models $\mathcal{G}$ will benefit the evaluation of the performance robustness;
- identification procedure to estimate a set of models $\mathcal{G}$. Similar procedures are needed in steps (a) and (c) by considering respectively the feedback controllers K_i and K_{i+1} and the estimation procedure should take into account the control design application of the set $\mathcal{G}$;
- robust control design method. The design of a controller on the basis of a set of models $\mathcal{G}$ in step (b) is needed to ensure (12.2).

In the remaining part of this chapter, the iterative scheme of model set estimation and robust control design will be described and applied to the motion control system of the wafer stage. The items mentioned above will be outlined in separate sections to clearly indicate the different aspects.

12.4 Elements of the Iterative Scheme

12.4.1 Performance and Control Objective

The mapping from the reference signals (r_2, r_1) to the output and input signals (y, u) of the plant in Figure 12.4 is given by the transfer function matrix $T(G_0, K_i)$ with

$$T(G_0, K_i) := \begin{bmatrix} G_0 \\ I \end{bmatrix} (I + K_i G_0)^{-1} \begin{bmatrix} K_i & I \end{bmatrix} \tag{12.6}$$

As a result, the data obtained from the feedback connection $\mathcal{T}(G_0, K_i)$ of Figure 12.4 can be described by

$$\begin{bmatrix} y \\ u \end{bmatrix} = T(G_0, K_i) \begin{bmatrix} r_2 \\ r_1 \end{bmatrix} + \begin{bmatrix} I \\ -K_i \end{bmatrix} (I + G_0 K_i)^{-1} v \tag{12.7}$$

The transfer function matrix $T(G_0, K_i)$ characterizes all closed-loop properties of a feedback connection of plant G_0 and controller K_i. Note that a feedback connection $\mathcal{T}(G, K)$ is internally stable if and only if $T(G, K)$ is stable.

In order to incorporate control design specification for the map $T(G, K)$, the control objective function $J(G, K)$ is taken to be a weighted form of the matrix $T(G, K)$ given in (12.6) and is defined as

$$\|J(G, K)\|_\infty := \|U_2 T(G, K) U_1\|_\infty \tag{12.8}$$

where U_2 and U_1 are (square) weighting functions. These weighting functions are chosen in such a way that the bandwidth of the resulting feedback connection can be adjusted, which will increase the speed of decay of the resulting servo error depicted in Figure 12.6. Furthermore, the weighting functions can be used to design a controller K that allows for an additional suppression of the low frequent vibration of the servo error.

In this particular situation, the weighting functions are chosen to comply with a loop shaped situation. By choosing:

$$U_2 = \begin{bmatrix} U_l & 0 \\ 0 & U_r^{-1} \end{bmatrix} \quad U_1 = U_2^{-1}$$

the performance objective function J can be written as

$$J(G,K) = \begin{bmatrix} U_l G \\ U_r^{-1} I \end{bmatrix} [I + KG]^{-1} [KU_l^{-1} \quad U_r] \tag{12.9}$$

$$= T(G_{ls}, K_{ls}) \tag{12.10}$$

with $G_{ls} = U_l G U_r$ and $K = U_r K_{ls} U_l$.

The performance characterization (12.8) is fairly general and will be used for analysing performance robustness and designing servo controllers for the wafer stage.

12.4.2 Set of Models and Evaluation of Performance Robustness

The characterisation of a set of models allows one to capture the actual system $G_0 \in \mathcal{G}$. In this way, performance robustness can be monitored via a worst-case performance evaluation over the set of models $\mathcal{G}$, whereas performance robustness can be enforced via a robust controller design.

Employing the knowledge of the stabilising controller K_i, a set of models $\mathcal{G}_i$ can be characterised by using the algebraic theory of fractional model representations [152]. The uncertainty structure used in our procedure is based on a dual-Youla parameterisation:

$$\mathcal{G}_i = \{G \mid G = (\hat{N} + D_k \Delta)(\hat{D} - N_k \Delta)^{-1},$$
$$\text{with } \Delta \in \mathcal{RH}_\infty \text{ and } \|\hat{V} \Delta \hat{W}\|_\infty < \gamma_i^{-1}\} \tag{12.11}$$

where (N_k, D_k) denotes a right coprime factorisation (RCF) of the controller K_i implemented on the plant G_0 during closed-loop experiments. Similarly, $(\hat{N}, \hat{D})$ denotes an RCF of a nominal model $\hat{G}_i$ of the plant G_0 that satisfies $T(\hat{G}_i, K_i) \in \mathcal{RH}_\infty$. The (stable and stably invertible) weighting functions $\hat{V}$, $\hat{W}$ are used to normalise the upper bound on $\hat{V} \Delta \hat{W}$ to γ_i^{-1}. A schematic representation of the uncertainty structure is depicted in Figure 12.8.

The uncertainty structure in (12.11) has several attractive properties that makes it useful for closed-loop identification and worst-case performance analysis:

- all models $G \in \mathcal{G}_i$, including the unknown plant G_0, are guaranteed to be stabilised by the current controller K_i. This property indicates that the dual-Youla-based set of models exploits the information that the actual

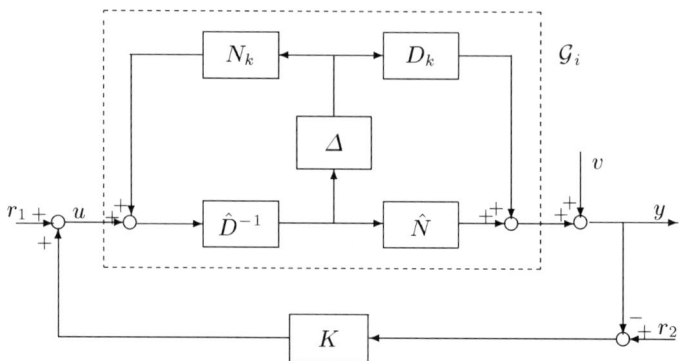

Fig. 12.8. Block scheme of dual-Youla-based uncertainty structure of models $G \in \mathcal{G}_i$ operating in closed-loop with a controller K

plant G_0 has been stabilised by the controller K_i during the closed-loop experiments;
- the calculation of the worst-case performance of a controller K over all models $G \in \mathcal{G}_i$ can be done relatively easy. In fact, the worst-case performance criterion reduces to an affine expression in Δ if the worst-case performance needs to be evaluated for the controller K_i used in characterising the model set $\mathcal{G}_i$ in (12.11).
- the uncertainty term Δ is accessible from closed-loop data. This will be utilised in the actual estimation of an upper bound for the uncertainty on the basis of closed-loop experiments.

In order to explain the favourable properties listed above, a linear fractional transformation (LFT) description of the set $\mathcal{G}_i$ in (12.11) is used to simplify the algebraic manipulations. An (upper) LFT $\mathcal{F}_u(Q, \Delta)$ is defined by

$$\mathcal{F}_u(Q, \Delta) := Q_{22} + Q_{21}\Delta(I - Q_{11}\Delta)^{-1}Q_{12} \qquad (12.12)$$

provided that $(I - Q_{11}\Delta)^{-1}$ exists. The set of models $\mathcal{G}_i$ in (12.11) can then be characterised by

$$\mathcal{G}_i = \{G \mid G = \mathcal{F}_u(Q_i, \Delta), \text{ with } \Delta \in \mathcal{RH}_\infty \text{ and } \|\Delta\|_\infty < \gamma_i^{-1}\}$$

where Δ indicates the same unknown (but bounded) uncertainty as in (12.11). The entries of the coefficient matrix Q_i in (12.12) dictate the way in which the set of models $\mathcal{G}$ is being structured. For the set $\mathcal{G}_i$ in (12.11), it can be verified that the entries of Q_i are given by

$$Q_i = \begin{bmatrix} \hat{W} & 0 \\ 0 & I \end{bmatrix} \begin{bmatrix} \hat{D}^{-1}N_k & \hat{D}^{-1} \\ (D_k + \hat{G}_i N_k) & \hat{G} \end{bmatrix} \begin{bmatrix} \hat{V}^{-1} & 0 \\ 0 & I \end{bmatrix}. \qquad (12.13)$$

As a special entry, one can recognise the nominal model $\hat{G}_i = \mathcal{F}(Q_i, 0) = Q_{22}$.

Since $\mathcal{G}_i$ is characterised by an LFT, the connection between any controller K and the set of models $\mathcal{G}_i$ can also be characterised by an LFT, thus enabling a relatively simple calculation of worst-case performance when K is applied to all models within $\mathcal{G}_i$. It appears that, for all models $G \in \mathcal{G}_i$, the control objective function $J(G, K)$ can be written as

$$J(G, K) = \mathcal{F}_u(M_i, \Delta)$$

with the entries of M_i given by (see [46]):

$$\begin{aligned}
M_{11} &= -\hat{W}^{-1}(\hat{D} + K\hat{N})^{-1}(K - K_i)D_k\hat{V}^{-1} \\
M_{12} &= \hat{W}^{-1}(\hat{D} + K\hat{N})^{-1}\begin{bmatrix} K & I \end{bmatrix}U_1 \\
M_{21} &= -U_2\begin{bmatrix} -I \\ K \end{bmatrix}(I + \hat{G}_iK)^{-1}(I + \hat{G}_iK_i)D_k\hat{V}^{-1} \\
M_{22} &= U_2\begin{bmatrix} \hat{N} \\ \hat{D} \end{bmatrix}(\hat{D} + K\hat{N})^{-1}\begin{bmatrix} K & I \end{bmatrix}U_1
\end{aligned} \quad (12.14)$$

Note that in this formulation, the controller K can be given by either the present controller $K = K_i$ referring to the situation of *posterior stability/performance assessment*, or by a newly-designed controller $K = K_{i+1}$, related to a *prior stability/performance robustness test*.

The entry M_{11} plays an important role in stability robustness analysis of a controller K applied to all models in the model set. For $K = K_i$, $M_{11} = 0$ is implying that all controller/model pairs resulting from $\mathcal{G}_i$ are stable, irrespective of the size of Δ. This property of the dual-Youla parameterisation has been indicated before. For $K = K_{i+1}$, stability robustness can be guaranteed by verifying that $\|M_{11}\|_\infty < \gamma_i$. This pertains to a test on a *lower* LFT: $\|\mathcal{F}_l(Q_i, -K)\|_\infty < \gamma_i$ (see [46]).

For performance evaluation, we distinguish again between the *posterior* and the *prior* situation.

Performance assessment for implemented controller

For posterior performance assessment (step (a) of the procedure), we have $K = K_i$, and the worst-case performance $\|J(G, K_i)\|_\infty$ over all models $G \in \mathcal{G}_i$ is evaluated by calculating

$$\|\mathcal{F}_u(M_i, \Delta)\|_\infty = \|M_{22} + M_{21}\Delta M_{12}\|_\infty$$

for all $\Delta \in \mathcal{RH}_\infty$ with $\|\Delta\|_\infty \leq \gamma_i^{-1}$. This test can be performed non-conservatively.

Robust performance test for designed (not yet implemented) controller

For a prior performance robustness test (step (b) of the procedure), we have $K = K_{i+1}$, and the necessary evaluation becomes

$$\|\mathcal{F}_u(M_i, \Delta)\|_\infty = \|M_{22} + M_{21}\Delta(I - M_{11}\Delta)^{-1}M_{12}\|_\infty \leq \gamma_i^{-1} \quad (12.15)$$

for all $\Delta \in \mathcal{RH}_\infty$ with $\|\Delta\|_\infty \leq \gamma_i^{-1}$. Note that evaluation of (12.15) can be done via the application of the main loop theorem and by computing an upper bound for the structured singular value of $\mu(M_i)$, see, e.g., Theorem 11.7 in [164]. With the result of the main loop theorem, $\mathcal{F}_u(M_i, \Delta)$ is well-posed, stable and $\|\mathcal{F}_u(M_i, \Delta)\|_\infty \leq \gamma_i^{-1}$ for all Δ with $\|\Delta\|_\infty < \gamma_i^{-1}$, if and only if

$$\mu(M_i) \leq \gamma_i \quad (12.16)$$

where the structured singular value $\mu(M_i)$ is computed with respect to the block diagonal structure of M_i given in (12.14). In case of a SISO plant with an unstructured dual-Youla based uncertainty Δ, $\mu(M_i)$ in (12.16) can be computed exactly. In the multivariable case, a tight overbound for $\mu(M_i)$ can be found via standard convex optimization techniques.

For step (c) of the procedure, the prior performance assessment test is used similar to the situation of step (a), albeit now with the *known* controller K_{i+1}, and an updated M_{i+1}.

12.4.3 Control-Relevant Estimation of a Set of Models

As indicated in (12.11), the structure of a set of models $\mathcal{G}$ is determined by an RCF $(\hat{N}, \hat{D})$ of a nominal model $\hat{G}$ and the weighting functions $(\hat{V}, \hat{W})$ that normalize the uncertainty or modelling error. The control-relevant estimation of a set of models $\mathcal{G}$ should address the minimisation

$$(\hat{N}, \hat{D}, \hat{V}, \hat{W}) = \arg \min_{N,D,V,W} \sup_{G \in \mathcal{G}} \|J(G, K)\|_\infty \quad (12.17)$$

subjected to both $G_0 \in \mathcal{G}$ and internal stability of the feedback connection $\mathcal{T}(\hat{G}, K)$. At the current state the minimisation of (12.17), using the variables $(\hat{N}, \hat{D}, \hat{V}, \hat{W})$ simultaneously, cannot be solved directly. Therefore, the control-relevant identification of a set of models in (12.17) is addressed by estimating the RCF $(\hat{N}, \hat{D})$ and the pair $(\hat{V}, \hat{W})$ separately:

Estimation of a nominal model

This involves the estimation of $\hat{G} = \hat{N}\hat{D}^{-1}$ such that (12.17) is being minimised using the RCF (N, D) only, subjected to internal stability of $\mathcal{T}(\hat{G}, K)$. The pair $(\hat{V}, \hat{W})$ is unknown and assumed to vary freely in order to satisfy $G_0 \in \mathcal{G}$.

Estimation of uncertainty

This consists of the characterisation of an upper bound on Δ in (12.11) via $(\hat{V}, \hat{W})$ such that (12.17) is being minimised using (V, W) only, subjected to $G_0 \in \mathcal{G}$. The RCF $(\hat{N}, \hat{D})$ is fixed to the estimate of the nominal model obtained previously.

By the separate identification of the RCF $(\hat{N}, \hat{D})$ and the weighting functions $(\hat{V}, \hat{W})$, only an upper bound on (12.17) can be minimised. However, it should be stressed that precise minimisation of (12.17) is not needed. If suffices to find a set of models that passes the *a posteriori* performance robustness test of (12.3) or (12.5). Furthermore, (standard) tools to estimate a nominal model and to characterize uncertainty can be applied as indicated in the application to the wafer stage. Finally, it can be noted that due to the separation being made, the attention can be focused on finding models of limited complexity [148]. The rationale is to avoid the computation of controllers on the basis of highly-complex models as much as possible, since this will lead to high-order controllers for which the computation may be badly conditioned.

12.4.4 Robust Control Design

A new and improved robust performing controller K_{i+1} can be designed on the basis of the estimated set of models $\mathcal{G}_i$ by the minimisation

$$K_{i+1} = \arg \min_K \sup_{G \in \mathcal{G}_i} \|J(G, K)\|_\infty \tag{12.18}$$

Basically, (12.18) is a robust performance control design, wherein a controller K_{i+1} is constructed such that the worst-case performance $J(G, K_{i+1})$ for all $G \in \mathcal{G}_i$ is being optimised. In order to use the available standard results on $\mathcal{H}_\infty$- and μ-controller synthesis, the transfer function M_i of the LFT $\mathcal{F}_u(M_i, \Delta)$ in (12.14) is represented as a lower fractional transformation $\mathcal{F}_l(P_i, K)$ as illustrated in Figure 12.9.

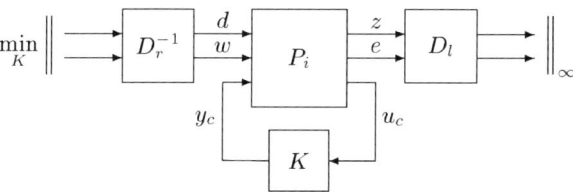

Fig. 12.9. Controller synthesis via $\mathcal{H}_\infty$ optimisation for fixed D-scaling

The entries of the standard plant P_i in Figure 12.9 can be found by extracting the controller K from the expression of M_i given in (12.14). Considering the map M_i given in (12.14), then $M_i = \mathcal{F}_l(P_i, K)$ where P_i is given by

$$P_i = \begin{bmatrix} \hat{W}^{-1} & 0 & 0 \\ 0 & U_2 & 0 \\ 0 & 0 & I \end{bmatrix} \left[\begin{array}{c|c|c} \hat{D}^{-1} N_k & \hat{D}^{-1} \; 0 & \hat{D}^{-1} \\ (D_k + \hat{G}_i N_k) & \hat{G}_i \; 0 & \hat{G}_i \\ 0 & I \; 0 & I \\ \hline -(D_k + \hat{G}_i N_k) & -\hat{G}_i \; I & -\hat{G}_i \end{array} \right] \begin{bmatrix} -\hat{V}^{-1} & 0 & 0 \\ 0 & U_1 & 0 \\ 0 & 0 & I \end{bmatrix}$$

The control design problem (12.18) can be tackled via a μ-synthesis that solves

$$\min_{K} \inf_{D_l, D_r \in \mathcal{D}} \left\| D_l \mathcal{F}_l(P_i, K) D_r^{-1} \right\|_\infty \tag{12.19}$$

iteratively for the scaling matrices D_l, D_r and the controller K, subjected to internal stability of the feedback connection of K and $\hat{G}_i$.

Finally, it can be noted that the control design used here is a generalisation of the robust controller synthesis as presented in, e.g., [34] or [114]. It can be verified that by ignoring the map from d onto z (representing the uncertainty), P reduces to

$$\begin{bmatrix} U_2 & 0 \\ 0 & I \end{bmatrix} \left[\begin{array}{c|c|c} \hat{G}_i & 0 & \hat{G}_i \\ I & 0 & I \\ \hline -\hat{G}_i & I & -\hat{G}_i \end{array} \right] \begin{bmatrix} U_1 & 0 \\ 0 & I \end{bmatrix}$$

and $M_i = \mathcal{F}_l(P_i, K) = U_2 T(\hat{G}_i, K) U_1$. In the special case of a *diagonal* weighting function $U = \mathrm{diag}(U_{in}, U_{out}^{-1})$ with $U_2 = U$ and $U_1 = U^{-1}$, the controller K_{i+1} that minimises $\|UT(\hat{G}_i, K_{i+1}) U^{-1}\|_\infty$ can be found by loop shaping techniques [34, pp. 107-108]. Explicit state-space formulae of the optimal controller for this special case can be found in [34] or [114].

12.5 Estimation of Model Set for the Wafer Stage

12.5.1 Access to Coprime Factorisations

The first step in the characterisation of the set of models $\mathcal{G}$, is the (approximate) identification of a stable nominal factorisation $(\hat{N}, \hat{D})$ of a (possibly unstable) nominal model $\hat{G}$. Access to a RCF of the system G_0 for identification purposes can be obtained by a simple filtering of the signals present in the feedback connection $\mathcal{T}(G_0, K_i)$.

As indicated in [150] or [45], a filtering of the reference signal $r := r_1 + K_i r_2 = u + K_i y$ via $x := Fr$ enables access to the various RCF of the system G_0 on the basis of closed-loop data. With (12.7), it can be seen that

$$x = F \begin{bmatrix} K_i & I \end{bmatrix} \begin{bmatrix} r_2 \\ r_1 \end{bmatrix} = F \begin{bmatrix} K_i & I \end{bmatrix} \begin{bmatrix} y \\ u \end{bmatrix} \tag{12.20}$$

and (12.7) reduces to

$$\begin{bmatrix} y \\ u \end{bmatrix} = \begin{bmatrix} G_0 S_{in} F^{-1} \\ S_{in} F^{-1} \end{bmatrix} x + \begin{bmatrix} (I + G_0 K_i)^{-1} \\ -K_i(I + G_0 K_i)^{-1} \end{bmatrix} v \qquad (12.21)$$

where $(G_0 S_{in} F^{-1}, S_{in} F^{-1})$ with $S_{in} = (I + K_i G_0)^{-1}$ can be considered to be a (right) factorisation of the system G_0. In order for this factorisation to be right coprime, the filter F in (12.20) is restricted to the form

$$F = [D_x + K_i N_x]^{-1} \qquad (12.22)$$

where (N_x, D_x) is an RCF of any auxiliary model G_x that is stabilised by the (initial) controller K_i used in the closed-loop experiments. For more details on this characterisation, see [150]. This includes choices for F that achieve normalisation of the factorisation $(N_{o,F}, D_{o,F})$, which has the additional advantage that redundant dynamics in the two factors is removed.

Consequently, a simple filtering (12.20) of the signals present in the feedback connection $\mathcal{T}(G_0, K_i)$ allows the access to an RCF of the system G_0. The system equation (12.7) can then be written in the form

$$\begin{bmatrix} y \\ u \end{bmatrix} = \begin{bmatrix} N_{o,F} \\ D_{o,F} \end{bmatrix} x + \begin{bmatrix} I \\ -K_i \end{bmatrix} [I + G_0 K_i]^{-1} v \qquad (12.23)$$

where x is given in (12.20), F is given in (12.22) and $(N_{o,F}, D_{o,F})$ is the RCF of the plant G_0 given by

$$\begin{bmatrix} N_{o,F} \\ D_{o,F} \end{bmatrix} = \begin{bmatrix} G_0 \\ I \end{bmatrix} [I + K_i G_0]^{-1} [I + K_i G_x] D_x. \qquad (12.24)$$

Since x in (12.20) is uncorrelated with v, (12.23) gives rise to an equivalent open-loop identification problem of the RCF $(N_{o,F}, D_{o,F})$ of the system G_0.

12.5.2 Feedback-relevant Estimation of Coprime Factorisations

In the estimation of the RCF $(\hat{N}, \hat{D})$, minimisation of (12.17) must be taken into account when estimating a nominal factorisation $(\hat{N}, \hat{D})$. Furthermore, $\hat{G} = \hat{N}\hat{D}^{-1}$ is subjected to internal stability of the feedback connection $\mathcal{T}(\hat{G}, K_i)$ in order to characterise the set of models $\mathcal{G}$ given in (12.11).

Clearly, at this stage, the set of models is unknown and (12.17) cannot be computed. In fact, the set of models is arbitrarily large as the norm bounded uncertainty Δ in (12.11) has not been characterised. Consequently, for any nominal model there exists a norm bounded uncertainty Δ that forms a set of models $\mathcal{G}$ for which $G_0 \in \mathcal{G}$. As $G_0 \in \mathcal{G}$, for any nominal model $G_0 \in \mathcal{G}$, the following upper bound for $\|J(\hat{G}, K_i)\|_\infty$ can be given.

$$\|J(\hat{G}, K_i)\|_\infty \leq \|J(G_0, K_i)\|_\infty + \|J(\hat{G}, K_i) - J(G_0, K_i)\|_\infty$$

As $\|J(G_0, K_i)\|_\infty$ in the above expression does not depend on the nominal model $\hat{G}$, the upper bound can be minimised by an estimated RCF $(\hat{N}, \hat{D})$ of a nominal model that minimises

$$\|J(\hat{G}, K_i) - J(G_0, K_i)\|_\infty \tag{12.25}$$

thus constituting a control-relevant identification criterion.

With the expressions introduced above, it can be shown [45] that

$$J(G_0, K_i) - J(\hat{G}, K_i) = U_2 \left(\begin{bmatrix} N_{o,F} \\ D_{o,F} \end{bmatrix} - \begin{bmatrix} \hat{N} \\ \hat{D} \end{bmatrix} \right) F \begin{bmatrix} K_i & I \end{bmatrix} U_1 \tag{12.26}$$

where $(\hat{N}, \hat{D})$ satisfies the constraint $\hat{D} + K_i \hat{N} = F^{-1}$. The estimation of a nominal factorisation for the positioning mechanism of the wafer stepper will be illustrated in the next section.

12.5.3 Estimation Results of Nominal Factorisations

To estimate a nominal factorisation $(\hat{N}, \hat{D})$, frequency domain measurements of the factorisation $N_{o,F}(\omega)$, $D_{o,F}(\omega)$ along a pre-specified frequency grid are used. The external signals r_1 and r_2 are designed accordingly to the experiment design discussed in Section 12.2.4.

Subsequently, the curve fitting procedure described in [44] is used to tackle the weighted minimisation of (12.25) frequency wise. As the curve fitting procedure is a non-linear optimisation, an initial estimate is required to start the optimisation. For that purpose, a multivariable least squares curve fitting procedure is used [43].

An amplitude Bode plot of the RCF $(\hat{N}, \hat{D})$ being estimated can be found in Figure 12.10. The resulting estimate of $col(\hat{N}, \hat{D})$ is an eighteenth order discrete time multivariable model having six inputs and three outputs. Computing $\hat{G}_i = \hat{N}\hat{D}^{-1}$ yields an eighteenth order nominal model, having three inputs and three outputs. The amplitude Bode plot of the model $\hat{G}_i$, along with the available frequency domain data computed via $N_{o,F}(\omega)D_{o,F}(\omega)^{-1}$ is depicted in Figure 12.11.

Although stability of the feedback connection $\mathcal{T}(\hat{G}_i, K_i)$ is not enforced during the estimation of the coprime factorisation $(\hat{N}, \hat{D})$, the model $\hat{G}_i$ is stabilised by K_i. This is mainly due to the fact that the control-relevant estimation of the coprime factors yields a nominal model $\hat{G}_i$ with a good fit of the control-relevant dynamical behaviour of the plant G_0 around the closed-loop relevant frequency area of 200 Hz.

With the knowledge of the eighteenth order curve-fitted nominal plant model $\hat{G}_i$, the weighting functions U_1 and U_2 for the control objective function $J(G, K)$ are designed. They are chosen to achieve decoupling of the multivariable plant at 90 Hz and to achieve a nominal bandwidth of approximately 90 Hz. Furthermore, two integrators are incorporated in each diagonal transfer of the loop-shaped plant to ensure low frequent disturbance rejection and tracking of the servo positioning mechanism.

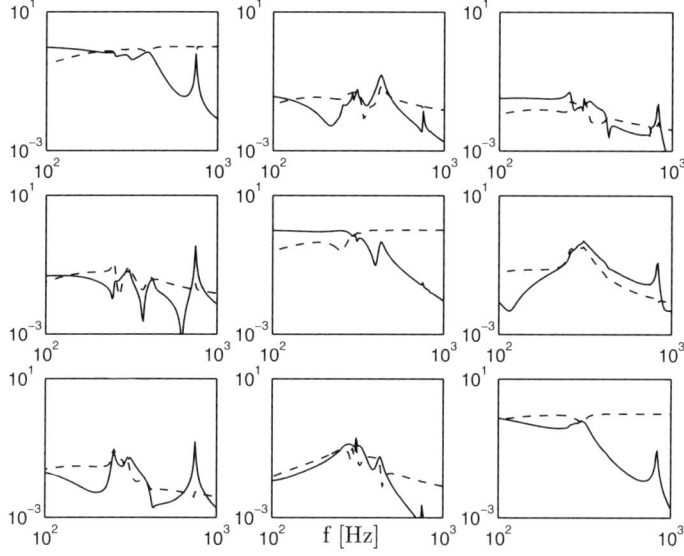

Fig. 12.10. Amplitude Bode plot of estimated coprime factors $\hat{N}$ (—) and $\hat{D}$ (- -)

12.5.4 Access to Model Uncertainty

Once an RCF $(\hat{N}, \hat{D})$ of the nominal model $\hat{G}_i$ is obtained, an estimation of the model(ling) uncertainty Δ in (12.11) can be performed. This involves the characterisation of an upper bound on Δ in (12.11) via the stable and stably invertible filters $(\hat{V}, \hat{W})$ such that (12.17) is being minimised and $G_0 \in \mathcal{G}_i$. For that purpose, first (an upper bound on) the allowable model perturbation Δ is determined by applying a model error bounding estimation technique. The uncertainty estimation routine described by [68] is used to obtain a frequency-dependent upper bound for Δ:

$$\|\Delta(\omega)\| \leq \delta(\omega) \text{ with probability } \geq \alpha \tag{12.27}$$

where α is a pre-chosen probability. In the multivariable case, the upper bound (12.27) can be obtained for each transfer function. Subsequently, stable and stably invertible weightings $\hat{V}$ and $\hat{W}$ can be determined that overbound the estimated upper bound $\delta(\omega)$.

In order to estimate a frequency-dependent upper bound on Δ, the map Δ must be accessible from data. This can be achieved by defining the filtered closed-loop signal

$$z := (D_k + \hat{G}_i N_k)^{-1} \begin{bmatrix} I & -\hat{G}_i \end{bmatrix} \begin{bmatrix} y \\ u \end{bmatrix} \tag{12.28}$$

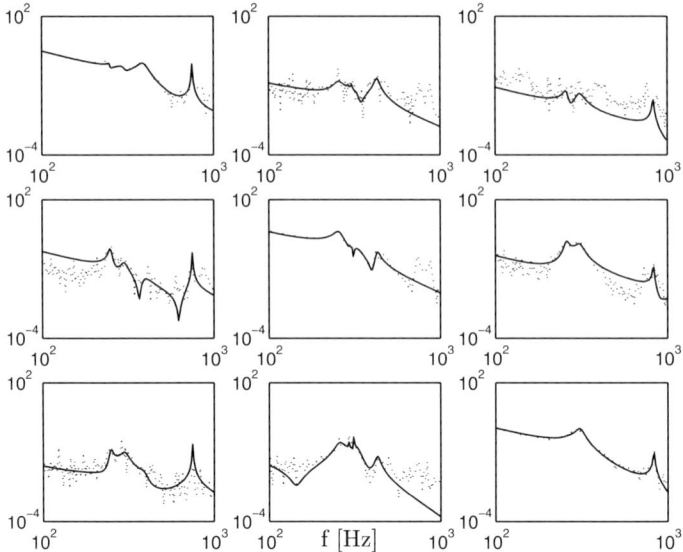

Fig. 12.11. Amplitude Bode plot of computed nominal model $\hat{G}_i$ (—) and frequency domain data ($\cdots$)

which can be shown to satisfy

$$z = \Delta x + D_k(I + G_0 K_i)^{-1} v \qquad (12.29)$$

As x is uncorrelated with v, this points to an open-loop bounded error identification problem to find an upper bound for a stable Δ. The estimated upper bound of Δ in (12.27) can then be used to complete the characterisation of the set of models $\mathcal{G}_i$.

12.5.5 Feedback-relevant Estimation of Model Uncertainty

It can be observed from (12.14) that knowledge of the controller K_i used during the closed-loop experiments is taken into account in the construction of the set of models. Application of the same controller $K = K_i$ to the uncertainty set for *a posteriori* robust performance evaluation greatly reduces the entries of the matrix M. Substitution of $K = K_i$ in (12.14) yields $M_{11} = 0$. This implies that when the controller K_i is applied to the estimated set of models $\mathcal{G}_i$, the upper LFT $\mathcal{F}_u(M, \Delta)$ modifies into

$$M_{22} + M_{21} \Delta M_{12} \qquad (12.30)$$

which is an affine expression in Δ.

Substituting M_{21} and M_{12} in (12.30) with $\Delta = \hat{V}\Delta\hat{W}$ yields the following expression

$$M_{22} + M_{21}\Delta M_{12} = M_{22} + W_2 \Delta W_1$$

where

$$W_2 = -U_2 \begin{bmatrix} -D_k \\ N_k \end{bmatrix}, \quad W_1 = \hat{D}^{-1}(I + K_i\hat{G}_i)^{-1}\begin{bmatrix} K_i & I \end{bmatrix} U_1 \quad (12.31)$$

Consequently, the effect of replacing an accurate (and high-order estimate) of the upper bound Δ by a low-order upper bound approximation $\tilde{\Delta}$ on the (robust) performance $\|J(G_0, K_i)\| = \|M_{22} + W_2 \Delta W_1\|$ can be bounded by the following triangular inequality

$$\|M_{22} + W_2 \Delta W_1\| \leq \|M_{22} + W_2 \tilde{\Delta} W_1\| + \|W_2(\Delta - \tilde{\Delta})W_1\| \quad (12.32)$$

From (12.32), it can be observed that, similar to identification of a control-relevant and low-complexity factorisation of a nominal model, a weighted difference between the actual and highly-complex uncertainty Δ and the low-complexity approximation $\tilde{\Delta}$ must be taken into account for a control relevant approximation of the model uncertainty. The weightings W_2 and W_1 are given in (12.31) and are known, once a nominal factorisation $(\hat{N}, \hat{D})$ has been estimated. With the (frequency-dependent) weightings W_1 and W_2, low frequent weighting filters $(\hat{V}, \hat{W})$ can be used to parameterise and overbound the frequency-dependent bound of the modelling uncertainty.

12.5.6 Estimation Results of Model Uncertainty

Given the nominal factorisation $(\hat{N}, \hat{D})$ and a normalised RCF (N_k, D_k) of the controller K_i, an estimation of the allowable model perturbation Δ in (12.11) is performed. For that purpose, the uncertainty estimation as presented in [68] has been applied to estimate a frequency-dependent upper bound on Δ. A complete discussion on the uncertainty estimation procedure of [68] is beyond the scope of this chapter. Here, we will just point to its main characteristics:

- it combines a worst-case bounding of unmodelled dynamics with a probabilistic bound on the variance error;
- it employs linearly parameterised models (basis functions) for which least squares or instrumental variable estimates are constructed;
- uncertainty regions for frequencies in any user-chosen frequency grid are computed from bias and variance errors.

The result of the uncertainty bounding procedure is summarised in Figure 12.12. In Figure 12.12, the amplitude bodeplot of the frequency domain data of Δ in (12.29) is compared with the estimated upper bound $\delta(\omega)$. It can be observed from Figure 12.12 that the upper bound of the frequency

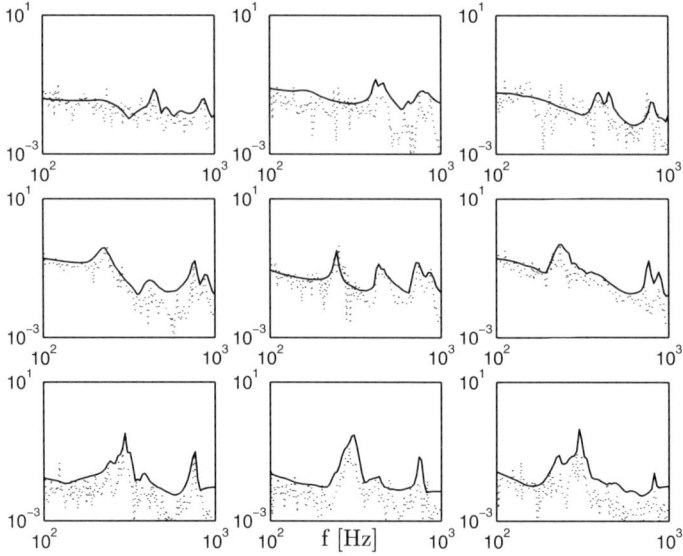

Fig. 12.12. Amplitude Bode plot of estimated uncertainty bound $\delta(\omega)$ (—) of Δ and frequency domain estimate of Δ ($\cdots$)

domain estimation of Δ is crossing the upper bound $\delta(\omega)$. Partly, this is due to the fact the upper bound only holds within a pre-specified probability of 95 per cent.

As $\delta(\omega)$ is only a frequency-dependent upper bound for Δ, low frequent weighting filters $(\hat{V}, \hat{W})$ are needed to parameterise and overbound the estimated uncertainty bound $\delta(\omega)$ depicted in Figure 12.12. In this way, the estimated upper bound can be taken into account during a robust controller design.

In the construction of $(\hat{V}, \hat{W})$, the weightings W_1 and W_2 given in (12.31) are used to emphasise the frequency range for the upper bounding of $\delta(\omega)$ by the parametric stable and stably invertible weightings $(\hat{V}, \hat{W})$. It can be observed from (12.31) that the input sensitivity $(I + K_i \hat{G}_i)^{-1}$, based on the nominal model $\hat{G}_i$, is incorporated in the weightings given in (12.31). As a consequence, the weightings emphasise (again) the closed-loop-relevant frequency area around 200 Hz.

12.6 Model Set-based Robust Control Design

On the basis of the identified set of models, a robust controller can be designed via μ-synthesis. As indicated in Section 12.4.4, a robust performance control design is used wherein a controller K_{i+1} is constructed such that the worst-case performance $J(G, K_{i+1})$ for all $G \in \mathcal{G}_i$ is being optimised. The control design is done via μ-synthesis using an alternating iteration between a $\mathcal{H}_\infty$ controller optimisation with fixed D-scalings and an adjustment of the scalings D for a fixed controller.

The μ-synthesis is invoked with second order D-scalings and four D-K iterations are performed to compute a robust performing controller. The μ-synthesis yields a high-order multivariable feedback controller and, in order to implement the controller, an additional closed-loop controller reduction [156] has been used to reduce the controller to a thirty-second order state-space realisation. A comparison between the initial controller K_i previously implemented on the wafer stage system and the newly-designed one K_{i+1} is given in terms of the amplitude Bode plot depicted in Figure 12.13.

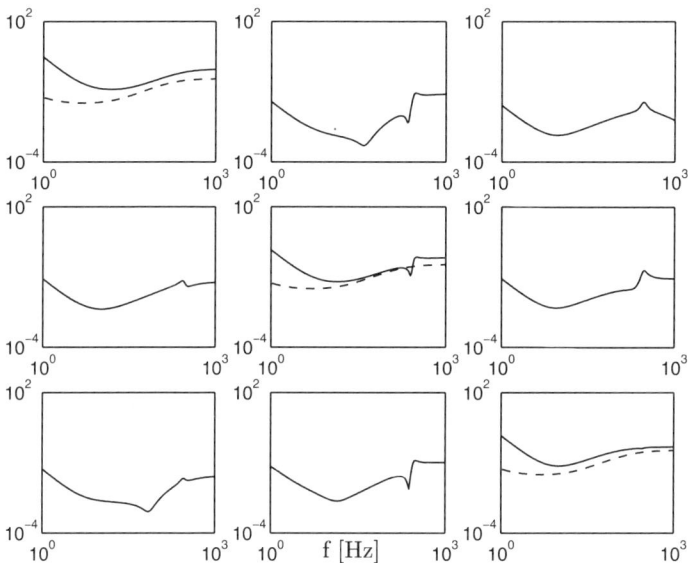

Fig. 12.13. Amplitude Bode plot of conventional parallel PID controller K_i (- -) and newly-designed thirty-second order multivariable controller K_{i+1} (—)

Compared to the initial controller K_i, it can be seen that the newly-designed K_{i+1} is a multivariable servo controller. Furthermore, it has additional dynamics to account for the modelled (uncertain) mechanical resonance modes of the wafer stage G_0.

Before implementing the new controller, the robust performance and stability needs to be evaluated. This can be done with the estimated set of models $\mathcal{G}_i$ by evaluating $\|J(G, K_{i+1})\|_\infty$ for all models $G \in \mathcal{G}_i$. Both stability and performance robustness of K_{i+1} can be evaluated with the structured singular value $\mu\{M\}$ as indicated in (12.16). In Figure 12.14, the structured singular value is plotted point wise over the frequency range between 10 Hz and 1000 Hz for both the initial K_i and the newly-designed feedback controller K_{i+1}. It can be observed that the newly-designed controller K_{i+1} has improved the performance robustly, as the maximum of the structured singular value $\mu\{M(e^{i\omega})\}$ has been lowered with a factor of approximately 4.

As a result, the performance index $\|J(G, K_{i+1})\|_\infty$ evaluated for all models $G \in \mathcal{G}_i$ with the new controller K_{i+1} is guaranteed to be four times better and, as a result, the performance of the closed-loop system has been improved robustly.

For presentation purposes, the weighting functions $\hat{V}_i$ and $\hat{W}_i$ that complete the set of models $\mathcal{G}_i$ in (12.11) are scaled to normalise the uncertainty Δ. As a result, *performance robustness* of the closed-loop system is guaranteed if $\max_\omega \mu\{M(e^{i\omega})\} < 1$. It can be seen from Figure 12.14 that this is not the case, but by adjusting the performance weighting functions U_1 and U_2 used in the performance characterisation (12.8), performance robustness can be guaranteed for a specific choice of the weighting functions U_1 and U_2 in the performance criterion $\|U_2 T(P, C) U_1\|_\infty$. Whether or not the performance weighting functions U_1 and U_2 are adjusted to guarantee performance robustness, the performance of the newly-designed controller K_{i+1} has been shown to be improved over the initial controller K_i.

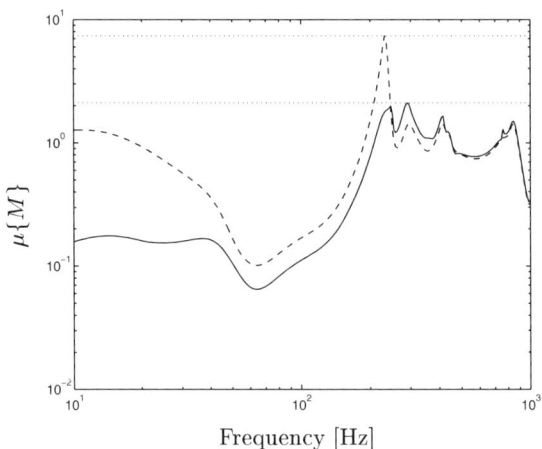

Fig. 12.14. Structured singular value $\mu\{M(e^{i\omega})\}$ for initial controller K_i (dashed) and newly-designed controller K_{i+1} (solid)

The newly-designed controller K_{i+1} has been implemented on the positioning mechanism of the wafer stage and the performance robustness improvement can be clearly seen from the time domain plots. In order to illustrate the improved performance of the positioning control, the reference signals r_1 and r_2 depicted in Figure 12.5 are put on the newly-designed feedback connection $\mathcal{T}(G_0, K_{i+1})$. A comparison with the servo error of Figure 12.6 obtained with the initial controller K_i is depicted in Figure 12.15. It can be seen from Figure 12.15 that both the speed and the accuracy of servo positioning have been improved successfully.

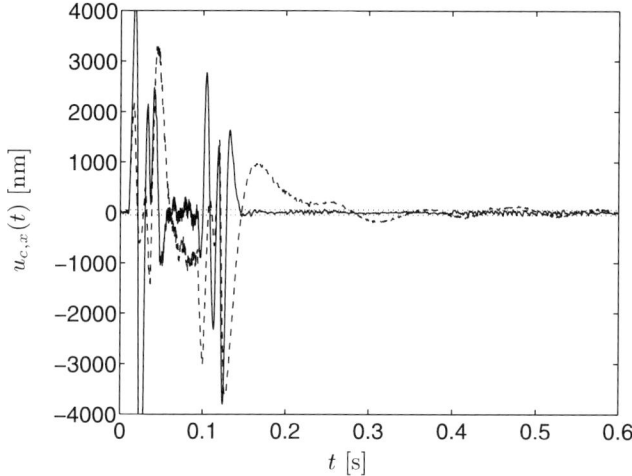

Fig. 12.15. Servo error response to a step in x-direction with initial parallel PID controller K_i (- -) and newly-designed multivariable controller K_{i+1} (—)

12.7 Summary

A wafer stage is part of a wafer stepper and used in chip manufacturing processes for accurate positioning of the silicon wafer on which the chips are to be produced. The wafer stage discussed in this application is a three degree-of-freedom (3-DOF) positioning mechanism and the accurate positioning requirements of the wafer stage demand a robust and high-performance multivariable servo controller that enables a fast and accurate positioning of the wafer stage in 3-DOF for a high throughput of silicon wafers.

This chapter illustrates the approximate and feedback-relevant parametric identification of a wafer stage and shows how models along with an uncertainty characterisation can be used to characterise a set of models. The set of models is constructed within a framework of algebraic model representations that allows the incorporation of the controller information into the construction of the model set. It is shown that the set of models is particularly useful for the closed-loop and control-relevant estimation of both the nominal model and the accompanying modelling error.

With the control-relevant estimation of a model set and the use of robust control synthesis tools, both stability and performance robustness of the model-based robust control design can be monitored. Additionally, monitoring of the performance robustness enables the possibility of guaranteeing performance improvement in an iterative scheme of a control-relevant estimation of a set of models followed by a robust control design. The procedure is shown to be suitable for the model-based robust control design of the wafer stage and is illustrated by a successful design and implementation of a robust controller.

Acknowledgements

The authors would like to acknowledge the support of Philips Research Laboratories (Eindhoven, the Netherlands) for providing the wafer stepper experimental set-up. Thanks also go to Okko Bosgra and Dick de Roover for fruitful discussions and their contributions to this paper.

13 Iterative Identification and Control: a Sugar Cane Crushing Mill

Robert R. Bitmead and Antonio Sala

Abstract. We deal with the "nuts and bolts" aspects of iterative system identification and control design on a specific process of a sugar cane crushing mill from North Queensland, Australia. Our emphasis is on the aspects in which the process knowledge is used to guide the myriad of design choices necessary *before* resorting to the use of special-purpose software for identification or control design. Of particular concern is the demonstration in concrete form of the principles developed in earlier chapters dealing with identification for control and model-based control design. We deliberately unfocus attention from the specific algorithms used in MATLAB® from the *System Identification* and the *(Robust) Control* Toolboxes. The objective is to connect the *a priori* knowledge, the design objective and the posing of inputs to the computational tools.

13.1 Introduction

This chapter details and develops some of the plant-specific aspects of the iterative identification and control design for a sugar cane crushing mill. Our aim is to expose the interplay between modelling and control design and process knowledge, as this knowledge evolves from purely *a priori* information to include information gathered from targeted experiments. This objective is to make more apparent by example some of the process-related choices during iterative design. We think of this as the *carpentry*[1] and *machinery* of the design, which binds the abstract stages of closed-loop data analysis, modelling and control design described in earlier chapters to the nuts-and-bolts implementation. The process, the modelling phase, the preparation of data for identification and the iterative identification and control steps will be described for the design of a controller for a sugar cane crushing mill. The plant is located at CSR Ltd Victoria Mill in Ingham, North Queensland (Australia). Our focus is to include sufficient detail to depict the role of process knowledge and objective in the eventual design. More completely detailed descriptions of the methods used and their results are given in [123].

The chapter tries to emphasise by example the importance that prior (empirical and not necessarily first principles) knowledge about the plant behaviour and the knowledge of control objectives has on the overall modelling

---

[1] We are indebted to Karl Johan Åström for coining this delightful metaphor linking the intellectual problem with the actual. He also suggested that, like carpentry, these methods cannot be learned well from a book. We hope, in this immediate instance, that he is not overly correct in this latter judgment.

and design process. The sugar cane crushing mill control problem is one of disturbance rejection. The plant performance is dominated by the variability of the sugar cane's mechanical properties. Because of the magnitude of disturbance energy and non-linearity in the process, a considerable degree of "carpentry" has to be carried out beforehand to design statistically meaningful experiments and to manipulate the data so that a model is obtained that is appropriate for the control purpose. Without this coupling of prior thinking to standard methods, it is highly doubtful that adequate performance might be achieved.

The structure of the chapter will be as follows. In the next section, the process will be reviewed together with the control objectives. This provides a starting point for the generation and validation of system models. Section 13.3 will describe the specific choices of the design variables for system identification and their connection with current system knowledge and with the control objectives. Data inspection and selection will be dealt with, in addition to data preparation for identification, using methods of digital signal processing to impose emphases on reflecting system knowledge. This is the core section, which demonstrates the principles of earlier chapters. Controller design will then be briefly addressed in Section 13.4, linking in the iterative control ideas of earlier. The iterative approach and its results will be presented in Section 13.5. The purpose of this chapter is to put some meat on the bones of many of the mainly in-principle developments so far.

13.2 The Sugar Cane Crushing Process

The purpose of a sugar cane crushing mill (or *sugar mill* for short) is to crush sugar cane to express the maximal amount of the sugar-bearing juice. A single mill is depicted in Figure 13.1 with the material entering from the left and exiting to the right. The *feed* to the mill is composed of shredded fibrous plant matter combined with juice (known as *bagasse*) with some juice remaining locked within plant cells. Normally, a sequence of four to six such mills is connected in cascade, with the exit material from Mill 2, say, forming the entry material for Mill 3, usually with the material pre-soaked in heated low-grade juice or water to aid extraction.

The sugar cane crushing process depicted in Figure 13.1 consists of the following stages:

- feed is delivered from upstream by a carrier system composed of a fixed channel with a moving carrier chain possessing *tines* to drag the bagasse up to the delivery into the chute. In these trials, the carrier chain is driven by a fixed-speed induction motor. The direction of motion is shown in the figure;
- the bagasse is gravity-fed through the *delivery chute*, which possesses a hydraulically actuated *chute flap* capable of altering the aperture at

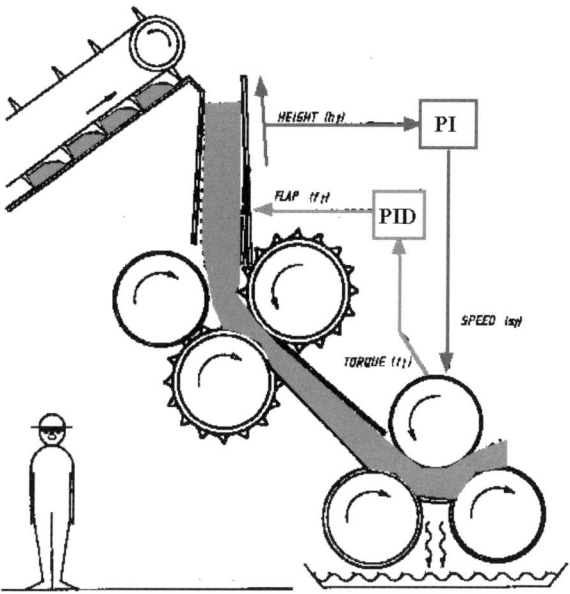

Fig. 13.1. Sugar cane crushing mill (single stage) depicting the initial PI/PID single loop controllers

the base of the chute. Gravity compaction of material is expected in the chute, and the flap helps regulate the flow from the base;
- the material is then forced by the upper set of *pressure feed rollers* into a channel of rectangular cross-section confined on all four sides. This further compacts the bagasse;
- the trio of grooved rollers at the bottom right perform the bulk of the extraction work as the bagasse is forced through the two narrow mill gaps. Expressed juice is collected primarily from this stage;
- once the bagasse leaves the mill, it moves either to a soaking trough from which it is lifted by tined carrier to the next mill or, if this is the final mill, it is taken by conveyor to be burned to provide energy (usually high pressure steam) for the factory and electricity for export to the local grid.

The objective of the case study is to show the procedure followed, to design a feedback controller for a mill using iterative identification and control. Power to the mill is provided by a steam turbine with a gear reduction unit or by direct hydraulic drive.

Each of the mills has two measured outputs:

- **Turbine torque:** computed from the turbine speed and turbine pressure using manufacturer's charts,
- **Chute height:** measured electrically by 16 conductivity probes on the sides of the chute, together with some logic circuitry,

and two manipulated inputs:

- **Turbine speed:** more exactly, the input signal to a turbine speed governor,
- **Chute flap:** position that restricts flow from the base of the chute into the pressure feed rollers.

As is evident from Figure 13.1, the conventional control strategy is to close a height-speed loop and, independently, a torque-flap loop. The reasoning is that excessive chute height might be managed by increasing throughput via speed, while over-torques might be controlled by reducing the material mass-flow rate. Empirically, these loops appeared to demonstrate the correct effect, although overall control performance was judged to be poor. This model will be revisited later.

Control is implemented using a factory-wide distributed control system. Sampling frequency is set to 1 Hz. Saturations are present in outputs and inputs, and they are frequently hit due to the severity of the disturbances. Rate limiters are also present in some of the control signal paths. Nevertheless, a linear discrete time, multi-input/multi-output model is used to capture the behaviour about the set points.

The control objective

The control objective is to maximise plant profitability by improved extraction efficiency. A translation of this goal into the measured plant variables is to maximise the torque, since this measures directly the energy expended in breaking the cane fibres to extract the juice. Overlaid on this objective are a number of constraints. Too high a torque (greater than 1.4 MNm, say) will lead to machinery damage, notably gear-tooth spalling or cam-ring fracture over the long term. The chute height is believed to help in achieving better pre-mill compaction but is limited to 100 per cent full and 0 per cent empty. The whole factory operates on a tight daily specification of throughput, which also must be accommodated by the controller. The objective is to maintain torque near a high set point, say 1.1 MNm, and high chute height while respecting the constraints. The milling train with PI-control of five successive mills extracts around 97 per cent of the available juice. Eventual performance will be measured by improving this figure.

The principal difficulty in achieving adequate control of the process is the unforeseeable disturbance that occurs due to variations in the mechanical crushing properties of the sugar cane presented to the mill. Farmers, akin to millers, maximise their revenue as well. This is done by increasing tonnage and commercial content of sugar (CCS) and leads farmers to select plant varieties most suited to their local growing conditions. Variations in local microclimates, soil types, plant varieties, *etc.* translate into dramatic variability of mechanical properties of the cane:

> *"Crushing some cane is like crushing broomsticks. Crushing others is like trying to squeeze soggy oatmeal."*

These severe process disturbances are effectively unmeasurable ahead of time and even inter-mill as the cane might not demonstrate poor feeding properties through the mill until, say, the third or fourth mill.

The fundamental control objective is to yield consistently high extraction performance and satisfaction of the constraints in the face of process variation due to the disturbances. This should be achieved by permitting operation at less conservative set points.

13.2.1 *Gedanken* Experiments, Physics and Preliminary Model Properties

Some simple, first-principles thought-experiments allow us to decipher part of the elementary structural properties of the system.

Consider the operation of a Platonic ideal sugar mill in which the bagasse is of uniform mechanical properties, the speed is fixed, the flap position is fixed and the mill operates at a fixed value of chute height and turbine torque. Now, suppose we increase the speed by a step, while leaving the flap position fixed. What happens? The chute height will decrease linearly over time as more material is drawn through the mill. Similarly, the eventual torque will drop by a step as slippage increases and the power per unit cane decreases. Thus,

> *a step increase in speed leads ultimately to a negative ramp in chute height and a negative step in torque. So, the transfer function from speed to chute height contains an integrator and the transfer function from speed to torque contains no integrator.*

A step increase (inward) in flap position should lead to a ramp upwards in chute height and a step down in torque. Thus,

> *the transfer function from flap to chute height contains an integrator and the transfer function from flap to torque contains no integrator.*

Next, consider the same Platonic example mill with stationary operating variable and suppose that the ideal, uniform bagasse is instantaneously replaced by incoming feed of less mechanical resilience. Then, if the speed and flap are unchanged, the chute height will exhibit a ramp increase due to more poorly-feeding cane residing in the mill, and the torque will correspondingly drop by a step. Thus,

> *the disturbance to the sugar mill due to cane variety variation can be modelled by ramps in chute height and negative steps in torque.*

Hence, the uncontrolled process model should look somewhat similar to:
$$\begin{pmatrix} Torque \\ Chute \end{pmatrix} = \begin{pmatrix} -A(z) & -C(z) \\ \frac{1}{1-z^{-1}}B(z) & \frac{-1}{1-z^{-1}}D(z) \end{pmatrix} \begin{pmatrix} Flap \\ Speed \end{pmatrix} + \begin{pmatrix} E(z) \\ -\frac{1}{1-z^{-1}}F(z) \end{pmatrix} w_t,$$
where $A, \ldots, F$ are discrete time transfer functions with positive DC gain and *which are stable*. It is natural to remark on the concordance between this modelling of the plant and disturbance behaviour at $z = 1$, and the use of integrators in the existing PID control solution shown earlier, which was found by experimentation rather than via modelling.

The presence of the integrator means that prior to any further experimentation on the mill, a closed-loop control strategy is necessary to avoid chute over/underflows. Hence, all identification experiments need to be carried out in closed-loop. The initial stabilising controller is the set of two single-loop PID controllers shown in Figure 13.1. This controller tells us much about the subsequent data filtering needs of iterative identification and control design.

13.2.2 Preliminary Controller Requirements

Once a local, linear model is derived, it will be used as the input to a controller design phase as described in earlier chapters. This controller design should be robust to account for the possibly large model errors likely to arise from the use of a linear model for such an obviously non-linear system, but also to account for the rudimentary modelling used for the disturbance process. We require the following of the controller:

- accommodation of modelling errors arising from linearisation, oversimplification;
- disturbance rejection of steps, ramps and related low-pass signals;
- open-loop stability (for off-line testing of the controller and manageable response during start-up, gaps in feed and bumpless transfer between controllers);
- feasible gains for implementation, so that actuator bounds are respected;
- integral antiwindup to allow for actuator (and sensor) saturation and gaps in feed;
- performance improved over the PID first design;
- comprehensibility and maintainability of the controller by factory engineers.

In performance terms, we are seeking to improve the bandwidth of the controlled process. The human and economic issues are less directly quantitatively captured.

13.3 Identification Procedure

Successful control design using model-based methods (tautologically) requires dynamical models for process and disturbance. Like many industrial processes, first-principles physics-based models are not well understood, so a black box approach to modelling is needed, via parameter identification from experimental data. Here, we shall consider the experiment design of the closed-loop (recall that the open-loop system is unstable because of the integrators) data collection together with the processing of these signals to assist in extracting an adequate model for control design. This design of the identification experiment has to take into account the closed-loop characteristics of the initial set-up and the evident non-linearities, jointly with the severity of the disturbances.

For closed-loop identification ($y = Gu + He$, $u = Ky + r$), the bias of the identification procedure is given by the following two equivalent expressions [108], both of which were presented in earlier chapters:

$$\int_{-\pi}^{\pi} \left\{ \left| G(e^{j\omega}) - \hat{G}_\theta(e^{j\omega}) \right|^2 \Phi_r(\omega) + \left| 1 + \hat{G}_\theta(e^{j\omega}) K(e^{j\omega}) \right|^2 \Phi_v(\omega) \right\} \quad (13.1)$$

$$\times \frac{\left| L(e^{j\omega}) \right|^2}{\left| 1 + G(e^{j\omega}) K(e^{j\omega}) \right|^2 \left| \hat{H}_\theta(e^{j\omega}) \right|^2} \, d\omega$$

or, equivalently the integrand can be replaced by

$$\left\{ \left| G(e^{j\omega}) + B_\theta(e^{j\omega}) - \hat{G}_\omega(e^{j\omega}) \right|^2 \Phi_u(\omega) + \left| H(e^{j\omega}) - \hat{H}_\theta(e^{j\omega}) \right|^2 \Phi_e^r(\omega) \right\} \frac{|L(e^{j\omega})|^2}{|\hat{H}(e^{j\omega})|^2}$$

where the bias term is

$$B(e^{j\omega}) = \left(H(e^{j\omega}) - \hat{H}_\theta(e^{j\omega}) \right) \frac{\Phi_{eu}(\omega)}{\Phi_u(\omega)}$$

and

$$\Phi_{eu} = \lambda_0 - \frac{|\Phi_{ue}|^2}{\Phi_u}$$

hence, bias on the transfer function depends on the ability of the parameterisation to model the disturbance spectrum (and also on the controller, acting in the cross-spectra). The pre-filter L weights the fit across different frequencies.

For the sugar mill, we demand that we capture the dominant dynamical features of the plant, $G(z)$, separately from information about the noise process, $H(z)$. This is because plant-model variation has the capacity to destabilise the closed-loop system, while in a linear context destabilisation is not possible due to an additive output disturbance. Further, we believe that we already have a tractable and acceptable model for the disturbance. Thus, our focus is on the reduction of $G - \hat{G}$ error in frequency bands pertinent for control design.

The variables under our purview that are available for modification in the identification phase are the reference signal spectrum $\Phi_r(\omega)$, the data filter $L(z)$, the class of noise models $\{\hat{H}_\theta(z)\}$, and the model structure and complexity.

13.3.1 Excitation Signal Design

Independent excitation signals need to be injected into the loop to help overcome the disturbances and to expose the plant dynamics in spite of an omnipresent and overwhelming disturbance process. As the disturbance dominates the data, the external excitation, r_t, needs to be big enough to be clearly discernible in the signals without striking the non-linear limits. Furthermore, the excitation signals must have the appropriate frequency spectrum so that a good fit is achieved in the regions interesting for control design. Those regions are the frequencies over which significant plant disturbance occurs (robust performance), and/or other regions determined by the controller to ensure robust stability, such as around the gain and phase crossover points – or their multivariable equivalents.

In most plants, a natural limitation is placed on the excitation signal not to interfere too much with normal plant activity during experimentation. With the sugar mill, it is impossible to experiment during down time, because the cane forms an important part of the mechanical process. We were very fortunate that CSR permitted us to conduct a number of trials during operation with excitation at the significant level needed to overcome the disturbance effects.

The following factors determine the excitation applied to the closed loop:

- **disturbance spectrum:** the gedanken experiments (backed by observation in operation) show that the disturbances have strong presence in the low-frequency region. Further, (13.1) illustrates that, if we seek to identify from the data a reasonably accurate plant model $\hat{G}$, then the reference spectrum, $\Phi_r(\omega)$, needs to dominate the disturbance spectrum, $\Phi_v(\omega)$, in this region of interest;
- **ultimate control objective:** we now add to this the control objective that, as a final objective, we expect the eventual control bandwidth should be pushed to roughly 0.2 Hz from less than 0.1 Hz. This focuses attention in the frequency region up to, say, 0.4 Hz with emphasis in the lower frequencies;
- **avoidance of constraints:** we are seeking to fit a linear model to the data derived from a system exhibiting saturation effects in many variables. Thus, we need to temper the desire to dominate the disturbance by limitations on signal aggression to respect constraints.
- **independence from the disturbance signal:** the capacity to distinguish plant behaviour from disturbance derives from the statistical independence of the excitation reference signal and the disturbance, effectively allowing decoupling.

The excitation signal solution adopted was the following:

- **speed input:** stabilised doubly integrated white noise[2], saturated at 20 per cent full-scale deviation, to overcome ramp-like disturbances on the height.
- **flap input:** stabilised singly integrated white noise at 10 per cent full-scale deviation, to overcome step-like effects on the torque.

These excitation signals are added to the feedback control signal from normal operations. These excitation signals are plotted in Figure 13.2, compared next to the total applied signal.

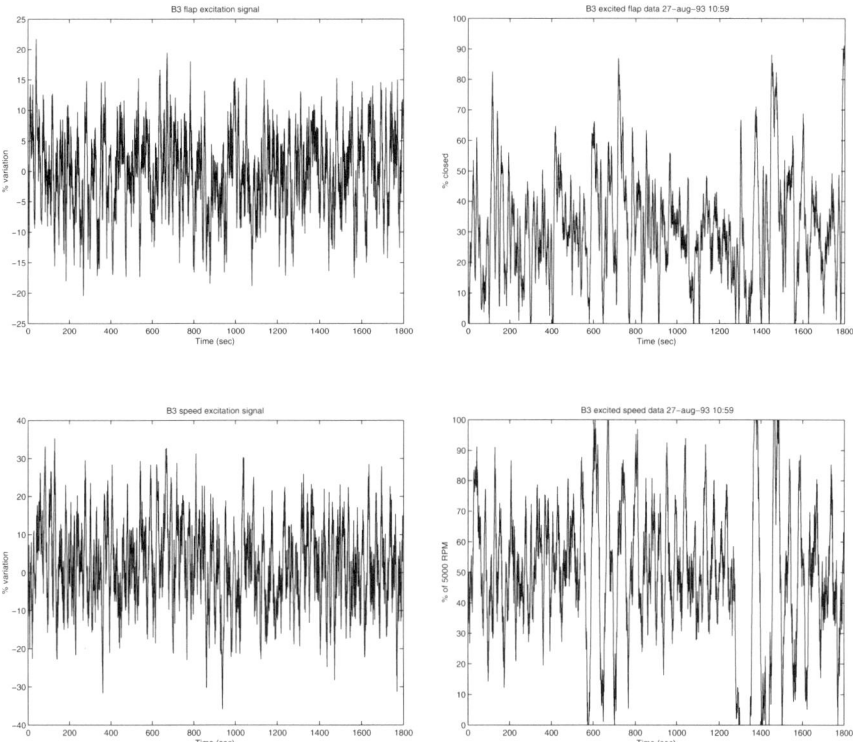

Fig. 13.2. Upper: flap excitation signal adjacent to total (excitation plus control) flap signal; lower: equivalent signals for speed excitation and speed

13.3.2 Data Preparation (Selection and Filtering)

The above excitation signal was applied to the process when operating under PI and PID feedback control as indicated in Figure 13.1. Closed-loop data

---

[2] That is, white noise passed through a stable linear system that has dominant low-pass behaviour resembling a double integrator but with finite DC-gain.

of excited speed, excited flap, response chute height and response turbine torque were collected over a thirty-minute trial at one-second sampling. The input data is displayed in Figure 13.2 and the corresponding 1800 samples of output data in Figure 13.3.

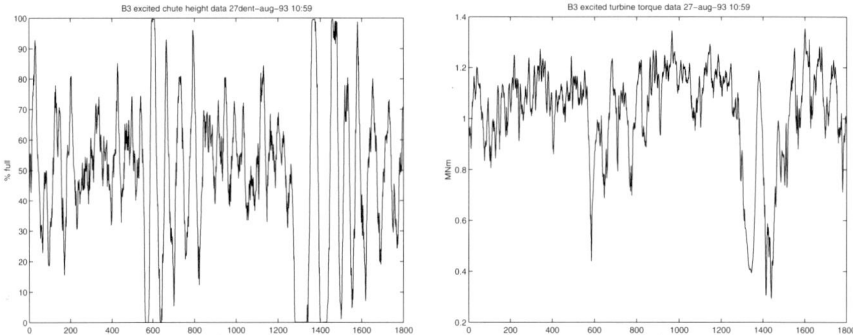

Fig. 13.3. Chute height and turbine torque closed-loop signals from thirty-minute experiment using excited PI/PID feedback

Once the excitation signals had been applied and records of closed-loop data obtained from the experiment, a number of critical decisions and operations remained to be made before actually calculating the parameters of the decided model structure using, say, the System Identification Toolbox of MATLAB®. This is a central point of this chapter – to indicate how process knowledge can influence these decisions in data selection and preparation. This is the *carpentry*.

Sample size and data selection

This has been discussed in Chapter 3 on data preparation, where an admonition was given about the need to select a data set of adequate information content and of suitable size from which to extract a usable model. Closed-loop data is, in general, beneficial for revealing the controlled properties of the plant. However, one needs to avoid the obviously non-linear effects of saturation, *etc.* in the data. Further, as developed earlier, there need not be any benefit in using very large data sets since, with approximation of plants, large amounts of data can lead to either over-confidence in the derived models or acceptance of overly complicated models for subsequent control design. These effects arise because of the misleading nature of statistical tests of acceptance – they need to presume that the model correctly describes the plant in order that their calculations hold. In the case of the sugar mill, the frequent presence of large excursions limits the data sample size.

Figure 13.2 displays the closed-loop input signals speed and flap. Figure 13.3 shows the closed-loop output signals. Visual inspection of the data shows a number of properties:

- there are periods in which the signals are dominated by the disturbances and regulation by the PI/PID controllers is poor;
- there are other periods where the signals are well-controlled into mid-range values and the input excitation can exert its effect;
- non-linear effects are very evident, notably in the chute height signal, where the saturation of the process is apparent. These parts of the signals illustrate chute overflows, gaps in feed, actuator saturation, *etc.*;
- a finer analysis would also show quantization effects in the chute height signal due to the discrete sensor;
- the signal environment is non-stationary, with the disturbance exhibiting sporadic presence. Thus, a stationary statistical model is also an approximation.

To improve the closed-loop identified model, we selected only data where the disturbances did not dominate, where the chute height and speed constraints were obeyed and the process demonstrated excited behaviour capable of being captured by a linear model. In spite of this dismissal of large parts of the data record, it was our belief that the resulting model would allow us to develop an improved control performance once we calculated an accurate plant model – relying on the gedanken experiments to characterise the disturbance process for the purposes of control design.

If the disturbance-dominated data had been used, then one would expect from an analysis of (13.1) that the identification with closed-loop data would concentrate its efforts on modelling the controllers rather than the plant itself.

In the data, significant saturations and disturbance dominance can be observed at samples 550 to 850 and 1250 onwards. We selected two data records: data 51 to 550 was used for evaluating an identified model, while data 851 to 1250 was used for validating this model on an independent data set. In actuality, we used both sets to fit models and the alternative set for validation and order selection.

The culling of data into sub-records of excitation dominance recognises that all data are not equivalent for system identification with the purpose of control design. We regard this part of the identification process as critical to the inclusion of system understanding and model objectives. Once the data had been selected, further pretreatment of them was needed to enhance the quality of model fit in frequency bands of interest where this fit was most needed.

Detrending, differencing and filtering

Zero-mean excitation was introduced and closed-loop data were gathered with control set points specified. In the case under consideration, the particular value of these set points on an open-loop model was, in some sense, useless. (The presence of integrators in the plant dynamics and controllers ensured this.) The controller we intend to design to reject step- and ramp-like disturbances from a plant containing integrators would not make use of the data

DC values. Accordingly, the first step in data processing was detrending, hence eliminating the DC component in the data. Note that if detrending had not been carried out, then (13.1) indicates that the modelling would be dominated by the system DC gains. We conducted detrending initially by removing the sample means from the input and output signals over the data window.

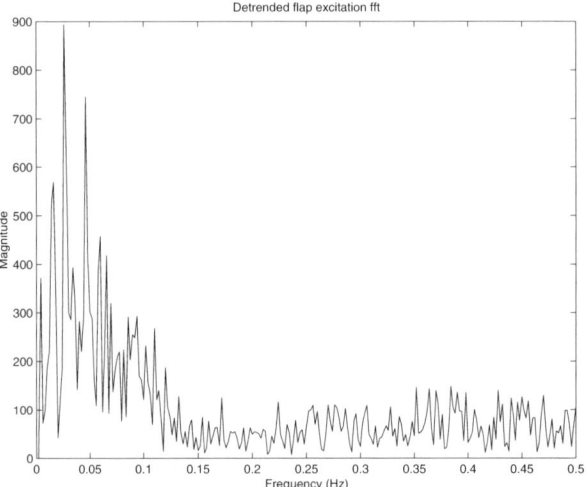

Fig. 13.4. FFT magnitude of flap excitation signal for the evaluation data set [51,550]

Consider the plot of the FFT magnitude of the detrended chute height signal for the evaluation data set in Figure 13.5. The corresponding FFT of the flap excitation signal is displayed in Figure 13.4. We make the following observations by inspection:

- **DC component:** the large DC component has been *notched* from the data.
- **Bandwidth:** the independent excitation signal shows the expected (designed) low-frequency dominance. But it does not include much energy in the frequency bands of likely controller design requirements past 0.2 Hz.
- **High frequency:** at high frequencies, there is some signal content although we are suspicious that this might be nothing more than measurement noise. We do not intend to use a model extending in validity to this frequency range.
- **Mysterious bump above 0.22 Hz:** in the chute height FFT plot (Figure 13.5), there appears an anomalous discernible bump between 0.2 Hz and 0.25 Hz. This is due to the carrier mechanism delivering the bagasse

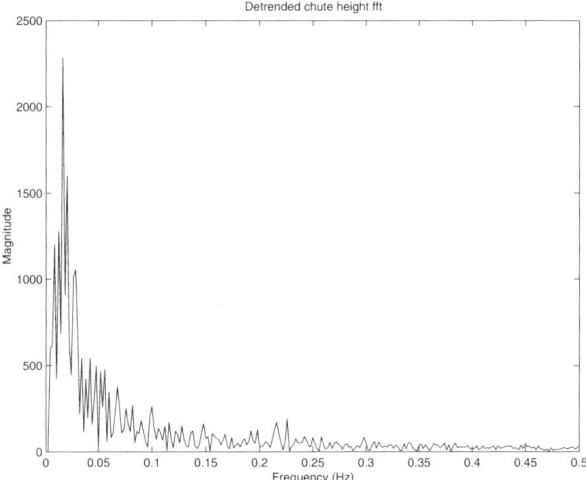

Fig. 13.5. FFT magnitude of detrended chute height for the evaluation data set [51,550]

to the top of the chute. The belt has regularly spaced tines that push the sugar cane, hence a periodic component is evident in the chute height.

Our response to these observations is to seek modification of the signal spectral content by filtering all the data through the same stable filter: $L(z)$ in (13.1). The filter we selected had three cascaded elements:

- **Differencer:** by taking differences between successive data samples (and thereby effectively differentiating the data) we are able to achieve several effects (as described earlier in the data preparation chapter): replacement of the detrending operation, since the differencer nulls out the DC terms; amplification of high-frequency bands with a gain linear with frequency thereby increasing spectral content at mid-to-high range at the expense of low frequency; amplification of very high-frequency information (poor parts of the signal);
- **Low pass filter:** by low-pass filtering the data past about 0.3 Hz, we were able to reverse the undesirable effects of the differencer in the very high-frequency range. We used a simple second order digital Butterworth filter;
- **Band-stop filter:** to remove the tine-induced periodic disturbance to the chute height, we introduced a sixth order digital Butterworth band-stop filter with cut-offs at 0.205 Hz and 0.240 Hz.

We note that, while our filtering did have some significant complexity in L, this would *not* be transferred to model complexity, provided that adequate data is used to eliminate transients from the filter. We remind the reader that all data records, chute, torque, flap and speed, were filtered through the same filter.

The frequency response of the composite data filter is illustrated in Figure 13.6. The FFT magnitude of the filtered chute height evaluation data is depicted in Figure 13.7.

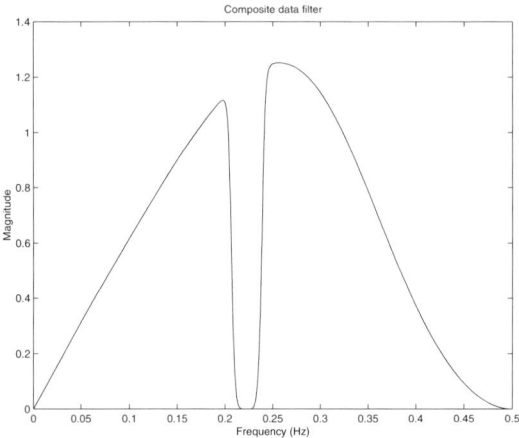

Fig. 13.6. Composite data filter frequency response illustrating differencing, notching and low-pass characteristics

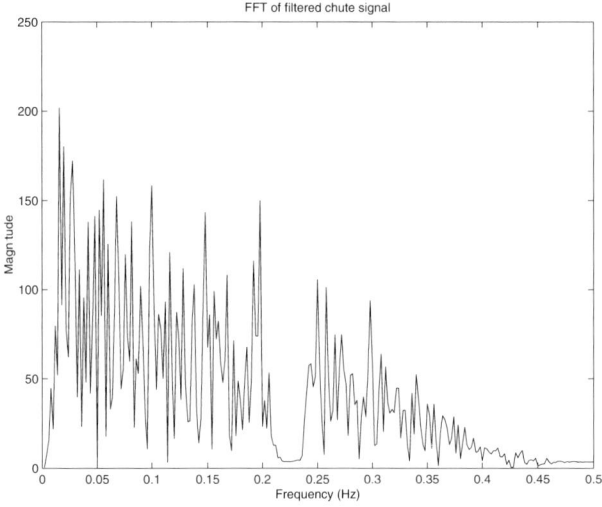

Fig. 13.7. FFT magnitude for filtered chute height signal. Compare this with Figure 13.5

We are led to the natural question of: what would have happened had we not filtered the data in this way? Again, we resort to (13.1) to assist in understanding the effects. Had the DC terms been left to dominate, then the fit of $\hat{G}$ to G would have been tight in the DC region, which is largely uninteresting for this control design problem. Had the differencer not been used, then the model fit would have emphasised the dominant low frequencies (<0.1 Hz) in the data, again which is not interesting for improved control. Lastly, had the tine filter not been used, this periodic component would have forced the high frequency part of the model to match this part exactly before dealing with the main fit. Since we are not interested in responding to the removal of the tine-based fluctuations in the data, fitting this component is not germane to the controller.

The chute height data used as the input to MATLAB® System Identification Toolbox is displayed in Figure 13.8 alongside the original evaluation data. (An offset and a scaling have been added to assist in comparison.) The

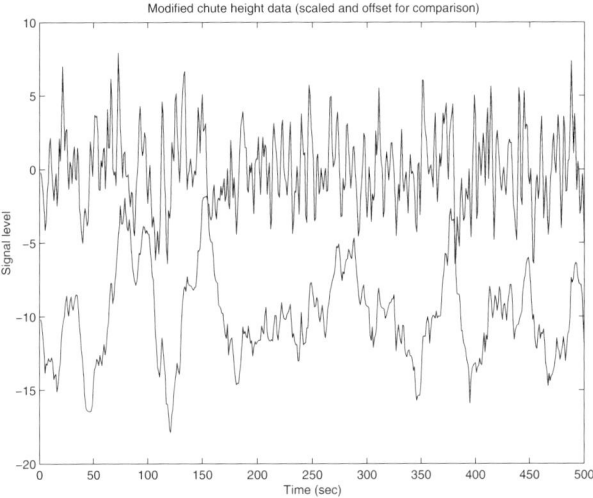

Fig. 13.8. Prepared, filtered chute height evaluation data (upper trace) prior to processing compared to original (lower trace). [Scaling and offset have been added.]

final filtering decision was related to control purposes: the objective was to build a controller with bandwidth value around 0.3 Hz, seeking an improvement over the PID setup. All the experimental excitation and output signals have a significant frequency content below 0.1 Hz. As the low-frequency dynamics could be, in principle, easily handled by integral or phase-lag elements in regulators, a better fit was sought at frequencies around the control bandwidth. The presence of the differencer was also justified by the presence of an integrator in the disturbance model.

13.3.3 Model Structure Selection

A low-order plant model is desired for control design. To try to achieve a low-bias model for the plant G using system identification with a low plant model order, a high-order noise model H should be used, even if it is going to be discarded later on in the controller design phase. Its role is to capture all those parts of the data not easily attributed to the low-order correlation between the inputs and the outputs. Increasing the capacity of the noise model, $\hat{H}$, to absorb these effects reduces the requirements of the plant model.

Plant model structure determination (order) has to be sorted out from the experimental data, because no prior first-principle model is available, although we do have expectations of eventual controller complexity. Identification was carried out over a first data set (the evaluation set) together with prediction error testing over a second independent (validation) data set with a collection of different candidate structures of varying order and delay. MATLAB® commands `arxstruc` and `selstruc` were used.

Figure 13.9 shows the output from the `selstruc` command. The picture consists of a number of crosses, one for each candidate model structure and its associated ARX model, fitted using the evaluation data set. The vertical axis measures the loss function: the "corrected" mean squared prediction error of the model, computed using the validation data set. The horizontal axis is the index of the number of parameters in each model structure, a measure of model complexity. Thus, the diagram indicates the predictive performance of each of many model structures when fitted using the evaluation set and measured on the independent validation set. By using two independent sets of data, there is no benefit in using complex models to fit random fluctuations in the data — they decrease predictive performance, as the experiment showed. There is also an information-theoretic need to penalize model complexity further and the mean square loss function is modified (or "corrected") with an additional complexity penalty such as the Akaike information criterion or the minimum description length.

It is our view that the penalty for model order for a model to be used for control design needs to be increased even further relative to that of a model to be used for prediction, say. This is because of the dramatically more difficult control design problem encountered for high-order models where these models do not capture significant dynamics in the band of interest.

13.3.4 Parameter Estimation

Once all the user-configurable decisions in the identification task had been taken, a Box-Jenkins model was fitted to the pre-filtered input/output data. The separate, independently parameterised, high-order noise model was discarded for later use. Detailed theory on identification algorithms can be found in [108]. In a sense, the interesting part of the system identification is embodied in the carpentry of data selection and preparation, model structure

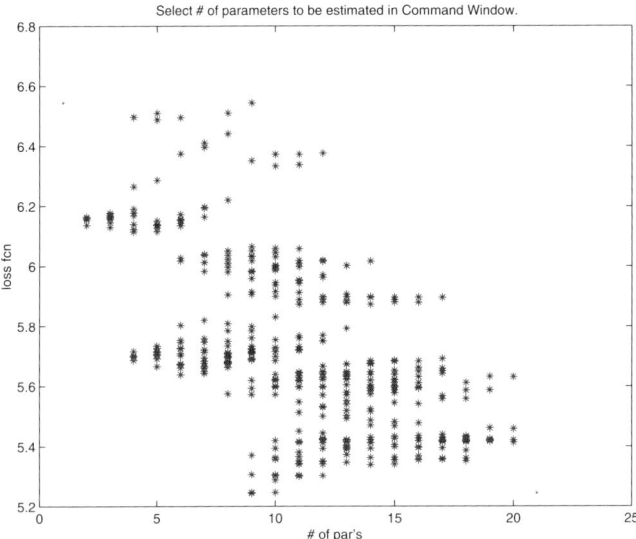

Fig. 13.9. Model order selection diagram from ID toolbox command *selstruc*

selection, *etc.*, where the problem specification and process knowledge enter the picture.

Model assessment

The resulting model obtained from parameter estimation has to be assessed and validated against the prior knowledge the engineer has about the plant. If any significant contradiction is present, a re-elaboration of the modelling steps is advised.

In particular, the model should be assessed in terms of adequacy of agreement with first-principles models, thought models, plant operation data or with other experiments. Areas for evaluation include:

- stability of the system. Since we will use the model for control design, it is normal to demand that the model and plant demonstrate similar open-loop stability properties;
- examination of frequency response plots for sensible bandwidth, gain, *etc.*;
- acceptable steady-state behaviour – correct signs and feasible gains;
- shape of the step response simulation;
- prediction error on validation sets;
- auto- and cross-correlation of prediction errors calculating up to indexes greater than the estimated order, to reveal unmodelled dynamics and known feedback;

- analysis of the estimated parameter covariance matrix to determine, for example, parameter confidence intervals, while being cautious of the validity of these calculations;
- evaluation of the suitability of the disturbance model vis-à-vis *a priori* information and models.

If a reasonable agreement with prior expectations is achieved, then the ensuing controller design phase can be carried out.

13.4 Controller Design

The algorithms that were chosen for control design in this particular case were frequency-weighted linear quadratic Gaussian (FW-LQG) and loop transfer recovery. FW-LQG can also be designed by incorporating input and sensor stable filters (weights) that change the process frequency response, and then designing a standard LQG controller for the augmented process. In the real implementation, those filters must be provided by the controller itself. Note that, in this way, the controller order increases, so model-reduction techniques need to be used subsequently. Details about the control design procedure can be found in [52, 123, 142].

The control criterion was chosen to be

$$J_{\text{LQG}} = \lim_{N \to \infty} \sum_{i=1}^{N} \left\{ (h_t - \bar{h})^2 + \lambda_t (t_t - \bar{t})^2 + \rho_s (s_t - \bar{s})^2 + \rho_f (f_t - \bar{f})^2 \right\}$$
(13.2)

Here, h_t, t_t, s_t and f_t represent the chute height, turbine torque, turbine speed and chute flap signal values. The barred quantities are their set point values. That is, we use a performance criterion in terms of the measured output variables only. This is because the model states from the identified system are not physically meaningful to us. It also allows us to make a link to loop transfer recovery LQG design. The design parameters λ_t, ρ_s and ρ_f are selected, in the first instance, to reflect what we know of the process, its behaviour and what we regard as important.

The parameter λ_t is used to balance the attention that we give to using the control signals s_t and f_t to manage the chute versus the importance we attach to controlling the chute height. Note that the units and ranges of these variables (0-100 per cent for chute height and 0.6–1.4 MNm for torque) need to be accommodated in selecting the value for λ_t – it needs to be at least 10^4 to weight the torque at all. Torque is the economically most important signal, which should dominate the criterion. The specific value selected was made by asking: *What is the maximum level of chute height disturbance we would accept for a 0.1 MNm improvement in torque?*. Similarly, we set ρ_s and ρ_f by asking the level to which we would permit speed or flap to alter in response to a 10 per cent change in torque. Subsequent tweaking of these gains then follows from experimental observation.

In the case study presented here, an open-loop stable regulator was a design requirement for the following reasons. In order to acquire acceptance for implementation of a controller, it was desirable to run it in open-loop for comparison with the existing strategy. In this way, confidence in operation could be developed by the plant engineers. Additionally, there is a requirement for a suitable response to saturation and windup issues arising when gaps in feed occur. This is more easily achieved with a stable control law.

In the same way that identified models have to be evaluated against prior knowledge, similar assessments should be made in practice with controllers prior to their implementation on the real plant. In this case, those validation chores included:

- checking Bode plots for:
  - sensible steady state behaviour,
  - correct frequencies of action;
- checking positions of poles and zeros;
- simulation runs, in particular, comparison to PID performance.

The conclusion is that prior knowledge (thinking and physical insight) is needed right from the initial modelling to the final control design phase, so that little autonomy is granted to ready-made automated identification or control design software.

13.5 Iterative Approach

The iterative scheme to be followed has its roots in the ideas described in the earlier chapters of this monograph and the references therein. The steps actually carried out in the sugar mill case study will be outlined here.

13.5.1 Closed-loop Identification

Given an operating controller, excitation signals are injected and closed-loop data measured to provide a new model capable of delivering an improved control performance. For the sugar mill, the excitation signal was selected as outlined earlier for all the modelling iterations. Likewise, the data filtering was maintained across the iterations.

The system identification approach varied from the so-called *direct method* described by (13.1) to the two-stage approach of Van den Hof and Schrama [147]. As we have remarked before, the objective in our system identification was to develop a good linear plant model $\hat{G}$ for G in spite of the the intrusion of the significant disturbance. The two-stage approach is designed to assist here. This capability, however, relies inherently on linearity which, in this case, makes it more sensitive to the selection of data.

The two-stage indirect approach

Direct parameter estimation via prediction error methods yields a bias of the estimates that depends upon the disturbance spectrum, as expressed in (13.1). The method proposed in [147] gives a tunable expression for the bias of the identified model. It does this by exploiting the independence of the excitation reference signal and the disturbance.

Stage 1 involves identifying a model for the transfer function from the reference signal, r_k, to the control signal, u_k.

$$u_k = \hat{S} r_k + \hat{V} \epsilon_k^u$$

where ϵ_k^u is the one-step ahead prediction error of u_k. This is analogous to a conventional open-loop estimation between the signals r_k and u_k. Because the reference r_k is independent of the process disturbance, there is no bias in the estimate of $\hat{S}(z)$ due to the disturbance, provided a high-order model is used.

Stage 2 generates (by simulation for the same data) the control action that would have been applied were no disturbance present, $u_k^r = \hat{S} r_k$. This signal inherits the independence from the disturbance enjoyed by r_k. Now, another model is fitted between u_k^r and plant output y_k.

$$y_k = \hat{G} r_k + \hat{H} \epsilon_k^y$$

where ϵ_k^y is the prediction error of y_k. Box-Jenkins estimation is used in both stages. The resulting plant model is again unbiased by the process disturbance because u_k^r is independent of the disturbance signal by construction.

With this approach, it is shown in the referenced work, by deriving an expression similar to (13.1), that if an accurate estimate of the sensitivity S in the first phase is achieved, then the plant bias in the plant model $\hat{G}$ is similar to the open-loop bias. In situations where a good open-loop plant model is desired, where excitation might not be adequate and disturbances are significant, the use of this closed-loop identification method can be advantageous.

13.5.2 Cautious Frequency Weighting LQG Design

In the iterative control design paradigm outlined earlier, one or several stages of model-free controller tuning is countenanced in which experimental closed-loop spectra are compared to anticipated design values to launch a redesign using frequency-weighted LQG without adjusting the process model used. In this formulation, the frequency weighting is $\|W\|^2 = \frac{\Phi_y}{\Phi_{y^c}}$, where Φ_y is the measured, achieved closed-loop spectrum of y_t and Φ_{y^c} is the designed value, both from data with the previous controller in operation.

In order to limit the magnitude of variation of the controllers ensuing, the proposed frequency weight is altered to

$$\|W\|^2 = \frac{\delta \Phi_y + (1 - \delta) \Phi_{y^c}}{\Phi_{y^c}}$$

In this way, given a previous LQG regulator responsible for the actual observed spectrum, frequency-weighted LQG design with $\delta = 0$ produces the same LQG regulator ($W = 1$) and, at the other limit, $\delta = 1$ produces a frequency weighting that maximises the use of the experimental data at the expense of cautious controller adjustment.

To accomplish the frequency-weighted LQG design, the weighting filter W is approximated by, at most, third order transfer functions (to capture the main characteristics without yielding a controller with unacceptably high order).

13.5.3 Modelling and Control Iterations

The iterative scheme for the A2 sugar mill was carried out as follows:

1. Closed-loop data were obtained with PID regulator (K_{-1}) and flap and speed external excitation.
2. A model $\hat{G}_0$ was identified.
3. A standard LQG regulator was designed based on that model ($K_0(\hat{G}_0)$) and tested without excitation.
4. Based on the different output and simulated output spectra, a cautious ($\delta = 0.6$) frequency-weighted regulator $K_1(\hat{G}_0)$, was designed and tested again.
5. The output spectra was recalculated, thus reweighting and redesigning regulator $K_*(\hat{G}_0)$. The obtained regulator was very similar to K_1, so no further frequency-compensation iterations were carried out for $\hat{G}_0$.
6. A new model $\hat{G}_1$ was identified, with K_1 operating in closed-loop with excitation. The identification method was a two-stage indirect approach.
7. A standard LQG regulator based on that model ($K_2(\hat{G}_1)$) was designed and tested.
8. Based on the output and simulated output spectra, a subsequent cautious frequency-weighted regulator was sought. Only for very small values of δ was a stable $K_3(\hat{G}_1)$ obtained, being it very similar to $K_2(\hat{G}_1)$. Hence, the frequency-compensation iterations were stopped.
9. A new experiment was conducted and the new model $\hat{G}_2$ was very similar to $\hat{G}_1$, hence modelling iterations were also stopped.

The final regulator to be applied was $K_2(\hat{G}_1)$.

Results

Figures 13.10 and 13.11 show the performance of a PID controller and the first LQG design based on direct identification $K_0(\hat{G}_0)$. The diagrams consist of time series data taken from the real process operating under each controller. The figures to the right are histograms of the data and indicate statistical measures of performance, notably mean and standard deviation.

292   R.R. Bitmead and A. Sala

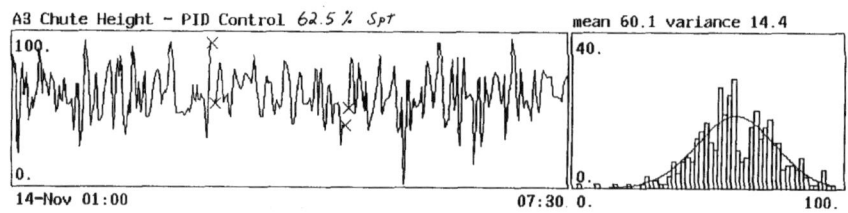
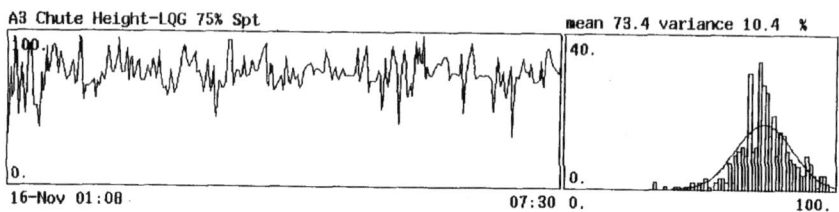

Fig. 13.10. PID versus LQG, chute height regulation

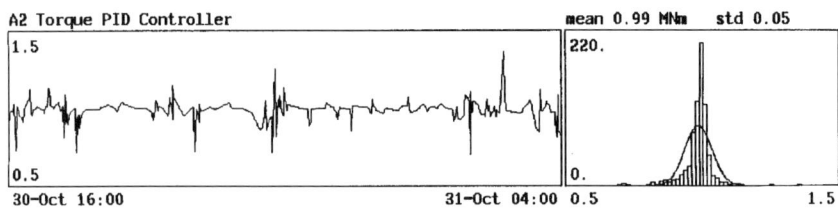
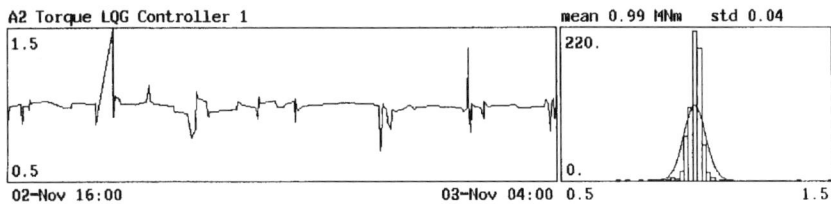

Fig. 13.11. PID versus LQG, turbine torque regulation

Better control should be discernible through a lower standard deviation of the torque. The first of the figures shows the chute height behaviour and the second one shows the torque regulation. Only a slightly lower variance was achieved in chute height and in torque regulation. The point to make is that the iterative scheme improved performance after subsequent closed-loop experiments were undertaken.

Figure 13.12 shows the performance of the final controller $K_2(\hat{G}_1)$. Torque variance was reduced to 60 per cent of the original PID value and height variance was reduced below 45 per cent of the initial value.

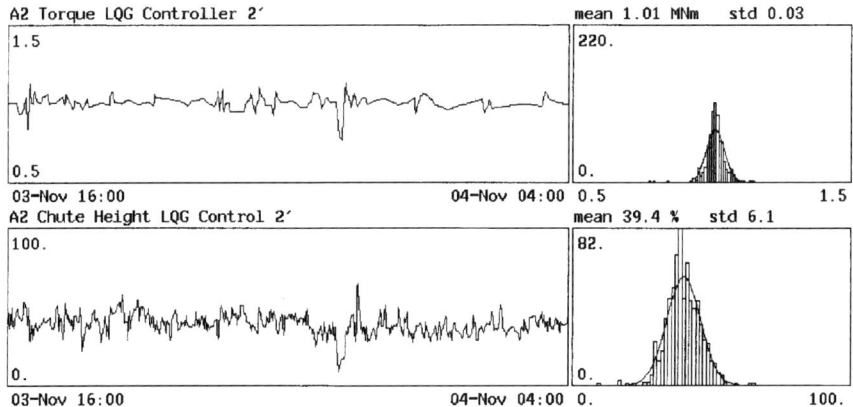

Fig. 13.12. Final LQG regulator after iterations

Energy savings were also achieved when using the new regulators. Table 13.1 below shows the results with respect to steam usage.

Iterative modelling and control achieved better performance in terms of variance and energy usage. Notwithstanding, a drawback lies at the heart of the technique (also applicable to any technique that tries to squeeze out performance via precise modelling): process variations need regulator retuning to keep performance levels optimal from season to season.

In fact, the identification experiments were carried out on Mill A2 during only one particular season. In that situation, the iterative approach showed the above improvements. In the following season, with different mill geometry and seasonal crop variation, the mill transfer function might have changed. In the next season's tests, performance with the same controller was no better than that of the original hand tuned PIDs. Thus, to achieve the benefits from the method in the subsequent year, a complete new redesign is needed.

Another test of the generality of the solution was its application to another of the mills, which is mechanically very similar. Starting with the final LQG controller for Mill A2, one additional iteration step to adapt to Mill A3 characteristics was carried out. Torque standard deviation decreased 10 per cent. Notwithstanding this apparent success in Mill A3, with some juicier cane varieties, large juice flows caused a torque decrease and the LQG controller overcompensated, producing very large fluctuations. So, operators

Table 13.1. Energy usage, LQG vs. PID

Mill	Control	Mean KW	Steam flow	% saving
A2	PID	391	9400	
	LQG	356	8800	6.4%
A3	PID	356	10600	
	LQG	323	10300	2.8%

often reverted back to PID control. Feedforward information about the state of previous mills was tried without great success.

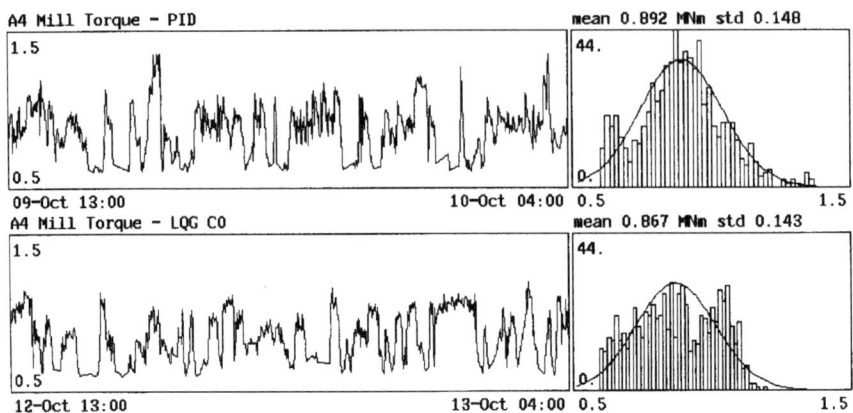

Fig. 13.13. Torque regulation in Mill 4

When trying this same philosophy of using the Mill 2 controller in Mill 4, no improvement was seen. Indeed, performance both with PID and with this controller was much worse than in other mills (see Figure 13.13). The poor performance suggests significant importance of non-linearities by this stage in the crushing process.

The conclusion if the experiments was that PIDs provided consistent (robust) performance for a wide range of cane varieties. Modelling and LQG control design could outperform them (in terms of variance and energy usage) only in the situations where experiments had been carried out and a controller carefully designed. Time variation due to mill geometry change and inter-season variation caused them to perform approximately the same as the original PID controllers. The overall project conclusion was that the process constraints (chute height limitations and high daily throughput targets) limited achievable control performance, so that improvements to extraction would hinge firstly on capital expenditure rather than controller redesign. This is, anyway, a valuable piece of information.

13.6 Conclusions

We have presented in this chapter an industrial example of iterative identification and control design applied to a sugar cane crushing mill. The important features of this analysis have been the connections drawn between the computational activities and the *a priori* process information. These two knowledge sources have acted hand in glove to guide the approach. Thus, experimentation is used both to generate data for the next model fit and also for testing the quality of control design and its departure from expected performance. The principles demonstrated in this example are those laid out with some generality in earlier chapters.

The iterative control design proved valuable in providing enhancement of model performance to match the controller needs, with relatively simple models and controllers. In the actual plant, it allowed for a reduced variance on outputs and savings in energy usage, although the same controllers applied to the subsequent mills or with altered mill geometries without redesign did not show much improvement.

Historically, this industrial plant of the CSR Victoria Sugar Mill was both an early development site and proving ground for iterative identification and control design [125, 161]. The motivating questions after the first round of model fitting and control design were: *"What do we do now that the new controller's performance is disappointing?" "What new information do we now have?"* and *"How can we use this new information to guide further actions?"* The new information was that the designed controller performed worse than expected.

There remain other underlying questions. As computers are introduced in all aspects of factory automation, it is relevant to research on developing methods so that readily accessible closed-loop data can be used to improve performance. It would be valuable to automate the information gathering phase, so that operational records covering significant time periods might be used to advantage. Most of the iterative ideas can be applied by using readily available software. However, current iterative approaches require good user supervision of design parameters. In a sense, this is an extension of adaptive control ideas with very slow adaptation and designer intervention into the loop. It would be useful to incorporate more of the aspects of adaptive control if possible.

Acknowledgements

The first author is grateful for the comments of Sam Crisafulli, Michel Gevers, Ari Partanen, Robert Peirce and Zhuquan Zang.
Research supported by CSR Herbert River Mills Group Australia and US National Science Foundation under Grant ECS-0070146.

References

1. P. Albertos. Modelling, identification and control design. In P. Albertos and A. Sala, editors, *Lecture Notes on Iterative Identification and Control design*, pages 1–13. Publicaciones Universidad Politecnica Valencia, 2000.
2. P. Albertos, A. Esparza, and J. Romero. Model-based iterative control design. In *Procs. American Control Conference*, pages 2578–2582, San Diego, California, Jun. 2000.
3. P. Albertos and J. Picó. Iterative controller design by frequency scale experimental decomposition. In *Procs. 32nd IEEE Conference on Decision and Control*, pages 2828–2832, San Antonio, Texas, Dec. 1993.
4. P. Albertos, A. Valera, J. Romero, and A. Esparza. Control of flexible robot manipulators: an iterative design. *Submitted and under reviewing*, 2002.
5. B.D.O. Anderson, R.R. Bitmead, C.R. Johnson Jr, P.V. Kokotovic, R.L. Kosut, I.M.Y. Mareels, L. Praly, and B.D. Riedle. *Stability of Adaptive Systems: Passivity and Averaging Analysis*. MIT Press, Cambridge, MA, 1986.
6. B.D.O. Anderson, X. Bombois, M. Gevers, and C. Kulcsár. Caution in iterative modeling and control design. In *IFAC Workshop on Adaptive Control and Signal Processing*, pages 13–19, Glasgow, 1998.
7. B.D.O. Anderson and M. Gevers. Fundamental problems in adaptive control. In D. Normand-Cyrot, editor, *Perspectives in Control Theory and Applications*, pages 9–21. Springer-Verlag, Berlin, 1998.
8. B.D.O. Anderson and J. B. Moore. *Optimal Control, Linear Quadratic Methods*. Prentice-Hall, Englewood Cliffs, N.J., 1989.
9. K.J. Åström. Matching criteria for control and identification. In *Proc. 2nd European Control Conference*, pages 248–251, Groningen, The Netherlands, 1993.
10. K.J. Åström. Fundamental limitations of control system performance. In A. Paulraj, C. Schaper, and V. Roychowdhury, editors, *Mathematical Engineering: A Kailath Festschrift*. Kluwer, 1996.
11. K.J. Åström. Limitations on control system performance. *European Journal of Control*, 6:1–19, 2000.
12. K.J. Åström and T. Bohlin. Numerical identification of linear dynamic systems from normal operating records. In *Proc. IFAC Symposium on Self-Adaptive Systems*, Teddington, UK, 1965.
13. K.J. Åström and B. Wittenmark. Problems of identification and control. *Journal of Mathematical Analysis and Applications*, 34:90–113, 1971.
14. K.J. Åström and B. Wittenmark. *Adaptive Control*. Addison-Wesley, second edition, 1995.
15. K.J. Åström and B. Wittenmark. *Computer-Controlled Systems: theory and design*. Prentice-Hall, NJ, third edition, 1997.
16. M. Athans and P.L. Falb. *Optimal Control*. McGraw-Hill, New York, 1966.
17. T. Başar and P. Bernhard. $\mathcal{H}_\infty$-*Optimal control and related minimax design problems - A Dynamic game approach*. Birkhauser, Boston, 1991.
18. R.L. Bagley and R.A. Calico. Fractional-order state equations for the control of viscoelastic damped structures. *J. Guidance, Control and Dynamics*, 14:304–311, 1991.

19. D.S. Bayard and R.Y. Chiang. A frequency domain approach to identification, uncertainty characterization and robust control design. In *Proc. 32nd IEEE Conference on Decision and Control*, pages 2266–2271, San Antonio, USA, 1993.
20. D.S. Bayard, Y. Yam, and E. Mettler. A criterion for joint optimization of identification and robust control. *IEEE Trans. Automatic Control*, 37:986–991, 1992.
21. R. Bellman. *Dynamic Programming*. Princeton University Press, Princeton, N.J., 1957.
22. R. Bellman, I. Glicksberg, and O. A. Gross. Some aspects of the mathematical theory of control processes. Technical Report R-313, The RAND Corporation, Santa Monica, Calif., 1958.
23. R.R. Bitmead. Iterative control design approaches. In *Proc. 12th IFAC World Congress*, volume 9, pages 381–384, Sydney, Australia, 1993.
24. R.R. Bitmead, B.D.O. Anderson, M. Gevers, and L.C. Kammer. Cautious controller tuning, submitted for publication. *Automatica*, 2000.
25. R.R. Bitmead, M. Gevers, and A.G. Partanen. Introducing caution in iterative controller design. In *11th IFAC Symp. on System Identification*, pages 1701–1706, Fukuoka, Japan, 1997.
26. H.S. Black. Stabilized feedback amplifiers. *Bell System Technical Journal*, 13:1–18, 1934.
27. H.W. Bode. Relations between attenuation and phase in feedback amplifier design. *Bell System Technical Journal*, 19:421–454, 1940.
28. H.W. Bode. *Network Analysis and Feedback Amplifier Design*. Van Nostrand, New York, 1945.
29. T. Bohlin. *Interactive System Identification: Prospects and Pitfalls*. Springer-Verlag, Berlin, 1991.
30. X. Bombois, B.D.O. Anderson, and M. Gevers. Mapping parametric confidence ellipsoids to Nyquist plane for linearly parametrized transfer functions. In G.C. Goodwin, editor, *Model Identification and Adaptive Control*, pages 53–71. Springer, 2001.
31. X. Bombois, M. Gevers, and G. Scorletti. A measure of robust stability for an identified set of parametrized transfer functions. *IEEE Trans. Automatic Control*, 45(11):2141–2145, 2000.
32. X. Bombois, M. Gevers, G. Scorletti, and B.D.O. Anderson. Robustness analysis tools for an uncertainty set obtained by prediction error identification. *Automatica*, 37(10):1629–1636, October 2001.
33. X. Bombois, G. Scorletti, B.D.O. Anderson, M. Gevers, and P.M.J. Van den Hof. A new robust control design procedure based on a prediction error identification uncertainty set. *Submitted to Automatica*, 2001.
34. P.M.M. Bongers. *Modeling and Identification of Flexible Wind Turbines and a Factorizational Approach to Robust Control*. Ph.D. thesis, Delft University of Technology, Delft, the Netherlands, 1994.
35. P.M.M. Bongers and O.H. Bosgra. Low order robust $\mathcal{H}_\infty$ controller synthesis. In *Proc. 29th IEEE Conf. Decision and Control*, pages 194–199, Honolulu, HI, 1990.
36. M. Campi, W.S. Lee, and B.D.O. Anderson. New filters for internal model control design. *Int. J. Robust and Nonlinear Control*, 4:757–775, 1994.
37. Y. Chen and C. Wen. *Iterative Learning Control: convergence, robustness and applications*. Springer LNCIS 248, London, 1999.

38. D.W. Clark, C. Mohtadi, and P.S. Tuffs. Generalised predictive control. *Automatica*, 23(2):137–148, 1987.
39. R. Clarke, J. J. Burken, J. T. Bosworth, and J.E. Bauer. X-29 flight control system - Lessons learned. *Int. Journal of Control*, 59(1):199–219, 1994.
40. J.C. Clegg. A nonlinear integrator for servomechanisms. *Trans. AIEE Part II*, 77:41–42, 1958.
41. Brian L. Cooley and Jay H. Lee. Control-relevant experiment design for multivariable systems described by expansions in othonormal bases. *Automatica*, 37(2):273–281, February 2001.
42. S. Crisafulli, R.R. Bitmead, G.J. Rey, C.R. Johnson Jr., and A.J. Connolly. System identification methods applied to a helicopter vibration control problem. In *IFAC Symposium on System Identification*, Copenhagen, Denmark, 1994.
43. R.A. de Callafon, D. Roover, and P.M.J. Van den Hof. Multivariable least squares frequency domain identification using polynomial matrix fraction descriptions. In *Proc. 35th IEEE Conference on Decision and Control*, pages 2030–2035, Kobe, Japan, 1996.
44. R.A. de Callafon and P.M.J. Van den Hof. Control relevant identification for $\mathcal{H}_\infty$-norm based performance specifications. In *Proc. 34th IEEE Conference on Decision and Control*, pages 3498–3503, New Orleans, USA, 1995.
45. R.A. de Callafon and P.M.J. Van den Hof. Filtering and parametrization issues in feedback relevant identification based on fractional model representations. In *Proc. 3rd European Control Conference*, volume 1, pages 441–446, Rome, Italy, 1995.
46. R.A. de Callafon and P.M.J. Van den Hof. Suboptimal feedback control by a scheme of iterative identification and control design. *Mathem. Modelling of Systems*, 3(1):77–101, 1997.
47. R.A. de Callafon, P.M.J. Van den Hof, and M. Steinbuch. Control-relevant identification of a compact disc pick-up mechanism. In *Proc. 1993 IEEE Conf. on Decision and Control*, pages 2050–2055, San Antonio, Texas, 1993.
48. D.K. de Vries and P.M.J. Van den Hof. Quantification of uncertainty in transfer function estimation: a mixed probabilistic - worst-case approach. *Automatica*, 31:543–558, 1995.
49. C.A. Desoer and M. Vidyasagar. *Feedback Systems: Input-Output Properties*. Academic Press, New York, 1975.
50. J.C. Doyle. Guaranteed margins for LQG regulators. *IEEE Trans. Automatic Control*, AC-23:756–757, 1978.
51. J.C. Doyle, B.A. Francis, and A.R. Tannenbaum. *Feedback Control Theory*. MacMillan, New York, 1992.
52. J.C. Doyle, K. Glover, P.P. Khargonekar, and B.A. Francis. State space solutions to standard $\mathcal{H}_2$ and $\mathcal{H}_\infty$ control problems. *IEEE Trans. Automatic Control*, 34:831–846, 1989.
53. J.C. Doyle and G.Stein. Multivariable feedback design: Concepts for a classical/modern synthesis. *IEEE Trans. Automatic Control*, 26:4–16, 1981.
54. J.C. Doyle and G. Stein. Robustness with observers. *IEEE Trans. Automatic Control*, AC-24:607–611, 1979.
55. W. Draijer, M. Steinbuch, and O.H. Bosgra. Adaptive control of the radial servo system of a compact disc player. *Automatica*, 28:455–462, 1992.
56. A.K. El-Sakkary. The gap metric: Robustness of stabilization of feedback systems. *IEEE Trans. Automatic Control*, AC-26:240–247, 1985.

57. U. Forssell and L. Ljung. Closed-loop identification revisited. *Automatica*, 35:1215–1241, 1999.
58. U. Forssell and L. Ljung. Some results on optimal experiment design. *Automatica*, 36:749–756, 2000.
59. G.F. Franklin, J.D. Powell, and A. Emami-Naeini. *Feedback Control of Dynamic Systems (third edition)*. Addison-Wesley, 1994.
60. M. Gevers. Towards a joint design of identification and control? In H.L. Trentelman and J.C. Willems, editors, *Essays on Control: Perspectives in the Theory and its Applications*, pages 111–151, New York, 1993. Birkhäuser.
61. M. Gevers. Modeling, identification and control. In *Communications, computing control and signal processing 2000.(dedicated to T. Kailath)*, pages 1–15. Kluver Academic, 1996.
62. M. Gevers, X. Bombois, B. Codrons, F. De Bruyne, and G. Scorletti. The role of experimental conditions in model validation for control. In A. Garulli, A. Tesi, and A. Vicino, editors, *Robustness in Identification and Control - Proc. of Siena Workshop, July 1998*, volume 245 of *Lecture Notes in Control and Information Sciences*, pages 72–86. Springer-Verlag, 1999.
63. M. Gevers and L. Ljung. Optimal experiment designs with respect to the intended model application. *Automatica*, 22:543–554, 1986.
64. G.C. Goodwin, M. Gevers, and B. Ninness. Quantifying the error in estimated transfer functions with application to model order selection. *IEEE Trans. Automatic Control*, 37:913–928, 1992.
65. G.C. Goodwin and K.S. Sin. *Adaptive Filtering, Prediction and Control*. Prentice-Hall, Englewood Cliffs, N.J., 1984.
66. M. Green and D.J.N. Limebeer. *Linear Robust Control*. Prentice-Hall, Englewood Cliffs, 1995.
67. G. Gu and P.P. Khargonekar. A class of algorithms for identification in $\mathcal{H}_\infty$. *Automatica*, 28:299–312, 1992.
68. R.G. Hakvoort. *System Identification for Robust Process Control: Nominal Models and Error Bounds*. Ph.D. thesis, Delft University of Technology, Delft, the Netherlands, 1994.
69. R.G. Hakvoort, R.J.P. Schrama, and P.M.J. Van den Hof. Approximate identification with closed-loop performance criterion and application to LQG feedback design. *Automatica*, 30(4):679–690, 1994.
70. R.G. Hakvoort and P.M.J. Van den Hof. Identification of probabilistic system uncertainty regions by explicit evaluation of bias and variance errors. *IEEE Trans. Automatic Control*, 42(11):1516–1528, 1997.
71. A.C. Hall. Application of circuit theory to the design of servomechanisms. *Journal of the Franklin Institute*, 242:279–307, 1946.
72. F.R. Hansen. *A fractional representation approach to closed-loop system identification and experiment design*. Ph.D. thesis, Stanford University, 1989.
73. F.R. Hansen, G. Franklin, and R. Kosut. Closed-loop identification via the fractional representation: Experiment design. *Proc. American Control Conference*, 1989.
74. A.J. Helmicki, C.A. Jacobson, and C.N. Nett. Control oriented system identification: A worst-case/deterministic approach in $\mathcal{H}_\infty$. *IEEE Trans. Automatic Control*, 36:1163–1176, 1991.
75. J.W. Helton and O. Merino. *Classical control using $\mathcal{H}_\infty$ Methods*. SIAM, Philadelphia, 1999.

76. R. Hildebrand and M. Gevers. Identification for control: optimal input design with respect to a worst-case ν-gap cost function. In *10th ERNSI Workshop on System Identification*, Cambridge, UK, 2001.
77. H. Hjalmarsson, M. Gevers, and F. De Bruyne. For model-based control design, closed-loop identification gives better performance. *Automatica*, 32:1659–1673, 1996.
78. H. Hjalmarsson, M. Gevers, S. Gunnarsson, and O. Lequin. Iterative Feedback Tuning: theory and applications. *IEEE Control Systems Magazine*, 18:26–41, August 1998.
79. H. Hjalmarsson, S. Gunnarsson, and M. Gevers. Optimality and suboptimality of iterative identification and control design schemes. In *Proc. American Control Conference*, volume 4, pages 2559–2563, Seattle, Washington, 1995.
80. B.L. Ho and R.E. Kalman. Effective construction of linear state-variable models from input-output functions. In *Proc. Third Allerton Conf. on Communication, Control, and Computing*, pages 449–459, Urbana, IL, 1965. University of Illinois.
81. U. Holmberg, S. Valentinotti, and D. Bonvin. An identification-for-control procedure with robust performance. *Control Engineering Practice*, 8:1107–1117, 2000.
82. I.M. Horowitz. *Synthesis of Feedback Systems*. Academic Press, New York, 1963.
83. I.M. Horowitz. *Quantitative Feedback Design Theory (QFT)*. QFT Publications, Boulder, Colorado, 1993.
84. I.M. Horowitz and U. Shaked. Superiority of transfer function over state-variable methods in linear time-invariant feedback system design. *IEEE Trans. Automatic Control*, AC-20:84–97, 1975.
85. T. Hu and Z. Lin. *Control Systems With Actuator Saturation : Analysis and Design*. Birkhäuser, 2001.
86. P.A. Ioannou and J. Sun. *Robust Adaptive Control*. Prentice-Hall Press, Upper Saddle River, NJ, 1996.
87. R. Isermann. *Digital Control Systems*. Springer, second edition, 1989.
88. H.M. James, N.B. Nichols, and R.S. Phillips. *Theory of Servomechanisms*. McGraw-Hill, New York, 1947.
89. L. Johnston and M. Gevers. A comparison of model order reduction approaches. *Submitted for publication*, 2001.
90. R.E. Kalman. Contributions to the theory of optimal control. *Boletín de la Sociedad Matemática Mexicana*, 5:102–119, 1960.
91. R.E. Kalman. A new approach to linear filtering and prediction problems. *Transactions ASME, Series D, J. Basic Eng.*, 82:34–45, 1960.
92. R.E. Kalman. When is a linear control system optimal? *Trans. ASME (D): J. Basic Engineering*, 86:1–10, March 1964.
93. R.E. Kalman and R.S. Bucy. New results in linear filtering and prediction theory. *Trans ASME (D): J. Basic Engineering*, 83 D:95–108, 1961.
94. R.E. Kalman, Y. Ho, and K.S. Narendra. *Controllability of Linear Dynamical Systems*, volume 1 of *Contributions to Differential Equations*. John Wiley & Sons, Inc., New York, 1963.
95. L.C. Kammer, R.R. Bitmead, and P.L. Bartlett. Direct iterative tuning via spectral analysis. *Automatica*, 36:1301–1307, 2000.

96. R.E. Klein. Using bicycles to teach system dynamics. *IEEE Control Systems Magazine*, 9(3):4–9, 1989.
97. R.L. Kosut and B.D.O. Anderson. Least-squares parameter set estimation for robust control design. In *Proc. American Control Conference*, pages 3002–3006, Baltimore, MD, 1994.
98. I.D. Landau. *System Identification and Control Design*. Prentice Hall, Englewood Clifffs, NJ, 1990.
99. I.D. Landau. From robust control to adaptive control. *Control Engineering Practice*, 7(9):1113–1124, 1999.
100. I.D. Landau. Identification in closed-loop: a powerful design tool (better design models, simple controllers). *Control Engineering Practice*, 9(1):51–65, 2001.
101. I.D. Landau and A. Karimi. Recursive algorithms for identification in closed loop: A unified approach and evaluation. *Automatica*, 33(8):1499–1523, 1997.
102. W.S. Lee, B.D.O. Anderson, R.L. Kosut, and I.M.Y. Mareels. A new approach to adaptive robust control. *Int. J. Adaptive Control and Signal Processing*, 7(3):183–211, 1993.
103. W.S. Lee, B.D.O. Anderson, R.L. Kosut, and I.M.Y. Mareels. On robust performance improvement through the windsurfer approach to adaptive robust control. In *Proc. of 32nd IEEE-CDC*, San Antonio, Texas, 1993.
104. W.S. Lee, B.D.O. Anderson, I.M.Y. Mareels, and R.L. Kosut. On some key issues in the windsurfer approach to adaptive robut control. *Automatica*, 31:1619–1636, 1995.
105. W.S. Lee, I. M. Y. Mareels, and B.D.O. Anderson. Iterative identification and two step control design for partially unknown unstable plants. *Int. Journal of Control*, 74:43–57, 2001.
106. K. Liu and R.E. Skelton. Closed loop identification and iterative controller design. *29th IEEE Conf on Decision and Control*, pages 482–487, 1990.
107. L. Ljung. Asymptotic variance expressions for identified black-box transfer function models. *IEEE Trans. Automatic Control*, AC-30:834–844, 1985.
108. L. Ljung. *System Identification: Theory for the User*. Prentice-Hall, Englewood Cliffs, NJ, 2nd edition, 1999.
109. L. Ljung and J. Sjöberg. A system identification perspective on neural nets. Technical report, Dept. of Electrical Eng., Linköping Univ., Sweden, May 1992.
110. L. Ljung and T. Soderström. *Theory and practice of recursive identification*. MIT Press, Cambridge, MA, 1983.
111. D.C. MacFarlane and K. Glover. *Robust controller design using normalized coprime factor plant descriptions*. Springer, New York, 1990.
112. J.M. Maciejowski. *Multivariable Feedback Design*. Addison-Wesley, Wokingham, England, 1989.
113. P.M Mäkilä, J.R. Partington, and T.K. Gustafsson. Worst-case control-relevant identification. *Automatica*, 31(12):1799–1819, 1995.
114. D. McFarlane and K. Glover. A loop shaping design procedure using $\mathcal{H}_\infty$ synthesis. *IEEE Trans. Automatic Control*, 37:759–769, 1992.
115. K.L. Moore. *Iterative Learning Control for Deterministic Systems*. Advances in Industrial Control. Springer-Verlag, Berlin, 1993.
116. M. Morari and E. Zafiriou. *Robust Process Control*. Prentice-Hall, Englewood Cliffs, N.J., 1989.
117. B. Ninness and G.C. Goodwin. Estimation of model quality. *Automatica*, 31(12):32–74, 1995.

118. H. Nyquist. Regeneration theory. *Bell System Technical Journal*, 11:126–147, 1932.
119. G. Obinata and B.D.O. Anderson. *Model Reduction for Control System Design*. Springer-Verlag, Berlin, 2001.
120. A. Oustaloup. *La Commande CRONE*. Hermès, Edition CNRS, Paris, 1991.
121. A. Oustaloup. *La Derivation Non Entière: Théorie, Synthèse et Applications*. Hermès, Paris, 1995.
122. H. Panagopoulos and K.J. Åström. PID control design and $\mathcal{H}_\infty$ loop shaping design of PI controllers based on non-convex optimization. In *Proceedings IEEE Int. Conf. Control Applications and Symp. Computer Aided Control Systems Design*, Kohala Coast, Hawaii, aug 1999.
123. A.G. Partanen. *Controller refinement with application to a sugar crane crushing mill*. Ph.D. thesis, Australian National University, 1995.
124. A.G. Partanen and R.R. Bitmead. Two-stage iterative identification/controller design and direct experimental controller refinement. In *Proc. 32nd IEEE Conf on Decision and Control*, pages 2833–2838, San Antonio, 1993.
125. A.G. Partanen and R.R. Bitmead. The application of an iterative identification and controller design to a sugar cane crushing mill. *Automatica*, 31:1547–1563, 1995.
126. L. Pernebo. An algebraic theory for the design of controllers for linear multivariable system - Part I: Structure matrices and feedforward design. *IEEE Trans. Automatic Control*, AC-26:171–182, 1981.
127. L. Pernebo. An algebraic theory for the design of controllers for linear multivariable system - Part II: Feeback realizations and feedback design. *IEEE Trans. Automatic Control*, AC-26:183–194, 1981.
128. R. Pintelon, P. Guillaume, Y. Rolain, J. Schoukens, and H. Van hamme. Parametric identification of transfer functions in the frequency domain - a survey. *IEEE Trans. Automatic Control*, 39:2245–2260, 1994.
129. I. Podlubny. *Fractional Differential Equations*. Academic Press, New York, N.Y., 1999.
130. I. Podlubny. Fractional-order systems and PID controllers. *IEEE Trans. Automatic Control*, AC-44:208–214, 1999.
131. L.S. Pontryagin, V.G. Boltyanskii, R.V. Gamkrelidze, and E.F. Mischenko. *The Mathematical Theory of Optimal Processes*. John Wiley, New York, 1962.
132. K. Poolla, P.P. Khargonekar, A. Tikku, J. Krause, and K. Nagpal. A time-domain approach to model validation. *IEEE Trans. Automatic Control*, 39:951–959, May 1994.
133. W.L. Rogers and D.J. Collins. X-29 $\mathcal{H}_\infty$ controller synthesis. *Journal of Guidance Control and Dynamics*, 15(4):962–967, 1992.
134. M.G. Safonov and M. Athans. Gain and phase margins for multiloop lqg regulators. *IEEE Trans. Automatic Control*, AC-22:173–179, 1977.
135. B.F. La Scala and R.R. Bitmead. A self-tuning regulator for vibration control. In *34th IEEE Conference on Decision and Control*, New Orleans, 1995.
136. R.J.P. Schrama. Accurate identification for control: The necessity of an iterative scheme. *IEEE Trans. Automatic Control*, 37:991–994, 1992.
137. R.J.P. Schrama. *Approximate Identification and Control Design*. Ph.D. thesis, Delft University of Technology, NL, 1992.

138. R.J.P. Schrama and O.H. Bosgra. Adaptive performance enhancement by iterative identification and control design. *Int. J. Adaptive Control and Signal Processing*, 7(5):475–487, 1993.
139. R.J.P. Schrama and P.M.J. Van den Hof. An iterative scheme for identification and control design based on coprime factorizations. In *Proc. American Control Conference*, pages 2842–2846, Chicago, IL, 1992.
140. M. M. Seron, J. H. Braslavsky, and G. C. Goodwin. *Fundamental limitations in filtering and control*. Springer-Verlag, Berlin, 1997.
141. R.E. Skelton. Model error concepts in control design. *Int. Journal of Control*, 49(5):1725–1753, 1989.
142. S. Skogestad and I. Postlethwaite. *Multivariable Feedback Control: analysis and design*. John Wiley & Sons, UK, 1996.
143. R.S. Smith and J.C. Doyle. Model validation: a connection between robust control and identification. *IEEE Trans. Automatic Control*, 37:942–952, 1992.
144. T. Söderström and P. Stoica. *System Identification*. Prentice-Hall International, London, UK, 1989.
145. G. Stein. Respect the unstable - the 1995 bode lecture. In *30th IEEE Conference on Decision and Control*, Honolulu, Hawai, December 1990.
146. M. Steinbuch, G. Schootstra, and O.H. Bosgra. Robust control of a compact disc player. In *Proc. 31st Conf. Decis. and Control*, pages 2596–2600, Tucson, AZ, USA, 1992.
147. P.M.J. Van den Hof and R.J.P. Schrama. An indirect method for transfer function estimation from closed loop data. *Automatica, 29*, 6:1523–1527, 1993.
148. P.M.J. Van den Hof and R.J.P. Schrama. Identification and control - closed-loop issues. *Automatica*, 31:1751–1770, 1995.
149. P.M.J. Van den Hof, R.J.P. Schrama, and P.M.M. Bongers. On nominal models, model uncertainty and iterative methods in identification and control design. *Lecture Notes in Control and Information Sciences*, 192:39–50, 1994.
150. P.M.J. Van den Hof, R.J.P. Schrama, R.A. de Callafon, and O.H. Bosgra. Identification of normalized coprime plant factors from closed-loop experimental data. *European Journal of Control*, 1:62–74, 1995.
151. M Vidyasagar. The graph metric for unstable plants and robustness estimates for feedback stability. *IEEE Trans. Automatic Control*, AC-29:403–417, 1984.
152. M. Vidyasagar. *Control System Synthesis: A Factorization Approach*. MIT Press, Cambridge, MA, 1985.
153. G. Vinnicombe. Frequency domain uncertainty and the graph topology. *IEEE Trans. Automatic Control*, 38:1371–1383, 1993.
154. G. Vinnicombe. The robustness of feedback systems with bounded complexity controllers. *IEEE Trans. Automatic Control*, 41:795–803, 1996.
155. G. Vinnicombe. *Uncertainty and Feedback - $\mathcal{H}_\infty$ loop-shaping and the ν-gap metric*. Imperial College Press, 2000.
156. P.M.R. Wortelboer. *Frequency Weighted Balanced Reduction of Closed-Loop Mechanical Servo-Systems: Theory and Tools*. Ph.D. thesis, Delft University of Technology, Netherlands, 1993.
157. Y. Yamamoto. An overview on repetitive control: what are the issues, and where does it lead to? In *Proc. IFAC Workshop on Periodic Control*, pages 135–140, Cernobio, 2001. Elsevier.
158. G. Zames. Feedback and optimal sensitivity: Model reference transformations, multiplicative seminorms, and approximative inverse. *IEEE Trans. Automatic Control*, AC-26(2):301–320, 1981.

159. G. Zames and A.K. El-Sakkary. Unstable systems and feedback: The gap metric. In *Proc. 18th Allerton Conf. on Communication, Control, and Computing*, pages 380–385, Urbana, IL, 1980. University of Illinois.
160. Z. Zang, R.R. Bitmead, and M. Gevers. Iterative model refinement and control robustness enhancement. In *Procs. 30th IEEE Conference on Decision and Control*, pages 279–284, Brighton, UK, Dec. 1991.
161. Z. Zang, R.R. Bitmead, and M. Gevers. Iterative weighted least-squares identification and weighted LQG control design. *Automatica*, 31(11):1577–1594, 1995.
162. Y. Zhang, P.G. Mehta, R.R. Bitmead, and C.R. Johnson Jr. Direct adaptive control for tonal disturbance rejection. In *American Control Conference*, pages 1480–1482, Philadelphia PA, 1998.
163. K. Zhou and J.C. Doyle. *Essentials of Robust Control*. Prentice Hall, NJ, 1998.
164. K. Zhou, J.C. Doyle, and K. Glover. *Robust and Optimal Control*. Prentice Hall, NJ, 1995.
165. J. Ziegler and N.B. Nichols. Optimum settings for automatic controllers. *Trans ASME*, 64:759–768, 1942.

Index

a priori knowledge, 46
adaptive control, 6, 183
adaptive control design, 145
algorithm
– iterative, 176, 291
approximation, 41, 182
autoregressive filters, 170

bandwidth
– closed-loop, 122
bias, 26
bias error, 4, 186
bias formula, 277
– closed-loop, 44
– open-loop, 44
Bode, 63, 72, 73, 76
Bode diagram, 140
Bode's Integral, 70
Bode's relations, 71, 72

cancellation of poles and zeros, 87, 88
caution, 168, 172, 177, 196, 290
– controllers K_α, 179
certainty equivalence, 201
change of experimental conditions, 146
closed-loop identification, 42, 47, 50, 152, 233, 289
compact disc player, 225
complementary sensitivity, 67, 68, 70, 92, 94, 97
constraints, 278
control
– disturbance rejection, 101
– feedforward, 102
– minimum-variance, 202
– model reference, 12
– model-based, 121, 142
– open-loop, 102
– reference tracking, 101
– robust, 4, 169, 212, 259
– servo, 225
– two degrees of freedom, 102
control design
– model-based, 8

control error, 14
control objective function, 253
control-relevant identification, 262
controller
– cancellation, 124, 125, 128, 132, 135
– edge, 123, 132
– generalised predictive, 123, 136
– identification, 48
– PID, 121
– PID optimised, 121
– tuning, 169, 183
– validation, 289
controller validation, 253
coprime factor
– identification, 233
– normalised, 192
coprime factor identification, 233, 260
coprime factorisation, 199, 200, 231
– normalised, 190, 231
cost function, 25
criterion
– control, 167
– global, 167, 169, 170, 173, 274
– identification, 167, 173, 175
– local, 167, 170, 173, 175
critical point, 67–69

data filtering, 44, 55–57, 218, 279, 281
– successive, 59
data preparation, 41, 55
data selection, 44, 49, 51, 279
detrending, 55–58, 281
differencer, 283
disturbance rejection, 213, 272, 274
dual control, 6, 194
dual-Youla uncertainty structure, 255

excitation, 44, 47, 48
– design, 278
– sinusoidal, 216
experiment design, 42, 44, 47
– optimal, 201
experimental conditions, 15

feedback amplifier, 63, 65, 72

feedback amplifiers, 63
feedback-relevant identification, 235
feedforward control, 102
fractional system, 73–75, 97
fractional transfer function, 73
frequency scale, 123

gang of four, 67
gedanken experiment, 275
generalised sensitivity function, 178
generalised stability margin, 95–97, 190
grey-box modelling, 47

$\mathcal{H}_\infty$ control, 64, 86, 95–97, 238, 259
$\mathcal{H}_\infty$-identification, 61
$\mathcal{H}_\infty$ loop shaping, *see* loop shaping
Hankel norm, 190
Hansen scheme, 182
harmonic distortion, 218
helicopter vibration control, 211

ideal loop transfer function, 72, 76
identification, 18
– algorithms, 27
– bias, 26, 47
– closed-loop, 31, 50, 122, 152, 194, 289
– direct, 174, 289
– feedback-relevant, 235
– frequency response, 127
– least squares, 25, 43
– open-loop, 27
– parameter estimation, 26, 219, 286
– two-stage, 174, 181, 289
– uncertainty, 239
– variance, 26
identification for control, 7, 11, 14, 195
– optimal, 202
identification/control design cycle, 149
IMC design for unstable plants, 161
impractical control objectives, 146
internal model control, 150
internal stability, 88
iterative design, 7
Iterative learning control
– MIMO, 110, 114
iterative learning control (ILC), 105
iterative procedure, 252

least squares identification, 25

LFT, 256
limitations to control, 80
loop
– achieved, 167
– design, 167
loop shaping, 64, 73, 95–97, 238
– $\mathcal{H}_\infty$, 86, 238
LQG control, 167, 215
– frequency weighted, 221
– frequency-weighted, 171, 288
– scaling, 288

maximum complementary sensitivity
    M_t, 68, 96
maximum sensitivity M_s, 63, 68, 70, 96
minimum-variance control, 202
model, 17
– ARMAX, 30
– ARX, 29
– Box-Jenkins, 24
– output error, 28, 153, 220
– parameterised, 20
– regressors, 24
– structures, 19
model assessment, 287
model reduction, 5
model reference control, 12
model set estimation, 252
model structure, 44, 45, 55
– selection, 286
model uncertainty, 64, 65, 76, 95
– bounding, 239, 259, 263, 265
model validation, 54, 281, 286, 287
modelling
– experimental, 18
– physical, 18
MOOG INC, 211
motion control, 243
μ-synthesis, 259

noise model, 46, 49, 175, 276, 278
non-stationarity, 53, 55, 57
normalised coprime factorisation, 231
notch filter, 283
ν-gap metric, 91, 164, 177, 178, 195
ν-metric, *see* ν-gap metric
Nyquist, 65, 76
Nyquist's stability criterion, 67

Index   309

open-loop control, *see* two degrees of freedom
optimal controller, 201
oscillating modes
– hidden, 121
oscillator, 213, 214
overshoot, 124

parameter estimation, 26, 116, 219, 286
parameters
– bias, 26, 44
– – closed-loop, 47
– variance, 26
Parseval's bias formula, *see* bias formula
parsimony, 46, 54
performance
– assessment, 252, 257
– robustness, 253
– worst-case, 191
performance degradation, 201, 204
performance index, 122, 136
performance robustness, 193, *see* robust performance
phase compensation, 214
power spectral analysis, 124, 130, 133
power spectrum analysis, 126
prediction, 22
– error, 25
prediction error identification, 4
prediction error method, 42
– least squares, 43
process variations, 68, 69

quantisation, 51
quantitative feedback theory, QFT, 76

rate limiter, 216
reduced complexity, 45
reference model, 123
reference signal, 47
robot manipulators, 142
robust
– control, 4, 169, 212, 259
– performance, 173, 175, 178, 181, 182, 253, 258

– stability, 168, 178, 182
robust stability, 192, 238
robust stability test, 240, *see* stability margin

sampling period
– selection, 123
sanitised data, 219
saturation, 274
sensitivity, 67–69, 71, 92, 96, 97
servo control design, 225, 243
– open-loop, 102
simultaneous stabilisation, 180
stabilising controller, 198
stability, 65, 67–70, 87–89, 93
– guarantees, 15
– margin, 67, 95–97, 178, 198
– robustness, 238
step response, 11, 89, 127, 131–133, 137, 160, 287
subspace identification, 3
sugar cane crushing process, 272
swept-sine test, 215

time–bandwidth product, 61
timescale, 121
transient instability, 147
triangle inequality, 173, 175
two degrees of freedom (2-DOF), 75–78, 87, 101–104
two-stage identification, 174, 181, 289

uncertainty identification, 239, 259

validation via correlation technique, 157
variance error, 4
Vinnicombe, 91–92, 96, 165, 177–179, 190, 196

wafer stepper, 243
water bed effect, 71
windsurfing approach, 143

Youla-Kucera parameterisation, 179